Fortschritte der Chemie organischer Naturstoffe

Progress in the Chemistry of Organic Natural Products

59

Founded by L. Zechmeister
Edited by W. Herz, G. W. Kirby, R. E. Moore,
W. Steglich, and Ch. Tamm

Authors:
A.-M. Eklund, S.-I. Hatanaka, I. Wahlberg

Springer-Verlag
Wien New York 1992

Prof. W. HERZ, Department of Chemistry,
The Florida State University, Tallahassee, Florida, U.S.A.

Prof. G. W. KIRBY, Chemistry Department,
The University, Glasgow, Scotland

Prof. R. E. MOORE, Department of Chemistry,
University of Hawaii at Manoa, Honolulu, Hawaii, U.S.A.

Prof. Dr. W. STEGLICH, Institut für Organische Chemie und Biochemie der Universität
Bonn, Bonn, Federal Republic of Germany

Prof. Dr. CH. TAMM, Institut für Organische Chemie der Universität Basel,
Basel, Switzerland

© 1992 by Springer-Verlag/Wien
Softcover reprint of the hardcover 1st edition 1992

Library of Congress Catalog Card Number AC 39-1015

Typesetting: Macmillan India Ltd., Bangalore-25

Printed on acid free paper

With 1 Figure

ISSN 0071-7886
ISBN-13: 978-3-7091-9152-1   e-ISBN-13: 978-3-7091-9150-7
DOI: 10.1007/978-3-7091-9150-7

Contents

List of Contributors

EKLUND, Dr. A.-M., Reserca AB, Box 17007, S-104 62 Stockholm, Sweden.

HATANAKA, Prof. S.-I., Department of Biology, University of Tokyo, Komaba 3-8-1, Meguro-ku, Tokyo 153, Japan.

WAHLBERG, Dr. I., Reserca AB, Box 17007, S-104 62 Stockholm, Sweden.

Amino Acids from Mushrooms*

SHIN-ICHI HATANAKA, Department of Biology, College of Arts and
Sciences, The University of Tokyo, Tokyo, Japan

With 1 Figure

Contents

* Dedicated to the late Professor H. GRISEBACH.

I. Introduction

Two dimensional paper chromatography or automated amino acid analysis of the "amino acid-fraction" prepared from ethanol extracts of plants and fungi reveals the existence of many ninhydrin-positive substances. These are mostly protein amino acids in the free state, but also include several non-protein amino acids, i.e., α-aminoadipic acid, ornithine, citrulline, β-alanine, γ-aminobutanoic acid and cysteic acid as well as ethanolamine.

Higher plants and fungi often contain one or more specific non-protein amino acids. Some of these are species-, section-, or genus-specific, although species lacking these specific amino acids are also known within such groups. Furthermore, many non-protein amino acids are distributed sporadically at higher levels of classification. That the distribution patterns of non-protein amino acids may have a bearing on the phylogenetic relationships among various groups of plants and fungi has been discussed.

Non-protein amino acids are structurally diverse, just like protein amino acids, and may be aliphatic, neutral, acidic or basic. A large number of non-protein amino acids can be regarded as derivatives of the twenty different protein amino acids. Naturally occurring non-protein amino acids are mostly α-amino acids, whose configuration at C-2 is S (L-form) except for a few cases. This suggests that most of the non-protein amino acids have their origin in the protein amino acids. In fact, it has been demonstrated that many are indeed biosynthesized from the corresponding parent protein amino acids. However, most biosynthetic pathways leading to most of them have not been studied and biosynthetic routes to many of the non-protein amino acids are not obvious. Coexistence of several structurally related non-protein amino acids in a single species sometimes gives strong clues as to their biosynthetic pathways.

Some of the non-protein amino acids have been isolated from fungi as their poisonous principles or more generally as biologically active substances. A large number of non-protein amino acids are known as antibiotics or as components of antibiotics. Amino acid antagonists can usually affect primary metabolism, transport, or permeation of protein amino acids. In some cases protein synthesis is strongly inhibited by the presence of non-protein amino acids. Like the usual amino acids, non-protein amino acids may occur as their γ-glutamylpeptides. The formation of γ-glutamylamino acids affects the stability of the free amino acid. Examples are known where, after enzymatic liberation of the γ-glutamyl moiety, the free amino acids display biological functions or yield odorous substances after sequential enzymatic reactions.

There are several reasons for studying non-protein amino acids. First, the isolation and structural clarification of physiologically active principles or chemically interesting natural products is important. Second, if a specific amino acid is discovered in one taxon, it is usually of interest from a chemotaxonomical point of view to investigate its distribution in other closely related taxa. Third, it is important to know chemical constituents of foods, beverages, and animal feeds in detail, even if many such substances are present in very small amounts.

In the last two decades, BELL (24, 23) and FOWDEN (108, 112, 109) have published several reviews dealing with non-protein amino acids in plants and fungi. These contributions also included discussions of their interesting phylogenetic and chemo-ecological properties. ROSENTHAL's book

(*363*) is the only one so far in which the toxicological properties of the non-protein amino acids have been emphasized. Handbook style publications on non-protein amino acids edited in 1985 by DAVIS (*75*) and HUNT in the same year (*194*) are also very useful. Numerous non-protein amino acids are listed as antibiotics in a dictionary of antibiotics edited by BYCROFT (*188*). Articles on toxic non-protein amino acids by FOWDEN and LEA (*111*) and ROSENTHAL and BELL (*364*) in the book "Herbivores" are also valuable. Comprehensive treatises on fungal metabolites were published by TURNER in 1971 (*431*) and by TURNER and ALDRIDGE (*432*) in 1983.

Except for the amino acids occurring as components of *Amanita* toxins, little attention has been paid to the free non-protein amino acids of edible or non-toxic mushrooms. Scientific studies of specific amino acids of mushrooms therefore started much later than studies of amino acids produced by higher plants. TH. WIELAND published an excellent book on the peptides of poisonous Amanitas in 1986 (*453*). BENEDICT reviewed mushroom toxins other than those in *Amanita* species in 1972 (*27*). In 1982, CHILTON (*50*) published a very useful review, the first such on secondary amino acids of mushrooms. After the valuable handbook on toxic and hallucinogenic mushroom by LINCOFF and MITCHEL (*267*), BRESINSKY and BESL (*40*) recently published a comprehensive treatise on poisonous mushrooms, their identification, symptoms, chemistry, biochemistry, pharmacology, therapy, etc., with many recent references. Reviews on γ-glutamylpeptides in plants and mushrooms by KASAI and LARSEN (*226*) and on active constituents of *Amanita* by EUGSTER (*94*) have appeared in this series.

A Specialist Periodical Report "Amino Acids and Peptides" published by Royal Society of Chemistry contains a survey of newly discovered non-protein amino acids and is helpful.

Non-protein amino acids in fungi cannot be considered to be essentially different from those of higher plants. However, many non-protein amino acids were isolated from mushrooms that have not yet been reported in higher plants. The reverse is also true. It must be noted, however, that our knowledge is very limited and fragmentary.

This review deals with the chemical structures, occurrence, and, in some cases, biosyntheses or possible biological significance of non-protein amino acids and their γ-glutamylpeptides known from mushrooms. When necessary, closely related natural non-protein amino acids of non-fungal origin are also referred to.

Scientific names of some fungi have been revised since the first publication of papers on their amino acids. However, in this review, the names used in the first publication have been adopted.

This is not a comprehensive review, and emphasis will be placed on neutral aliphatic amino acids. Many of these seem to be characteristic of mushrooms.

Almost all naturally occurring amino acids belong to the L-series. When the prefixes D- or L- are not shown, the absolute configuration at C-2 should be understood to be L or (2S).

II. General Methods

In the identification of species used for analyses, it is essential to consult taxonomic specialists and to deposit voucher specimens in herbaria. Additionally, free amino acid patterns of the same species may vary from sample to sample collected in various localities and habitats. Inclusion of this information in research papers may also be helpful to researchers.

Although fruitbodies can be dried and stored before analysis, attention should be paid to the possibility that amino acids may decompose at high temperatures. Through freeze-drying, the amino acids of fresh fruitbodies can be preserved and the extraction is easier and more efficient.

Free amino acids as well as γ-glutamylpeptides of mushrooms are usually extracted by homogenizing with aqueous ethanol or methanol. When the extract is passed directly through a column of cation exchange resin, (e.g. Amberlite IR-120, CG-120, Dowex 50, or ZeoKarb 225 of H$^+$-form) and the retained amino acids are displaced with ammonia-water, an "amino acid-fraction" is obtained.

Besides automated methods such as the use of an amino acid analyzer or high performance liquid chromatography, conventional two-dimensional paper chromatography is still widely used for screening free amino acids. The last named technique can give valuable information on the properties or sometimes even the partial structure of unidentified or unknown amino acids. Thus, their location on the chromatograms suggest their acidic, basic or neutral properties and the specific color reaction with ninhydrin may indicate whether the amino acid is unsaturated, hydroxylated or heterocyclic. In many cases ninhydrin colors on paper are quite different from those in an amino acid analyzer which is performed at a constant pH. Copper complexes of α-monoamino acids are ninhydrin-negative, a property that is often used for an α-amino acid test (256). The combination of paper chromatography with paper electrophoresis may also give good separations of many amino acids.

References, pp. 117–140

Identification of amino acid based only on the elution position or retention time can sometimes lead to error. It is clear that two or more specific amino acids might be eluted at the same time, because the standard systems for protein hydrolysates or even for physiological fluids have been designed to separate and determine only a limited number and often only the most frequently occurring amino and imino acids. OGAWA *et al.* (*326*) recently reported some examples of which specific non-protein amino acids overlapped with common protein amino acids on an automated analyzer.

Preparative fractionation as well as purification of amino acid and their γ-glutamyl dipeptides is carried out using a column of ion exchange resin or microcrystalline cellulose powder. Acidic amino acids are separated efficiently on anion exchanger Dowex 1 in acetate-form by eluting them successively with diluted acetic acid. There are also reports that polysaccharides with ionizable radicals, such as DEAE-Sephadex in acetate-form have been applied successfully. For neutral and basic amino acids, cation exchangers (Amberlite CG-120 or Dowex 50 in H$^+$- or Na$^+$-form) are useful; the amino acids are separated from each other by elution with dilute HCl, NH$_4$OH, pyridine or an appropriate buffer. Because they are zwitterions, amino acids are generally easy to crystallize. Basic ones are usually separated as stable hydrochlorides. Recent advances in various instrumental analyses have made it possible to elucidate quickly the chemical structures of small molecules in natural products. Non-protein amino acids from mushroom are not exceptional. Difficulties lie sometimes only in the determination of stereochemistry. Although the absolute configuration at C-2 is usually determined from the positive shift of optical rotation values with increasing acidity based on Clough-Lutz-Jirgensons rule, oxidation with L- or D-amino acid oxidase gives more exact results, particularly in the case of racemates or partial racemates.

III. Structure, Occurrence and Biochemistry

1. Neutral Aliphatic Amino Acids

1.1. Hydroxyvaline, Hydroxyleucines, and Hydroxyisoleucines

Hydroxylation is one of the simplest and most direct modifications of protein amino acids; it results in the formation of unique non-protein amino acids in the free or bound state. Hydroxylated forms of acidic amino acids will be treated in the next chapter. The parent non-hydroxylated acids may be protein or non-protein amino acids. Intensive studies of hydroxyleucines and hydroxyisoleucines have been carried out, partially because some are components in the deadly toxic cyclopeptides of *Amanita phalloides* (Vaill.) Secr. and related species.

In this chapter hydroxyvaline, hydroxyleucines, hydroxyisoleucines will be reviewed first; subsequently other hydroxyamino acids will be noted. Some reports on their distribution in living organisms other than the Amanitas will also be cited.

Proline, leucine and isoleucine in the cyclopeptides of *Amanita phalloides* are hydroxylated, probably by a reaction similar to that in which hydroxyproline and hydroxylysine are formed in polypeptides. Various kind of hydroxyamino acids are also known as components of

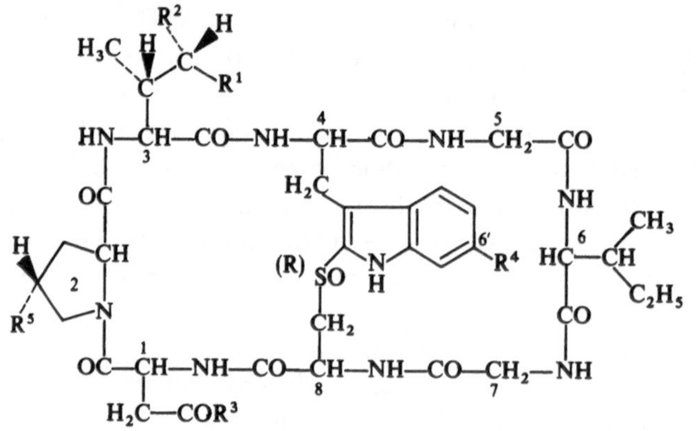

Name	R^1	R^2	R^3	R^4	R^5
α-Amanitin	CH_2OH	OH	NH_2	OH	OH
β-Amanitin	CH_2OH	OH	OH	OH	OH
γ-Amanitin	CH_3	OH	NH_2	OH	OH
ε-Amanitin	CH_3	OH	OH	OH	OH
Amanin	CH_2OH	OH	OH	H	OH
Amanin amide[a]	CH_2OH	OH	NH_2	H	OH
Amanullin	CH_3	H	NH_2	OH	OH
Amanullinic acid	CH_3	H	OH	OH	OH
Proamanullin	CH_3	H	NH_2	OH	H

[a] In *A. virosa* only.

(1)

General formula of the amatoxins (from *454*)

antibiotics. Chemistry, biochemistry, and molecular biological aspects of the toxic cyclic peptides in *Amanita* were recently reviewed by WIELAND (*454*). Chemical structures, including stereochemistry and conformational analysis of hydroxyamino acids in these toxins were reported mainly during the 1960's to 1970's from WIELAND laboratory. Structures and biological functions of amatoxins, which were known at that time, were concisely reviewed by WIELAND (*453*).

The amatoxins (**1**) are bicyclic octapeptides with a bridge of 6'-hydroxytryptathionine-(*R*)-sulfoxide and consist exclusively of L-amino acids. Their physiological action is known to be due to the inhibition of DNA-dependent RNA-polymerase II. The 4-hydroxyl group in the side chain 3 seems to be related to the toxicity. The phallotoxins (**2**), bicyclic heptapeptides, have a thioether bridge, and their toxic effect is probably caused by their tight binding to F-actin. Except for prophalloin, they contain (2*S*,4*S*)-4-hydroxyproline. Acidic phallotoxins contain (2*R*,3*R*)-3-hydroxyaspartic acid, whereas neutral ones contain D-threonine. All

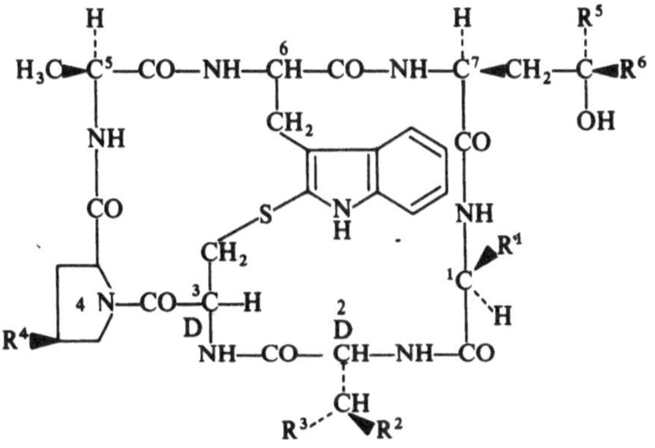

Name	R^1	R^2	R^3	R^4	R^5	R^6
Phalloin	CH_3	CH_3	OH	OH	CH_3	CH_3
Phalloidin	CH_3	CH_3	OH	OH	CH_2OH	CH_3
Phallisin	CH_3	CH_3	OH	OH	CH_2OH	CH_2OH
Prophalloin	CH_3	CH_3	OH	H	CH_3	CH_3
Phallacin	$CH(CH_3)_2$	OH	CO_2H	OH	CH_3	CH_3
Phallacidin	$CH(CH_3)_2$	OH	CO_2H	OH	CH_2OH	CH_3
Phallisacin	$CH(CH_3)_2$	OH	CO_2H	OH	CH_2OH	CH_2OH

(**2**)

General formula of the phallotoxins (from *454*)

hydroxylated leucine and isoleucines in the amatoxins are located in position 3 and those in the phallotoxins in position 7. Determination of the absolute configurations of the 4-hydroxyleucines and isoleucines, particularly those present in bound form, is not easy, because the interconversion takes place at C-2 as well as at C-4. In the case of the hydroxyisoleucines, the configuration at C-3 is also changed. It is worth noting that, during chemical study of various hydroxyamino acids in the cyclic peptides of *A. phalloides*, much information was obtained on the synthesis, resolution, reactivity and particularly on the interconversion between optical isomers or diastereomers.

The closely related *Amanita verna* (Bull.) Pers. also contains amatoxins and phallotoxins and is deadly toxic. Concentrations of various amatoxins and phallotoxins in this species are very similar to those in *Amanita phalloides*. Amatoxins and phallotoxins have been detected by thin-layer chromatography in several other *Amanita* species belonging to the subgenus *Lepidella*. BEUTLER and DER MARDEROSIAN (*30*) identified cyclopeptides, together with various tryptamines, in various species of

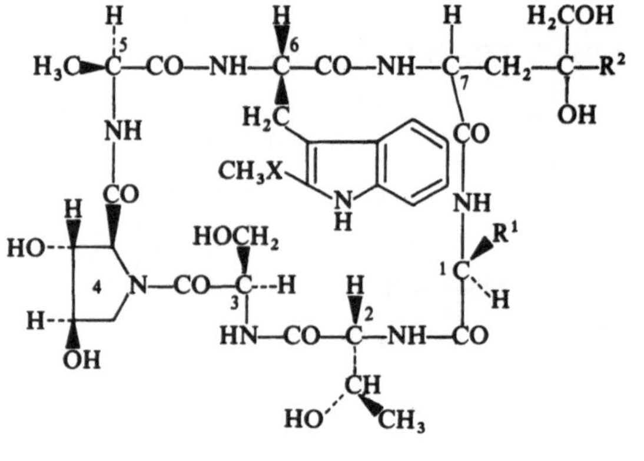

Name	X	R¹	R²
Viroidin	SO_2	$CH(CH_3)_2$	CH_3
Desoxoviroidin	SO	$CH(CH_3)_2$	CH_3
Ala¹-viroidin	SO_2	CH_3	CH_3
Ala¹-desoxoviroidin	SO	CH_3	CH_3
Viroisin	SO_2	$CH(CH_3)_2$	CH_2OH
Desoxoviroisin	SO	$CH(CH_3)_2$	CH_2OH

(3)

General formula of the virotoxins (from *454*)

Amanita which were collected in different localities. Fruitbodies of *A. ocreata* which was known to contain amatoxins also appeared to contain some phallotoxins. This species is closely related to *A. phalloides*. Interestingly, amatoxins are known also in species of other genera and in other families, namely, *Galerina* (Cortinariaceae), *Lepiota* (Agaricaceae) and *Pholiota* (Bolbitiaceae) (*40*). Phallotoxins have never been detected outside the genus *Amanita*.

In addition to amatoxins and phallotoxins, a third group of toxic cyclic peptides, the virotoxins (*100*) (**3**), is known in *Amanita virosa* Lam. ex Secr. Virotoxins are monocyclic heptapeptides which lack the sulfur-bridge, contain D-threonine, D-serine, and more markedly (2*S*,3*R*,4*R*)-3,4-dihydroxyproline (2,3-trans-3,4-trans-3,4-dihydroxy-L-proline), instead of (2*S*,4*S*)-4-hydroxyproline in position 4 (*42*). 3,4-Dihydroxyproline is known from cell walls of diatoms (*311, 224*). In this case, however, the configuration is (2*S*,3*S*,4*S*) (*311, 224*). They were the first records of the natural occurrence of dihydroxyprolines. Recently, the (2*S*,3*R*,4*R*)-dihydroxyproline proved to be a potent and specific inhibitor of β-D-glucuronidase (*367*). The composition of the toxic peptides in *A. virosa* is known to differ depending on localities of the samples. This may also be the case in other mushrooms.

It is highly interesting that the fruitbodies of *Amanita phalloides* produce a cyclic decapeptide, antamanide (**4**), which nullifies the toxic effect of phalloidin (*460*). Antamanide is composed of only protein amino acids of L-form.

```
 ┌─L-Pro—L-Phe—L-Phe—L-Val—L-Pro─┐
 └─L-Pro—L-Phe—L-Phe—L-Ala—L-Pro─┘
```

(**4**)

Antamanide

A considerable number of species of the genus *Cortinarius* is known to be toxic. In 1984, TEBBETT and CADDY (*419*) reported the isolation of cortinarin A, B, and C from *C. speciossimus* and suggested possible structures. All are very similar to each other and are bicyclic peptides composed of ten L-amino acids, D-threonine and a tryptophan bridge. Results of the screening of about 60 different species of *Cortinarius* showed that all contain cortinarin A and C, except for *C. violaceus* which lacks cortinarin A. *C. orelanus*, *C. orellanoides*, and *C. speciossimus*, also contain cortinarin B. According to TEBBETT and CADDY, cortinarin A and B are toxic because of their structural similarity to vasopressin. Cortinarin C seems not to be toxic.

Many hydroxyamino acids give characteristic brownish-violet spots with ninhydrin on paper chromatograms, the color changing to the usual violet or purple with time. This property is often useful for identification of novel amino acids.

1.1.1. 3-Hydroxyvaline (5)

Recently AOYAGI and SUGAHARA (*15*) identified (2*S*)-3-hydroxyvaline (5) from fruitbodies of *Pleurocybella porrigens*. Although this amino acid was completely overlapped by threonine in an elution profile in an amino acid analyzer, the two were distinguished by TLC with a Li-citrate buffer system. With the use of Amberlite IR-120 (H$^+$-form), Dowex 1, (acetate-form), Dowex 50 (pyridinium-form), Avicel cellulose, and finally with Wako C-200 silica gel column, the pure amino acid was obtained. Identification was carried out by elementary analysis, SIMS, ^{1}H-NMR, reduction with HI and red phosphorus, degradation by Ba(OH)$_2$ giving glycine, and conclusively by direct comparison with a synthetic sample. Positive shift of the value of the optical rotation in a more acidic solution indicated that the natural form is the L-form.

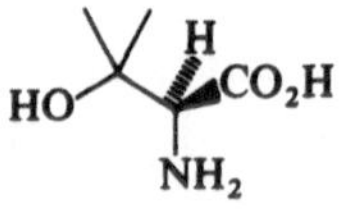

(5)

(2*S*)-3-Hydroxyvaline

3-Hydroxyvaline is present as a constituent of the antibiotic YA-56 (*207*, *327*). Free 3-hydroxyvaline is not known to possess biological activity.

4-Hydroxyvaline was isolated from crown gall tumors of *Kalanchoe daigremontiana* (*351*). However, whether it has the (2*S*,3*S*) or the (2*S*,3*R*) configuration was not clear.

1.1.2. 4-Hydroxyleucine (6)

This amino acid is a component of three phallotoxins; phalloin (*461*), prophalloin and phallacin. It has also been found in hydrolysates of casein (*65*) and gelatin (*456*). (2*S*)-4-Hydroxyleucine has been synthesized in good yield from (2*S*)-leucine by photochlorination (*237*, *101*).

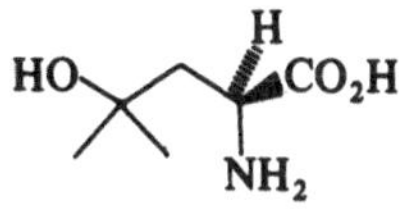

$$HO \diagdown \overset{\overset{\text{H}}{|}}{\diagup} \diagup \diagdown CO_2H$$
$$\underset{NH_2}{}$$

(6)

(2S)-4-Hydroxyleucine

(2S,3R)-3-Hydroxyleucine (L-*threo*-form) is known to occur in a plant, *Deutzia gracilis* (*213*), and the (2S,3S)-isomer (L-*erythro*-) is a component of the antibiotic telomycin (*387*). The absolute configuration (2S) was determined by oxidation with periodate to give 2-aminolevulinic acid; this was subjected to the iodoform reaction with iodine under strongly alkaline conditions. The aspartic acid obtained proved to be the (2S)-isomer (L) using stereospecific aminotransferase (*342*).

1.1.3. 4,5-Dihydroxyleucine (7)

Phalloidin, phallacidin, viroisin, desoxoviroidin, Ala[1]-viroisin and Ala[1]-desoxoviroidin contain 4,5-dihydroxyleucine whose absolute configuration is (2S,4R). Because of the presence of two centers of chirality, four isomers are possible. Synthesis yielded a mixture of all four isomers which could all be resolved (*459*). The synthetic mixture of lactones was fractionally crystallized into (2S,4S; 2R,4R) and (2S,4R; 2R,4S). Each of the ditoluyl tartrates was then enzymatically resolved. The configuration at C-2 was very likely (2S) from its optical rotation, and the (4R) configuration was suggested by Hudson's lactone rule. This was further confirmed by analysis and comparison of the ^{1}H-NMR spectra of the natural and resolved synthetic amino acid. In 1968, WIELAND *et al.*, reported that in the lactone of the natural (4R)-isomer, hydroxymethyl group can rotate freely; however; in the diastereomer, because of a hydrogen bond the geminal 5-protons are not equivalent (*458*).

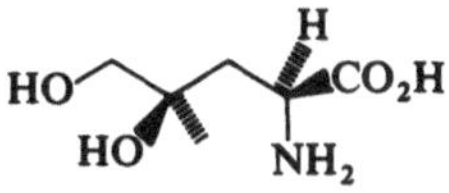

$$HO \diagup \overset{\overset{\text{H}}{|}}{\diagdown} \diagup \diagdown CO_2H$$
$$HO \diagdown \qquad \underset{NH_2}{}$$

(7)

(2S,4R)-4,5-Dihydroxyleucine

1.1.4. 4,5,5'-Trihydroxyleucine (8)

This amino acid was discovered in 1967 in a phallisin hydrolysate (*134*) and later also in phallisacin, viroisin and desoxoviroidin. The

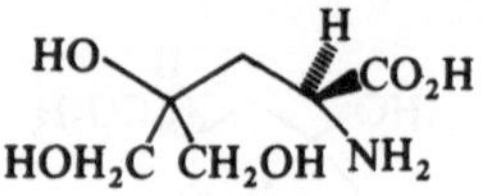

(8)

(2S)-4,5,5'-Trihydroxyleucine

structure was elucidated by means of periodate degradation by WEYGAND and MAYER in 1968 (451). In this case there exist no stereochemical problems other than that associated with C-2.

1.1.5. 4-Hydroxyisoleucine (9)

This compound was first isolated as a new amino acid lactone from γ-amanitin hydrolysate (462) and later from ε-amanitin as well (455). In order to determine the absolute configuration, FAULSTICH et al. (101) synthesized the diastereomeric mixture of (2S)-4-hydroxyisoleucines by photochlorination of (2S)-isoleucine. The two diastereomers, (2S,4R) and (2S,4S), could be separated by crystallization; one of these was identical with the natural 4-hydroxyisoleucine. Consequently, the question was only as to the configuration at C-4. From an analysis of the ^{1}H-NMR spectrum by means of the nuclear Overhauser effect it was originally considered to be (4R) (458). Several years later, GIEREN et al. (140) unambiguously established it to be (4S) by X-ray analysis. This configuration was further supported by re-examination of the ^{1}H-NMR spectrum and Hudson's lactone rule.

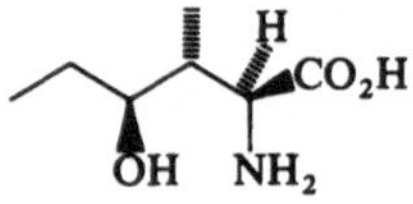

(9)

(2S, 3R, 4S)-4-Hydroxyisoleucine

In 1973, FOWDEN et al. (116) reported the isolation and identification of free 4-hydroxyisoleucine from the seeds of fenugreek, *Trigonella foenum-graecum* L. (Leguminosae). The amino acid represents 30–50% of the total free amino acids and 13 g of its lactone was obtained from 14 kg of seed meal. Results of the elementary analysis agreed with $C_6H_{13}NO_3$. Prolonged reduction over Adams Pt catalyst gave isoleucine and a small amount of alloisoleucine. The structure was further clarified from its ^{1}H-NMR spectrum and the absolute configuration at C-2 was shown to be S; that at C-4 was determined by direct comparison with the 4-hydroxyisoleucine isolated from γ-amanitin. Their results suggested that the 4-hydroxyisoleucines in fenugreek and in γ-amanitin have the same absolute configurations at three chiral carbon atoms and they concluded that the configuration was $(2S,3R,4R)$. However, because this report by FOWDEN et al. (116) appeared before GIEREN et al. (140) had correctly determined the configuration at C-4 of the 4-hydroxyisoleucine of γ-amanitin mentioned above, the hydroxyamino acid in fenugreek should also be $(2S,3R,4S)$.

Quantitative data on the epimerization of 4-hydroxyisoleucine by heating with 5 N $Ba(OH)_2$ or 6 N HCl have been presented (116). Under the first condition, epimerization was observed only at C-2 on a small scale, but with 6N HCl, the $(2S,3R,4S)$ epimer was converted to a mixture of $(2S,3R,4S)$, $(2S,3R,4R)$ and $(2S,3S,4S)$ in the ratio of 1:1:3, i.e., epimerization at C-3 and C-4 also occurred.

In view of the poor reproducibility in the isolation of a particular lactone after photochlorination and because of the epimerization of the lactones, ALCOCK et al. (6) carefully re-examined the ^{1}H-NMR spectral data (including NOE) and the CD data of the lactone hydrochloride isolated from fenugreek. They then carried out an X-ray crystal structure determination. All of the results were consistent and indicated conclusively that the absolute configuration of 4-hydroxyisoleucine from fenugreek is $(2S,3R,4S)$.

In fenugreek, $(2R,3R,4S)$-4-hydroxyisoleucine is also present as a minor component. Judging from the mild isolation procedure, it is unlikely to be an artifact (116).

As a result of their chromatographic survey of non-protein amino acids of mushrooms, DARDENNE et al. (69, 8) reported the presence of two amino acids giving a brown ninhydrin-coloration from fruitbodies of Lactarius camphoratus (Fr.) Fr. On a paper chromatogram, they were located side by side near valine in phenol saturated with 0.08 M citrate-phosphate buffer, pH 4.2. The faster moving one was isolated by use of preparative paper chromatography and Dowex 1 (OH$^-$-form). Elementary analysis, ^{1}H-NMR spectrum, reduction with hydriodic acid and red phosphorus, and hydrogenation on relatively large amounts of Adams Pt catalyst showed it to be 4-hydroxyisoleucine. Furthermore, oxidation by L-amino acid oxidase prepared from Habu snake (a poisonous snake from Okinawa Prefecture) venom (312) and an NOE experiment on the N-acetylated lactone established the absolute configuration as $(2S,3R,4S)$. Another slow-moving amino acid on the chromatogram was assumed to be the lactone of the above amino acid, i.e., $(2S,3R,4S)$-2-amino-3-methyl-4-pentanolide. Experimentally, the interconversion of the open chain 4-hydroxyamino acid and its lactone with weak alkali and acid was ascertained. The hydroxyisoleucines from three different biological sources, γ-and ε-amanitins, fenugreek, and *Lactarius camphoratus* thus have the same absolute configuration $(2S,3R,4S)$.

Strikingly, fruitbodies of *Lactarius camphoratus,* purified 4-hydroxy-isoleucine, and fenugreek meal all have a characteristic curry-like odor. Here we recall that fenugreek is used to prepare curry powder, a mixture of spices. Quite recently, GIRARDON *et al.* (*143*) identified the odoriferous compound of fenugreek as 3-hydroxy-4,5-dimethyl-2(5*H*)-furanone. This compound was first synthesized by SULSER *et al.* (*401*) as an aroma substance of bouillon; it was later isolated from aged Sake (Japanese rice wine) by TAKAHASHI *et al.* (*408*). Furthermore, TOKIMOTO *et al.* (*423*) pointed out that the sugary aroma of raw sugar is mainly caused by the same compound which they named "sotolon" (soto—raw sugar in Japanese). It is interesting that purified 4-hydroxyisoleucine alone gives the same aroma. This suggests that spontaneous formation of the odoriferous substance occurs and that it can be detected as an aroma even if present only in minute amounts.

1.1.6. 4,5-Dihydroxyisoleucine (10)

This amino acid was found in the hydrolysates of α-amanitin and β-amanitin. A diastereomeric mixture was synthesized by GEORGI and WIELAND (*138*). Dihydroxyisoleucine in α-amanitin was reduced to 4-hydroxyisoleucine *via* an epoxide without touching the chirality center at C-4. The obtained 4-hydroxyisoleucine was identical with that of $(2S,3R,4R)$ (*457*). The absolute configuration was confirmed by X-ray analysis (*140*). The $(4S)$ configuration of the 4-hydroxyisoleucine of γ-amanitin should be designated as (R), when CH_2OH is replaced with CH_3.

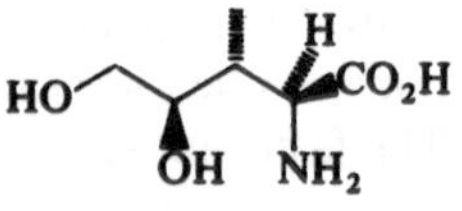

(10)

(2S, 3R, 4R)-4,5-Dihydroxyisoleucine

1.2. Other Hydroxyamino Acids

1.2.1. 2-Amino-3,4-dihydroxybutanoic Acid (11)

OGAWA and his associates several years ago reported isolation and identification of two unique hydroxyamino acids from edible fruitbodies

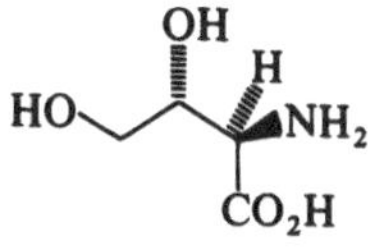

(11)

(2R, 3S)-2-Amino-3,4-dihydroxybutanoic acid

of *Lyophyllum ulmarium*. This edible mushroom is now cultivated in Japan on a commercial scale. The first amino acid reported was the D_s-*erythro*-3,4-dihydroxy-isomer of 2-aminobutanoic acid (*324*) (11). 2-Aminobutanoic acid itself is only seldom encountered in plant and fungal materials. Conventional column chromatography with ion exchangers and preparative paper chromatography were used. The yield was astonishingly high: 105 mg from 300 g fresh material. Results of elementary analysis and field desorption mass spectrometry (FDMS) showed the chemical formula $C_4H_9NO_4$. Presence of two hydroxyl groups on the methine and methylene carbon atoms was suggested by GC-MS analysis of the trimethylsilyl derivative. The chemical structure was clarified by a combination of two kinds of chemical degradation (see Scheme 1), and stereochemical properties of the products were compared with those of authentic samples. The naturally occurring amino acid produced D_g-glyceric acid and D_s-homoserine. The stereochemistry was further confirmed by the chromatographic method which is used to separate L_s-leucylpeptides of optical isomers. Natural occurrence of a free D-amino acid is infrequent and its origin and metabolism are of great interest. Experimental evidence was presented that 2-amino-3,4-dihydroxybutanoic acid in this mushroom seems to be present in only one isomeric form.

Another isomer, the L_s-*threo*-form, is known in a strain of *Streptomyces* (*450*) and in actinomycin Z_1 (*229*); a related amino sugar, 2-amino-2-deoxy-D_g-erythrose, is found in the glycoprotein fraction of the cell wall of the mushroom, *Agaricus bisporus* (*126*).

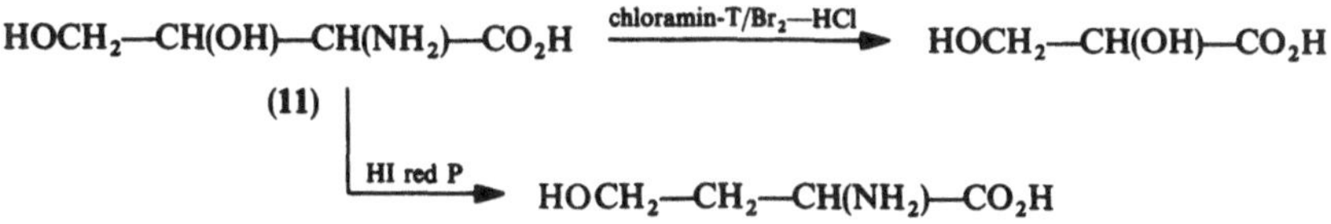

$$HOCH_2—CH(OH)—CH(NH_2)—CO_2H \xrightarrow{\text{chloramin-T/Br}_2—HCl} HOCH_2—CH(OH)—CO_2H$$

(11) $\xrightarrow{\text{HI red P}}$ $HOCH_2—CH_2—CH(NH_2)—CO_2H$

Scheme 1. Chemical decomposition of 2-amino-3,4-dihydroxybutanoic acid (from *324*)

1.2.2. 2-Amino-4-ethoxy-3-hydroxybutanoic Acid (12)

Subsequently OGAWA's group reported isolation of the closely related 2-amino-4-ethoxy-3-hydroxybutanoic acid from *L. ulmarium* (*325*). Its

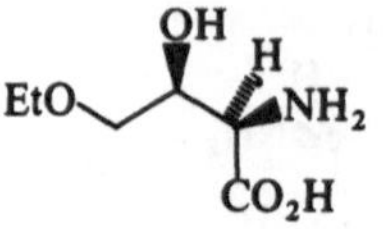

$$\text{EtO} \quad \overset{\text{OH}}{\underset{\text{CO}_2\text{H}}{\bigwedge}} \overset{\text{H}}{\underset{}{\text{NH}_2}}$$

(12)

(2*R*, 3*R*)-2-Amino-4-ethoxy-3-hydroxybutanoic acid

stereochemistry proved again to be that of the D_s-*erythro* form. Its concentration was much lower than that of the first amino acid and they obtained ca 40 mg colorless residue from 1 kg of mushrooms. Similar methods were applied for the separation, and the IR spectrum (1120 cm^{-1}) indicated the presence of an ether linkage in the molecule. Reduction with hydriodic acid and red phosphorous gave 2-aminobutanoic acid and 2-amino-4-hydroxybutanoic acid, and together with the results of ^{1}H-and ^{13}C-NMR spectrometry and GC-MS of its TMSi derivative, the structure was clarified. The absolute configuration was determined by comparing the reduction products with those of authentic compounds. This amino acid seemed to be 98% optically pure and this is the first report on its natural occurrence.

1.2.3. 4-Hydroxynorvaline (13)

Norvaline can be hydroxylated at C-3, C-4, and C-5, and all of these possibilities are known to occur in nature. Unquestionable evidence for the presence of norvaline itself in the free state has not yet been obtained.

The antibiotic cycloheptamycin has been reported to contain (2*S*, 3*S*)-hydroxynorvaline (*149*) while (2*S*)-5-hydroxynorvaline occurs in the free state in the legumes jack bean (*Canavalia ensiformis*) and kidney bean (*Phaseolus vulgaris*) (*422*).

Only 4- and 5-hydroxynorvaline are known in higher fungi as well as in plants. FOWDEN (*107*) first reported isolation and characterization of the former amino acid from seeds of sweet pea, *Lathyrus odoratus*. The isolate gave two spots on paper electrophoresis at pH 2.0, migrating identically with a synthetic mixture of diastereomers. FOWDEN assumed that partial epimerization had occurred at the 4-carbon atom during

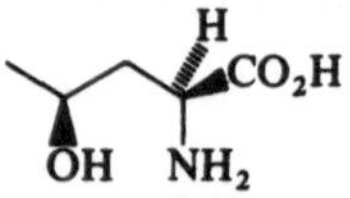

$$\overset{\text{H}}{\underset{\text{OH} \quad \text{NH}_2}{\bigwedge}} \text{CO}_2\text{H}$$

(13)

(2*S*, 4*S*)-4-Hydroxynorvaline

lactone formation in the isolation procedure. He thought that the natural form was the fast moving diastereomer 4R, by analogy with the mobilities of threonine and allothreonine. However it was later shown that this analogy is not applicable in this case (see below). 4-Hydroxynorvaline as well as the 5-hydroxy isomer were found to be distributed rather widely in seeds of *Astragalus* (Leguminosae) (*87*). (2S,4R)-4-Hydroxynorvaline is produced by *Streptomyces griseosporeus* as a weak antiviral agent (*313*).

The genus *Boletus* contains a relatively large number of species and on the basis of our screening many species still remain to be explored for non-protein amino acids. MATZINGER *et al.* (*283*) reported isolation of (2S,4S)- and (2S,4R)-4-hydroxynorvaline (**13**) from fruitbodies of *Boletus satanus* Lenz. Successive use of Dowex 50-, Dowex 1- and cellulose-column chromatography followed by final purification with preparative paper chromatography gave surprisingly high yields: 305 mg of the (2S,4S)-epimer and 340 mg of the (2S,4R)-epimer, respectively, from 426 g of fresh material. Comprehensive ^{1}H-NMR- and CD-spectra of the acid and lactone forms were presented and careful searches for the possible coexistence of the enantiomer in each isolate and for the occurrence of epimerization were carried out. They concluded that their "(2S,4S)-isolate" was a partial racemate of (2S,4S): (2S,4R) in the ratio of 3:2, and that the optical purity of the "(2S,4R)-isolate" was 88%. They confirmed that a change in configuration at C-2 and/or C-4 could not have occurred during the isolation procedure. According to their results, the fast moving diastereomer was (4S) (*erythro*-isomer), and they corrected the tentative assignment made by FOWDEN for the absolute configuration of the 4-hydroxynorvaline isolated from sweet pea. 5-Hydroxynorvaline is known from seeds of *Crotalaria juncea* (Leguminosae) (*337, 350*).

1.2.4. 2-Amino-7-hydroxyoctanoic Acid (**14**)

A hydroxyamino acid with an extraordinary long and straight carbon chain, 2-amino-7-hydroxyoctanoic acid, was reported by AOYAGI *et al.* (*9*) from fruitbodies of *Russula cyanoxantha* (Schw.) Fr. After gradient

(14)

(2S)-2-Amino-7-hydroxyoctanoic acid

elution by pyridine-acetic acid buffer from a column of Dowex 50 (pyridinium-form) and cellulose column chromatography, they obtained 220 mg of pure crystals from 190 g of dried mushroom. Elementary analysis and SIMS showed that the molecular formula was $C_8H_{17}NO_3$, and the formation of a copper complex indicated an α-amino acid. The carbon skeleton was deduced from the ^{1}H- and ^{13}C-NMR spectra and the position of the hydroxyl was determined on the basis of the formation of 2-amino-7-oxooctanoic acid on oxidation with acidic potassium permanganate. Positive shifts of the values of the optical rotation in solutions with stronger acidities suggested that the natural amino acid belonged to the L-series. The configuration at C-7 is unknown. The likely precursor, 2-aminooctanoic acid, has not yet been reported as a natural product.

1.3. Unsaturated Leucine and Isoleucine Homologues

Unsaturated isoleucines and structurally similar amino acids composed of seven carbon atoms, are known from taxonomically distant higher fungi.

1.3.1. 2-Amino-5-methyl-4-hexenoic Acid (15)

An unsaturated higher homologue of leucine, 2-amino-5-methyl-4-hexenoic acid, was reported by DARDENNE et al. (70) from fruitbodies of *Leucocortinarius bulbiger* (Alb. et Schwein) Singer. Column chromatography with Amberlite CG-120 and Dowex 1 afforded 388 mg of pure crystals from 1.1 kg of mushrooms. This amino acid gave a yellow ninhydrin color. On catalytic hydrogenation it was converted to the known 2-amino-5-methylhexanoic acid, and it was oxidized to aspartic acid and acetone by potassium permanganate. Positive shift of the optical rotation on acidification of the solution suggested the (2S) configuration. The racemate of this amino acid was synthesized by condensation of 1-bromo-3-methyl-2-butene and diethyl dimethylallyl acetamidocyano-acetate followed by hydrolysis with NaOH. The synthetic amino acid was

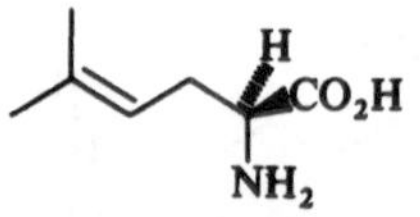

(15)

(2S)-2-Amino-5-methyl-4-hexenoic acid

similar in every respect to the natural one. The IR spectrum of a racemate prepared from the natural form was identical with that of the synthetic sample. Racemization was achieved in 10 N NaOH by heating at 195° for 26 hr.

1.3.2. 2-Amino-4-methyl-5-hexenoic Acid (16)

A similar unsaturated amino acid with 7 carbon atoms and a methyl group at C-4 was reported from the allegedly hallucinogenic fruitbodies of a *Boletus* sp. (section *Ixocomus* group Nudi); the substance constituted 0.04% of the dry weight (*136*). The structure was verified by elementary analysis, chelate formation with Cu^{2+}, catalytic hydrogenation, periodate-$KMnO_4$ oxidation yielding formaldehyde, ^1H-NMR spectrum and mass spectrometry. The configuration at C-2 was assumed to be (2S) from the optical rotation at different acidities. This was consistent with the order of retention behavior of diastereomeric trifluoroacetyl-L-prolyl-DL-amino acid esters (*366*). The absolute configuration at C-4 was later determined by chemical synthesis (*137*). (2S,4S)-2-Amino-4-methylhexanoic acid was prepared from 2(S)-methylbutan-1-ol and acetylaminomalonate, followed by optical resolution of the acetyl derivative by hog kidney acylase. The synthetic (2S,4S)-form was identical in every respect with the dihydroamino acid obtained from the natural amino acid by catalytic hydrogenation.

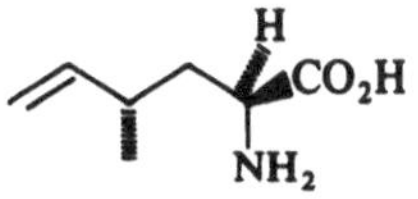

(16)

(2S, 4S)-2-Amino-4-methyl-5-hexenoic acid

The fruitbodies of the *Boletus* sp. mentioned above were collected in New Guinea. They differed from other hallucinogenic mushrooms in causing multiple or inverted vision about one hour after ingestion. The symptoms last for several hours and full mental control is not recovered for several days (*136*). Pharmacological tests for the detection of LSD_{25}, mescaline, or psilocybin type were all negative. Any biological activity at the level of metabolism has not been reported for this amino acid.

A more strongly unsaturated form, 2-amino-4-methyl-5-hexynoic acid, and its 4-hydroxylated derivative, 2-amino-4-hydroxymethyl-5-hexynoic acid, together with a straight chained 2-amino-4-hydroxy-6-heptynoic acid, are also known from seeds of *Euphoria longan* (*403*).

An isomeric amino acid with respect to the position of unsaturation, 2-amino-4-methyl-4-hexenoic acid, its saturated form, 2-amino-4-methylhexanoic acid (homoisoleucine), and a small amount each of 2-amino-4-methyl-6-hydroxy-4-hexenoic acid and 3-(methylenecyclopropyl)-3-methylalanine were reported from seeds of *Aesculus californica* (Hyppocastanaceae) by FOWDEN and his associates (*117, 289, 114*). Subsequently, they also characterized *cis*-2-(carboxycyclopropyl)glycine and exo(*cis*)-3,4-methanoproline from akee (*Blighia sapida*) (*118*). The hypoglycemic activity of unripe akee is well known to be caused by the related substance hypoglycin A, 3-(methylenecyclopropyl)alanine, and B, the γ-glutamyl derivative of hypoglycin A. A few years later, they reported isolation and characterization of 2-amino-5-methyl-6-hydroxy-4-hexenoic acid from *Blighia unijugata* (Sapindaceae) (*113*).

Considering the distribution pattern of these characteristic amino acids and their possible biosynthetic relationships, FOWDEN and SMITH pointed out the phytochemical similarity of the genera *Aesculus* and *Blighia* (*113*).

Feeding experiments with ^{14}C-precursors were undertaken (*114*), with isoleucine being the most effective precursor. The biosynthesis of the unsaturated C_7 amino acids or their possible interconversions, however, are still not clear.

The specific amino acids occurring in these plants have not yet been reported from higher fungi. However, considering the limited number of fungal species surveyed and the known occurrence of related amino acids in fungi, it is very likely that these amino acids will be found also in fungi.

1.3.3. 3-Methylenenorvaline (2-Amino-3-methylenepentanoic Acid) (17)

Fresh fruitbodies of *Lactarius helvus* have a mild odor similar to that of *Lactarius camphoratus* mentioned previously, but when dried, the odor is much stronger. They are called "Maggipilz" in Germany and a small

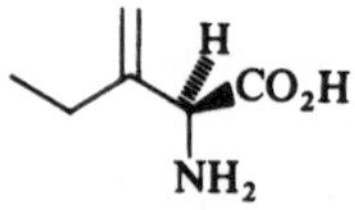

(17)

(2S)-3-Methylenenorvaline

quantity of dried powder is used as a spice. A larger amount, however, can cause nausea, vomiting or dizziness (286).

Two dimensional paper chromatography revealed several unidentified ninhydrin-positive substances in L. helvus (49). In 1968 LEVENBERG (264) reported isolation and characterization of a new amino acid, 2-amino-3-methylenepentanoic acid. By conventional use of Amberlite IR-120, followed by preparative paper chromatography, he obtained 52 mg of pure crystals from 8 pounds of mushrooms. The configuration at C-2 was determined by oxidation with L- and D-amino acid oxidase. Cleavage by periodate-$KMnO_4$ reagent yielded formaldehyde, propionic acid and ammonia. Catalytic hydrogenation on Adams catalyst gave an equimolecular mixture of isoleucine and alloisoleucine. The racemate of this amino acid was prepared by a Strecker synthesis procedure.

The specific rotation, $[\alpha]_D = +197°$ in water, is exceptionally high compared with the value given for the majority of amino acids (155) and it will be discussed in the next section in relation to the chemical structure. A slight but significant shift of the optical rotation in more acidic solution as well as the results of oxidation with L-amino acid oxidase suggested that the natural form belonged to the L-series of amino acids.

Several unidentified amino acids have been detected in fruitbodies of Lactarius subzonarius Hongo, one of which was isolated in our laboratory (180). The initial yellow coloration with ninhydrin changed to yellowish brown, grayish blue and finally reddish violet. This color change was nearly the same as reported for 2-amino-3-methylenepentanoic acid by LEVENBERG. The identification of our isolate was made by analysis of hydrogenation products on Adams catalyst, the ^{1}H-NMR spectrum and conclusively by direct comparison of the IR spectrum with that of the amino acid of Lactarius helvus. An odoriferous substance similar to the 3-hydroxy-4,5-dimethyl-2(5H)-furanone produced from 4-hydroxyisoleucine seems to be formed from this amino acid. Fruitbodies of Lactarius helvus, L. subzonarius, and the isolated 2-amino-3-methylenepentanoic acid all have similar curry-like odors, but there are slight differences in the odors among the three. Lactarius subzonarius is neither edible nor useful as a spice. In addition to the three species mentioned above, a few other Lactarius species have similar odors, i.e., L. cimicarius (Secr.) Gill. and L. serifluus (Fr.) Fr. (339).

γ-Glutamyl derivatives of (2S)-2-amino-3-methylenepentanoic acid and its more strongly unsaturated form, 2-amino-3-methylenepentenoic acid were found by CAMPOS et al. in the plant Phyladelphus coronarius (Saxifragaceae) (47). Presence of the last-named amino acid in the free state was also demonstrated in this plant (47).

1.3.4. 2-Amino-3-hydroxymethyl-3-pentenoic Acid (18) and 2-Amino-3-formyl-3-pentenoic Acid (19)

DOYLE and LEVENBERG (84) reported two structurally closely related amino acids from fruitbodies of *Bankera fuligineoalba*. In a manner similar to that used for isolation of 2-amino-3-methylenepentanoic acid, they obtained 85 mg of 2-amino-3-hydroxymethyl-3-pentenoic acid from 4 lb of materials. The 3-formyl compound was unstable and could not be crystallized. The ninhydrin reaction color of the first amino acid was initially pale yellow, changing after a short time to olive-green and finally to blue. The second amino acid developed first a light yellow-brown that changed to stable sepia in a very short time. On catalytic hydrogenation using a relatively large amount of Adams catalyst, the first amino acid gave isoleucine and alloisoleucine in equal amounts, while with a small amount of the catalyst 2-amino-3-methyl-3-pentenoic acid was detected among the reduction products. Oxidation with periodate-$KMnO_4$ reagent produced formaldehyde and acetaldehyde. The ^{1}H-NMR spectrum confirmed the proposed structure while the magnitude of long-range coupling suggested a *cis* relationship between the methyl and hydroxymethyl group about the double bond. The formylamino acid showed a UV absorption curve similar to that of tiglaldehyde with a maximum at ca. 223 nm. With alkaline borohydride it was reduced stoichiometrically to the first amino acid, 2-amino-3-hydroxymethyl-3-pentenoic acid. The first amino acid exhibited an extremely high specific optical rotation, $[\alpha]_D = +182$ in water, while, as stated before, the value for 2-amino-3-methylenepentanoic acid was $+197°$. These high values are probably

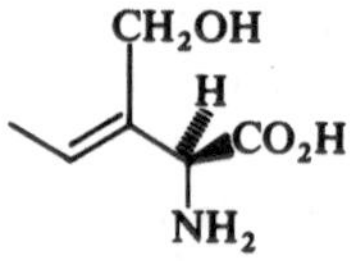

(18)

(2*S*, *Z*)-2-Amino-3-hydroxymethyl-3-pentenoic acid

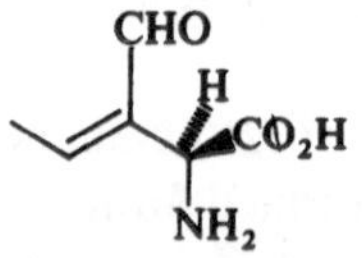

(19)

(2*S*, *Z*)-2-Amino-3-formyl-3-pentenoic acid

caused by 3,4-unsaturation. Further examples are (2S)-2-amino-3-methylenehexanoic acid of the next section, $[\alpha]_D = +149°$ (437), and (2S,Z)-2-amino-3-methyl-3-pentenoic acid, $+251°$, (171).

1.3.5. 2-Amino-3-methylenehexanoic Acid (20)

A higher homologue of 3-methylenenorvaline, 2-amino-3-methylenehexanoic acid, was reported by VERNIER and CASIMIR (437) from fruitbodies of *Amanita vaginata* var. *fulva*. This amino acid yielded a yellow spot on paper with ninhydrin and this changed to violet with time. For the separation, ion exchange chromatography with Dowex-1 and -50 was used. Various chemical degradations, elementary analysis, optical rotation, ^{1}H-NMR spectrum, and catalytic hydrogenation showed the structure to be (2S)-2-amino-3-methylenehexanoic acid.

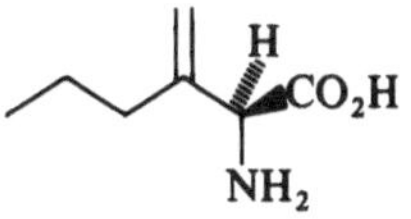

(20)

(2S)-2-Amino-3-methylenehexanoic acid

A number of unsaturated analogues of leucine, isoleucine, and alloisoleucine have been synthesized and were found to be effective competitive inhibitors of natural amino acids [e.g. (338)].

1.4. Unsaturated Norvalines and Norleucines, and Related Amino Acids

It is rather peculiar that three neutral aliphatic amino acids in proteins, i.e. valine, leucine, and isoleucine, all contain branched chain carbon skeletons. There should be some evolutionary reason(s) for the absence from proteins of neutral amino acids with straight chain carbon skeletons such as 2-amino-n-butanoic acid, norvaline and norleucine. Occurrence of these amino acids, bound or free, in biological materials seems to be extremely restricted. We recall here the famous and classic simulation experiments by MILLER and his associates on the prebiotic formation of amino acids by electric discharge in a gaseous mixture of ammonia, methane and hydrogen (287, 359). Alanine, glycine and 2-amino-n-butanoic acid were the three major amino acids formed, while the concentration of norvaline produced as a minor product was still over three times as high as that of valine. The amounts of norleucine and alloisoleucine formed were less than that of leucine but comparable to that of isoleucine. Furthermore, it has been demonstrated that all of the non-protein amino acids found in the Murchison meteorite can be produced by a similar electric discharge (467).

It is therefore very likely that 2-aminobutanoic acid, norvaline, and norleucine contributed to the formation of the prebiotic proteins or proteinoids. However, these amino acids probably were somewhat disadvantageous as protein constituents. First, the electronic properties of the straight carbon chain might not bring about the hydrophobicity necessary to satisfy the biological functions of proteins. As is well known, inter- and intramolecular properties and hence the biological functions of proteins are based to a great extent on the hydrophobicity of their amino acid residues. Secondly, the capability of straight chain amino acids to recognize another molecule or molecules is probably smaller than that of amino acids with side chains of more stereospecific structure. Molecular recognition is also one of the most important properties of protein molecules. For these reasons, it is very likely that the neutral straight chain amino acids which have no terminal functional group, such as an amino or carboxyl group, have been excluded as protein components in the course of chemical or molecular evolution. Information on this matter can surely be obtained in more detail by biophysical and organic chemical approaches.

In connection with the above facts, it is highly interesting that the two fungal genera *Amanita* and *Tricholomopsis* contain species which produce various amino acids having straight carbon skeletons, and further, carbon-carbon double bond(s) or, more rarely, a triple bond. It is also interesting that an unsaturated norvaline and several norleucines found in *Amanita*-species contain a chlorine atom.

As far as we know, most of the unsaturated amino acids with straight carbon chains are growth inhibitors of bacteria or fungi. If some fungal species have developed the capability to modify the straight chain amino acids and to produce the toxic amino acids, those species can possibly utilize them effectively as a deterrent to herbivores.

Depending on the chemical structures, most unsaturated amino acids give various color reactions with ninhydrin, including light yellow, dark yellow, khaki yellow, yellowish brown, brick brown or brown. A unique shade of colors is characteristic for each unsaturated amino acid while an amino acid with a straight carbon chain and a terminal triple bond, 2-amino-5-hexynoic acid, yields the usual violet color. As in the case of many hydroxyamino acids, these yellow or brown colors generally change to the usual violet with time, but some do not, and these are also intrinsic to each amino acid. Color reactions with ninhydrin of additional unsaturated acetylenic amino acids with seven carbon atoms have been reported (405). These properties may constitute a valuable tool for the identification of the unsaturated amino acids.

1.4.1. 2-Amino-3-butenoic Acid (Vinylglycine) (21)

The smallest straight chain neutral amino acid with a double bond, 2-amino-3-butenoic acid, was isolated by DARDENNE et al. (71) from fruitbodies of *Rhodophyllus nidorosus* (Fr.) Quél. As with other unsaturated amino acids, it gave a yellow spot on paper which on further heating changed to brown and finally to the usual violet. Isolation and purification were carried out with Amberlite CG 120 (H⁺-form) and Dowex 50 (chloropyridinium-form), successively. Formation of a copper complex on paper (256), empirical formula, and production of 2-amino-4-butanoic acid on catalytic hydrogenation indicated immediately the structure which was supported also by comparing the IR and ¹H-NMR spectra with those of a synthetic sample.

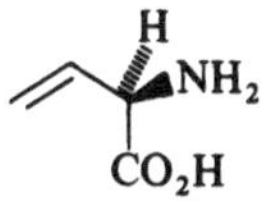

(21)

(2R)-2-Amino-3-butenoic acid

As regards the optical properties of the natural amino acid, DARDENNE et al. reported an important and instructive experiment by repeatedly recrystallizing the sample from ethanol-water and determining the optical rotation of the crystals and residues with each successive treatment. For example, the optical rotation of the first purified sample was $[\alpha]_D = -6°$, and that of the crystals obtained from the first filtrate of the crystallization, $-22°$. From the residues of the above recrystallization, the crystals were again separated by ethanol-water treatment. The specific optical rotations of the crystals and residues were $-58°$ and $-55°$, respectively. In parallel, racemic 2-amino-3-butenoic acid was synthesized, resolved, and the specific optical rotation of the D-isomer was determined to be $-94°$ (123). On this basis, DARDENNE et al. assumed that the natural form consisted of 55% of the D- and 45% of the L-form; on recrystallization, the D-amino acid separated more readily in crystalline form than the L-antipode. The last crystalline sample, however, would still have contained 20% of the L-form.

In most cases, optical activity is determined for the most "pure isolate". Therefore, there is a danger that interesting optical properties of the natural amino acids may sometimes be overlooked. Because the natural amino acids are often isolated only in very small amounts and it is not always possible to determine the optical rotations many times, it is recommended to use also the appropriate L-amino acid oxidase (304).

Fruitbodies of *Rhodophyllus nidorosus* may cause light gastrointestinal damage. This might be related to the existence of the above amino acid, but its content in this mushroom was not reported. During recent studies on antibacterial substances from mushrooms, MATSUMOTO (281) isolated 2-amino-3-butenoic acid as an active substance from the edible mushroom, *Rhodophyllus crassipes*. The optical rotation indicated that this amino acid occurs in partially racemized form also in this fungus. 2-Amino-3-butenoic acid inhibits the growth of *Bacillus subtilis*, *Escherichia coli*, *Shigella frexneri*, and *Citrobacter freundii* at a concentration of 10 µg/ml.

The saturated form, 2-aminobutanoic acid was reported as a component of an antibiotic, staphylomycin (435). On screening with two dimensional paper chromatography for free amino acids of fungi, we have rarely detected spots corresponding to 2-amino-3-butenoic acid.

1.4.2. Allylglycine (2-Amino-4-pentenoic) (22) and Propargylglycine (2-Amino-4-pentynoic Acid) (23)

These two amino acids were isolated and identified in our laboratory from fruitbodies of *Amanita pseudoporphyria* Hongo, together with six and seven carbon homologues (173). The amino acid-fraction prepared

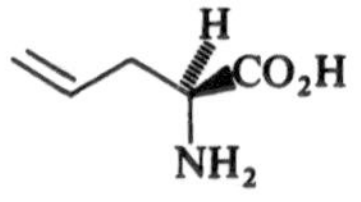

(22)

(2S)-2-Amino-4-pentenoic acid

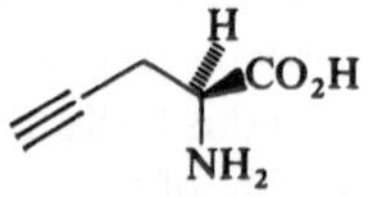

(23)

(2S)-2-Amino-4-pentynoic acid

from 6.18 kg of fruitbodies was fractionated repeatedly on cellulose columns using different solvent systems. We obtained 58 mg of allylglycine and 212 mg of propargylglycine. The results of elementary analyses were in good agreement with the formulae and their NMR spectra were also satisfactory.

The ninhydrin coloration of allylglycine was brown. On hydrogenation with Adams catalyst it was converted to norvaline. The observed specific rotation, $[\alpha]_D = -33.7°$, indicated that the isolate was pure L-enantiomer since the value for the optically resolved L-antipode of the synthetic sample was reported to be $[\alpha]_D = -37.1°$ (33). This was the first report on its natural occurrence.

Propargylglycine also gave a brownish ninhydrin coloration. It yielded norvaline over Adams catalyst while over Lindlar catalyst (270) it formed 2-amino-4-pentenoic acid, as expected. The IR spectrum showed the characteristic absorption at ca. 3250 cm^{-1} caused by the terminal acetylene group. In all respects, the properties of the isolate were identical with those of a synthetic sample. The optical purity also was as high as that of allylglycine since that was no significant difference between our observed value of $-31.6°$ and that of authentic material $-32.6°$ (382). SCANNELL and coworkers had already reported isolation and characterization of propargylglycine as an antimetabolite produced by a strain of Streptomyces (382).

These two amino acids have long been known as synthetic antibiotics. Thus, allylglycine inhibits the growth of Echerichia coli and Saccharomyces cerevisiae (78). Because of its potent inhibitory effect on glutamic acid decarboxylase which yields γ-aminobutanoic acid, allylglycine disturbs the GABA-mediated system in brain (5). Many investigations on the antimetabolic properties of propargylglycine have been reported. Similar to allylglycine, it inhibits the growth of E. coli and S. cerevisiae (139). There are also reports on the inhibition of amylase synthesis (221). Inactivations of vitamin B_6-dependent enzymes, e.g. crystathionine γ-lyase (1), cystathionine γ-synthase (421, 219), methionine γ-lyase (220), and aminotransferase (43) are also known.

Recently allyl- and propargylglycine were isolated together with 2-amino-4,5-hexadienoic acid [1.4.9.] and 2-amino-5-chloro-6-hydroxy-4-hexenoic acid [1.5.3.] from fruitbodies of Amanita abrupta Peck (470, 329). Because of a fatality in Japan clearly caused by the consumption of Amanita abrupta Peck, toxicological studies on the chemical constituents of this species were carried out by YAMAURA et al. (470). An aqueous extract was injected interperitoneally into mice and caused an acute decrease of serum glucose and liver glycogen after 6 hr. Activity of serum glutamic oxaloacetic transaminase and glutamic pyruvic transaminase increased markedly. Histological studies indicated remarkable liver cell necrosis and disappearance of glycogen granules in the liver. The same authors found the biochemical effects of (2S)-2-amino-4-pentynoic acid to be similar to those of the fruitbody extract (470). The natural occurrence of the saturated form of these two amino acids, norvaline (2-aminopentanoic acid), free or bound, is so far unknown. Careful scrutiny of its possible occurrence in *A. pseudoporphyria* gave inconclusive results. Although a tiny amount of norvaline was possibly detected, isolation was unsuccessful.

It would be interesting and important to determine whether these two antimetabolites play any chemo-ecological roles in natural habitats. There must also be some protective mechanism against the toxicities in the fungi producing them.

As stated before, non-protein amino acids such as 2-amino-4-chloro-4-pentenoic acid, inactivate B_6-enzymes. However, another mechanism, and probably a much more frequent one, is the inhibition of protein synthesis.

The prolyl-tRNA synthetase isolated from *Convallaria majalis*, which produces a large amount of azetidine-2-carboxylic acid, was shown to be unable to activate this amino acid. In contrast, the same enzyme from *Phaseolus aureus* was able to catalyze the formation of the corresponding aminoacyl-tRNA (340). That is, proline and azetidine-2-carboxylic acid were discriminated against in protein synthesis by *Convallaria majalis*, but not by *Phaseolus aureus*.

This line of study has been extended to a considerable number of non-protein amino acids in FOWDEN's laboratory (111). Interestingly, all non-protein amino acids which are coplanar at C-4 mimic phenylalanine as substrates of phenylalanine-tRNA synthetase of *Aesculus hippocastanum*. As one example, 2-amino-4,5-hexadienoic acid at 5 mM gave 23% of V_{max} retained by phenylalanine (7).

1.4.3. 2-Amino-3-hydroxy-4-pentynoic Acid (24)

Sclerotium rolfsii is a sclerotial Basidiomycotina anamorph and parasitic on various species of plants. Cultures of this fungus are toxic to

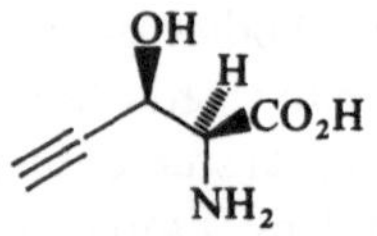

(24)

(2S, 3R)-2-Amino-3-hydroxy-4-pentynoic acid

poultry and domestic mammals, causing such acute symptoms as anorexia, muscular weakness, inordination of movements, paralysis and coma.

POTGIETER et al. (352) reported a toxic amino acid, (2S,3R)-2-amino-3-hydroxy-4-pentynoic acid, from sclerotia of this fungus. The amino acid which proved to be lethal to New Hampshire chickens (LD_{50}, 150 mg/kg) was very unstable in neutral and alkaline solution, but stable in acid. The color reaction with ninhydrin was bright yellow to brown, a reaction that was completely inhibited in the presence of Cu^{2+}, thus indicating that it was an α-amino acid. After passing the aqueous extract through a Dowex 50 column only 20% (250 mg) of the amount calculated from the toxicity were obtained from 500 g of dried sclerotia probably owing to instability of the substrate. Elementary analysis and mass spectrum were in good agreement with formula $C_5H_7NO_3$. This amino acid showed strong and sharp absorption IR bands at 3264 cm^{-1} and 2134 cm^{-1}, suggesting the presence of a terminal acetylenic group. The chemical structure, (2S,3R)-3-hydroxypropargylglycine, was clarified by analysis of the ^{1}H-NMR spectrum and was confirmed by hydrogenation. The sample yielded a mixture of (2S,3R)- and (2R,3S)-2-amino-3-hydroxypentanoic acid on hydrogenation over Pd/charcoal. The natural form however was considered to be (2S,3R) after oxidative deamination with D- and L-amino acid oxidase.

2-Amino-3-hydroxy-4-pentynoic acid was synthesized as a diastereomeric mixture from copper glycine and propiolaldehyde. The absolute configuration of the natural form was later confirmed by X-ray crystallography (59).

1.4.4. 2-Amino-5-hexenoic Acid (25)

Fruitbodies of species in the subgenus Lepidella of Amanita have characteristic free amino acid patterns that depend on the individual species. In particular, many interesting non-protein straight chain, unsaturated C-5 to C-7 amino acids, some containing chlorine, are known from the group. The subgenus Lepidella is a large group and a large number of species have been described (21). From the chemotaxonomical

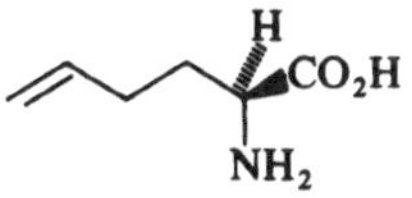

(25)

(2S)-2-Amino-5-hexenoic acid

point of view, it should be interesting to accumulate unequivocal data on specific non-protein amino acids of as many species as possible.

Recently, we had an opportunity to analyze free amino acids of the fruitbodies of *Amanita gymnopus* Corner & Bas. Although little material was available, four characteristic amino acids, 2-amino-5-hexenoic acid, 2-amino-5-hexynoic acid [1.4.6], 2-amino-4,5-hexadienoic acid [1.4.9] and a new chlorine-containing amino acid [1.5.5] were separated in relatively pure states (*179*). A mixture of neutral amino acids obtained from 78 g of six fresh fruitbodies was fractionated with ion pair reverse phase chromatography using LiChroprep RP-8. The chromatographic behavior of the first three amino acids on cellulose thin layers developed separately with four different solvent systems and the ^{1}H-NMR spectra were compared with those of the respective synthetic or authentic natural preparations. Configurations at C-2 of all these amino acids were determined to be (2S) by L-amino acid oxidase prepared from venom of Habu-snake (*312*). The natural occurrence of 2-amino-5-hexenoic acid is not known other than in *A. gymnopus*.

On the basis of the antagonistic effect of the (*S,E*)-isomer of 2-amino-4-hexenoic acid on the metabolism of methionine in *Escherichia coli*, the conformation of methionine on the site of its utilization was discussed (*393*). The terminal group and the β-methylene group of methionine probably assume a *cis*-like configuration similar to *cis* rather than *trans*-crotylglycine. The (*S,E*)-form of 2-amino-4-hexenoic acid was also characterized as a component of an antibiotic ilamycin produced by *Streptomyces islandicus* (*417, 418*).

1.4.5. 2-Amino-4-hexynoic Acid (26)

This amino acid was isolated in our laboratory first from fruitbodies of *Tricholomopsis rutilans* (Schaeff.: Fr.) Sing. (*174*). After removal of the basic and acidic amino acids with Dowex 1, fractionation was carried out on a cellulose column to give 19 mg of pure crystals from 3 kg of mushrooms. Chelation with Cu^{2+} and elementary analysis suggested that it was an unsaturated leucine or an isomer. Because the product of hydrogenation with Adams catalyst was norleucine and that of oxidation

(26)

(2S)-2-Amino-4-hexynoic acid

with potassium permanganate was aspartic acid, the isolated amino acid should be 2-amino-4-hexynoic acid or 2-amino-4,5-hexadienoic acid. The ^{1}H-NMR spectrum and hydrogenation over Lindlar catalyst (*270*) supported the former possibility. An acetylenic bond which is not terminal usually cannot be detected by any characteristic absorption bands in the IR spectrum. Racemic 2-amino-4-hexynoic acid was synthesized by condensation of diethyl formylaminomalonate (*131*) and 1-bromo-2-butyne (*62*) in the presence of sodium ethoxide. The product, diethyl 1-formylamino-3-pentyne-1,1-dicarboxylate was hydrolyzed successively with alkali and acid according to BLACK and LANDOR (*32*). The synthetic sample was in all respects identical with the natural amino acid. Optical rotation measurements indicated that the natural form probably belonged to the L-series. Ninhydrin reagent yielded a brick-brown coloration on paper. Subsequently, 2-amino-4-hexynoic acid was demonstrated to be present also in the fruitbodies of *Amanita pseudoporphyria* (*173*) and *A. miculifera* Bas & Hatanaka (*181*). The yield was 267 mg from 6.18 kg fruitbodies of the former and 50 mg from 580 g fruitbodies of the latter. *Amanita miculifera* Bas & Hatanaka was described during the studies of *Amanita*-amino acids as a new species (*22*). Identification of the amino acid was made by direct comparison of the various chemical and spectroscopic properties with those of the authentic preparation.

In view of the differing values of the specific optical rotation in water of 2-amino-4-hexynoic acid isolated from *T. rutilans, A. pseudoporphyria* and *A. miculifera*, $[\alpha]_D = -54.4°$ (c, 1), $-39.3°$ (c, 1) and $-16.8°$ (c, 1.4), respectively, it is most likely that the optical purity of this amino acid depends on the source. In this connection, it should be pointed out that a closely related 2-amino-4,5-hexadienoic acid is present probably as a racemate in *A. miculifera* (*vide infra*).

2-Amino-4-hexynoic acid is known to have inhibitory and regulatory effects as a structural analogue of L-methionine on ATP: L-methionine S-adenosyltransferase prepared from several organisms (*63*).

The distribution of this acetylenic amino acid and other related compounds will be described in [1.4.7.].

1.4.6. 2-Amino-5-hexynoic Acid (**27**)

In 1985, AOYAGI and SUGAHARA (*14*) reported isolation and identification of this amino acid from fruitbodies of *Cortinarius claricolor* var. *tenuipes* Hongo as a strong growth inhibitor against *Bacillus subtilis* B-50. After 169 g of the freeze-dried samples were defatted, their amino acids were extracted. Fractionation and purification of the amino acids were

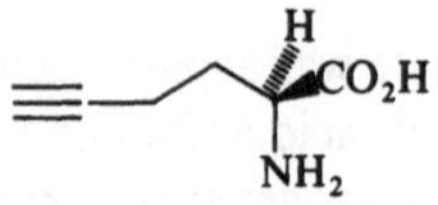

(27)

(2S)-2-Amino-5-hexynoic acid

carried out with Amberlite IR-120, activated charcoal, Dowex 1 (acetate-form) and Dowex 50 (pyridinium-form), to give 920 mg of pure crystals. Elementary analysis and FD mass spectroscopy showed the molecular formula as $C_6H_9NO_2$. Chelate formation with Cu^{+2} indicated an α-amino acid. On catalytic hydrogenation over Adams catalyst it consumed 2 moles of hydrogen forming norleucine as the sole ninhydrin positive product. Because oxidation with acidic $KMnO_4$ gave glutamic acid, the position of unsaturation should be C-5. The IR spectrum had bands at 3300 cm^{-1} and 2120 cm^{-1} characteristic of a terminal acetylene. According to the Clough-Lutz-Jirgensons rule, the configuration at C-2 was determined to be (2S). This acetylenic amino acid inhibited the growth of *Bacillus subtilis* B-50 in minimal medium at 0.1 μg/ml. This is the first report of the natural occurrence of 2-amino-5-hexynoic acid.

As stated in the previous section, the presence of this amino acid was verified more recently in fruitbodies of *Amanita gymnopus*. Also, in this fungus, the concentration was very high and we could obtain 10 mg of crystals from 78 g of fresh fruitbodies. It is worth noting that the color with ninhydrin is the usual violet, and 2-amino-5-hexenoic acid gives also a violet spot on paper under the same conditions.

1.4.7. 2-Amino-3-hydroxy-4-hexynoic Acid (28)

Shortly after the isolation and characterization of 2-amino-4-hexynoic acid from fruitbodies of *Tricholomopsis rutilans*, the coexistence of its 3-hydroxy form was reported (*175, 314*). The ^{1}H-NMR spectrum strongly suggested that the isolate was a mixture of two diastereomers.

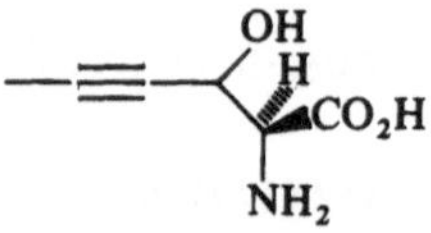

(28)

(2S, 3R)-, (2S, 3S)-2-Amino-3-hydroxy-4-hexynoic
acid

The two migrate at different speeds in cellulose columns developed with n-butanol-methylethyl ketone-ammonia-water (*198*) and were separated from each other in this way. This solvent has been used to separate the *threo-* and *erythro-*forms of several 2-amino-3-hydroxyacids (*198, 292, 135*). Results of elementary analysis and optical rotation measurements indicated that both amino acids had the elementary composition $C_6H_9NO_3$ with the (2S) configuration. Their chromatographic behavior suggested that the fast-moving was the *threo*, and the slow-moving the *erythro* isomer. Rf-values of the *threo* isomer of some known 2-amino-3-hydroxyacids using the above solvent were reported to be larger than those of corresponding the *erythro* isomer (*198, 292, 135*) and the order of Rf-values are reversed in the solvent, isopropanol-acetic acid-water (*86*). On the basis of the Karplus relationship in ^{1}H-NMR spectroscopy, a larger coupling constant $J_{2,3}$ is expected in the *erythro* than in the *threo* isomer; thus the value 3.5 Hz was obtained for the fast-moving and 4 Hz for the slow-moving amino acid. The presence of a triple bond was indicated by a weak absorption band at 2230 cm^{-1} and by catalytic hydrogenation with Lindlar catalyst.

Chemical synthesis of 2-amino-3-hydroxy-4-hexynoic acid was carried out by condensation of tetrolaldehyde and copper glycine, the product being separated into a fast-moving and slow-moving amino acid as for the natural products. MIX (*292*) had already reported the synthesis of the mixture of four isomers of 3-hydroxynorleucine, separation of the diastereomers and their stereochemical characterization. Comparison of the separated and hydrogenated products of the natural amino acids with those of the known synthetic samples was therefore possible and led conclusively to the structures. The naturally occurring fast-moving amino acid was (2S,3R)-2-amino-3-hydroxy-4-hexynoic acid and the slow-moving one was the (2S,3S)-form.

According to the taxonomic system adopted by SINGER (*391*), the genus *Tricholomopsis* is composed of two sections, *Tricholomopsis* and *Platyphyllae* (Table 1). The type species of the latter is *T. platyphylla* (Pers.:Fr.) Sing. The results of our paper chromatographic survey indicated clearly that 2-amino-4-hexynoic acid and/or its 3-hydroxylated form occur in all four species tested from the section *Tricholomopsis*.

Table 1. *Sections of the genus* Tricholomopsis (from *391*)

Tricholomopsis Sing.
    Sect. 1. *Tricholomopsis* (*Rutilantes* Sing.)
        2. *Platyphyllae* Sing.

Table 2. *Distribution of Specific Non-Protein Amino Acids in the Genus* Tricholomopsis

	2-Amino-4-hexynoic acid	2-Amino-3-hydroxy-4-hexynoic amino acid	γ-Glutamyl acetylenic acid	3-(3-Carboxy-4-furyl)alanine	Pipecolic acid
T. rutilans	+	+	+	+	−
T. sasae	+	+	+	−	−
T. decora	−	+	+	+	−
T. bambusina	+	−	−	+	−
T. platyphylla (Oudemansiella platyphylla, see Text)	−	−	−	−	+

However, *T. platyphylla* contained none of these specific non-protein amino acids (Table 2) (*317*). This species is now considered to be less related to other members of *Tricholomopsis* and the name was changed to *Oudemansiella platyphylla* (Pers.:Fr.) Mos. (*302*).

Concentrations of the acetylenic amino acids in the section *Tricholomopsis* might vary with localities to a great extent. Thus, CHILTON and OTT (*53*) failed to detect any of the unsaturated norleucines in the fruitbodies of *T. rutilans* and *T. decora* collected in Washington.

Several feeding experiments with labeled compounds were carried out on the formation of 2-amino-4-hexynoic acid using fruitbodies of *T. rutilans* (*315*). Although the occurrence of free norleucine could not be detected in the fruitbodies, [2-^{14}C]acetate injected into a fruitbody was incorporated into norleucine and, when cold norleucine was given at the same time, the incorporation markedly decreased. Rapid metabolism of norleucine was observed in the experiments in which radioactive norleucine was given. Radioactivity was incorporated first into 2-amino-4-hexenoic acid and then into 2-amino-4-hexynoic acid, suggesting strongly a stepwise dehydrogenation. Whether the former intermediate had *cis*- or *trans*-configuration could not be determined with certainty, but it behaved on paper chromatography just like the *cis*-form. The formation of 2-amino-4-hexynoic acid could be observed 48 hr after the injection. The process might be analogous to the desaturation mechanism in the case of fatty acids (*301*).

References, pp. 117–140

1.4.8. 2-Amino-6-hydroxy-4-hexynoic Acid (29)

The source of this amino acid was *Amanita miculifera* described previously (*181*). The yield was relatively high, 110 mg being obtained from 580 g of fresh fruitbodies. Elementary analysis indicated the formula $C_6H_9NO_3$. Its ninhydrin coloration was brown. The oxidation product with $KMnO_4$ was aspartic acid. Over Adams catalyst it was catalytically reduced with consumption of two mol/eqvts of hydrogen and the product was further reduced to norleucine with hydriodic acid and red phosphorus. Additionally, it absorbed one mole hydrogen over Lindlar catalyst, indicating the presence of one triple bond in the molecule. The structure was assumed from the MS and ^{1}H-NMR spectrum along with the above experimental results.

(29)

(2*R*)-2-Amino-6-hydroxy-4-hexynoic acid

Unexpectedly, the positive optical rotation of this amino acid became less positive in acid solution from $[\alpha]_D = +36.2°$ (c, 1; H_2O) to $+9.6°$ (c, 0.5; 3N HCl). From the Clough-Lutz-Jirgensons rule, it very likely belonged to the D-series of amino acids. This possibility and the chemical structure as well were verified by chemical synthesis, followed by enzymatic optical resolution.

Racemic 2-amino-6-hydroxy-4-hexynoic acid was synthesized by alkylation of formylaminomalonate with 1-chloro-2-butyne-4-ol (*60*) and its chloroacetyl derivative was prepared in the usual way. The hydrolysis product of the racemic chloroacetyl amino acid with renal acylase I was passed over a cation exchanger. The effluent was hydrolyzed to give the D-isomer and from 2N ammonia-eluate the L-antipode was obtained immediately after concentration. The optical rotations of a recrystallized sample of the former were $+38.0°$ and $+7.0°$, and those of the latter, $-35.0°$ and $-10.0°$, under the same conditions as for the natural amino acid.

This result showed that the natural 2-amino-6-hydroxy-4-hexynoic acid consisted of the pure D-enantiomer. It is one of the rare examples of the occurrence of a free amino acid in the pure D-form in nature. In view of the finding that 2-amino-4,5-hexadienoic acid in this fungus seems to be racemic, there might be an interesting mechanism for the enzymatic interconversion of the unsaturated amino acids in *A. miculifera*.

In higher plants, D-amino acids occur often in the form of γ-glutamyl-
(*127*), N-malonyl- (*323*), or other low molecular peptides (*234, 320, 121*).
D-Amino acids seem to be present also in the free state, and the knowledge
on their occurrence is increasing. Amino acids of D-series might play some
important roles in amino acid metabolism.

1.4.9. 2-Amino-4,5-hexadienoic Acid (30)

As far as natural products are concerned, this is the only known
amino acid with an allenic bond. In 1968, CHILTON and associates (*56*)
reported isolation and identification of 2-amino-4,5-hexadienoic acid
from fruitbodies of *Amanita solitaria* (Fr.) Secr. sensu D. E. Stuntz. The
fungus is now regarded as the same as *A. smithiana* (*50*). This amino acid
was separated by cellulose chromatography from the ethanol extract.
Reaction with ninhydrin reagent gave a brown coloration that did not
change with time. Its IR spectrum showed strong absorption at
$1930 \, \text{cm}^{-1}$ assigned to an allene and at $850 \, \text{cm}^{-1}$ assigned to the
methylene of a terminal allene, together with several bands characteristic
of amino acid zwitterion. On catalytic reduction, it consumed two moles
of hydrogen, giving norleucine which was identified as the N-acetyl
derivative by mass spectrometry and elementary analysis. This allenic
amino acid was optically inactive at the sodium D line. The norleucine
derived from the allenic amino acid, however, was dextrorotatory similar
to L-norleucine (2S) and led the authors to conclude that the absolute
configuration was (2S).

(30)

(2S)-2-Amino-4,5-hexadienoic acid

This compound was one of several allenic amino acids synthesized by
BLACK and LANDOR (*32*) in the same year. They used formylaminomalonate instead of acetylaminomalonate because of the greater ease with
which the formyl protecting group is hydrolyzed.

This amino acid has since been obtained also from fruitbodies of
several other species of *Amanita*; *A. pseudoporphyria* Hongo (*173*), *A.
neoovoidea* Hongo (*169*), *A. abrupta* Peck (*470, 329*), *A. gymnopus* Corner
& Bas (*179*) and the newly described *A. miculifera* Bas & Hatanaka (*181*).
It is very likely that the optical purity of this amino acid varies with fungal

species. As stated above, CHILTON et al. reported that it was optically inactive at the sodium D line. However, the values of the specific rotations of the amino acid isolated from A. pseudoporphyria were $[\alpha]_D = -52.1°$ (c, 0.8; H_2O), $-9.7°$ (c, 0.4; 3 N HCl) and those from A. neoovoidea were $-24.0°$ (c, 1; H_2O), $-4.0°$ (c, 0.5; 3N HCl). The same allenic amino acid obtained from A. abrupta was oxidized completely by L-amino acid oxidase, indicating that it consisted of only the (2S)-form. However, the norleucine derived from the allenic amino acid of A. miculifera was inactive in the region 350 nm to 600 nm. Under the same conditions, i.e., concentration, solvent and temperature, (2S)-norleucine was distinctly dextrorotatory. These results suggested strongly that the allenic amino acid in A. miculifera is a racemate.

This amino acid seems to be distributed widely in section Roanokenses of the genus Amanita (Table 3).

According to CHILTON et al. (55), synthetic 2-amino-4,5-hexadienoic acid is weakly toxic to guinea pigs. Test animals receiving 100 mg/kg showed a decrease in respiratory rate and depth, skin blanching and hypothermia. Death occurred 25–40 hr later. Guinea pigs receiving 10 mg/kg of the allenic norleucine did not show any significant symptoms.

1.4.10. 2-Aminohept-4-en-6-ynoic Acid (31)

This amino acid was isolated from fruitbodies of Amanita pseudoporphyria (173). The yield was 520 mg from 6.18 kg f.w. The ninhydrin coloration was orange brown. Elementary analysis gave the empirical formula $C_7H_9NO_2$. The IR spectrum showed an absorption band at 3260 cm^{-1} characteristic of a terminal acetylene and the UV spectrum was identical with that reported for the synthetic racemate (32). Over Adams catalyst it absorbed 3 mol/eqvts of H_2 and 2-aminoheptanoic acid was formed. One mol/eqvt. of H_2 was consumed over Lindlar catalyst. The major oxidation product with potassium permanganate was aspartic acid. The ^{1}H-NMR spectrum supported also the structure, 2-aminohept-4-en-6-ynoic acid. Neither natural occurrence in other or-

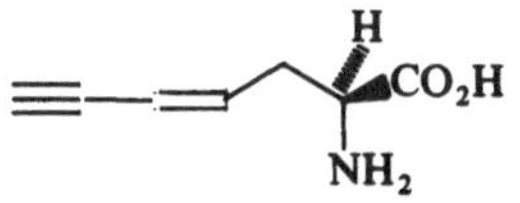

(31)

(2S)-2-Aminohept-4-en-6-ynoic acid

Table 3. *Structures and Distributions of Unsaturated Amino Acids of Norleucine-type*

Structures $R = -CH(NH_2)CO_2H$		Names	Occurrence
C_4	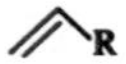	(2RS)-2-Amino-3-butenoic acid	*Rhodophyllus nidorosus* *R. crassipes*
C_5		(2S)-2-Amino-4-pentenoic acid	*Amanita pseudoporphyria*
		(2S)-2-Amino-4-pentynoic acid	*A. peudoporphyria* *A. abrupta*
	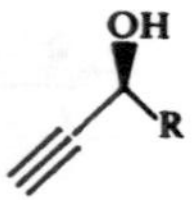	(2S,3R)-2-Amino-3-hydroxy-4-pentynoic acid	*Sclerotium rolfsii*
	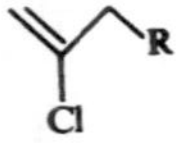	(2S)-2-Amino-4-chloro-4-pentenoic acid	*A. pseudoporphyria*
		(2S)-2-Amino-5-hexenoic acid	*Amanita gymnopus*
C_6		(2S)-2-Amino-5-hexynoic acid	*Cortinarius claricolor var. temuipes* *A. gymnopus*
		(2S)-2-Amino-4,5-hexadienoic acid	*A. smithiana* *A. pseudoporphyria* *A. neoovoidea* *A. miculifera* (2RS) *A. abrupta* *A. gymnopus*

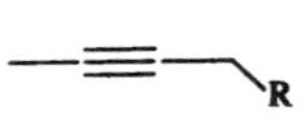

(2S)-2-Amino-4-hexynoic acid

Tricholomopsis rutilans
A. pseudoporphyria
A. miculifera

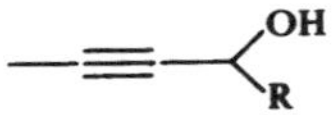

(2S,3R)-, (2S,3S)-2-Amino-3-hydroxy-4-hexynoic acid

T. rutilans

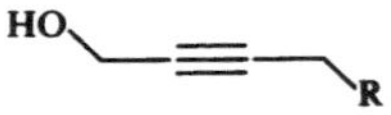

(2R)-2-Amino-6-hydroxy-4-hexynoic acid

A. miculifera

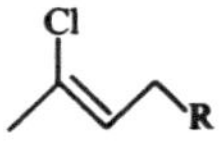

(2S,Z)-2-Amino-5-chloro-4-hexenoic acid

A. smithiana

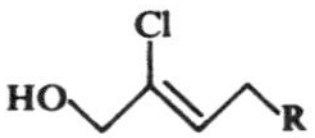

(2S,Z)-2-Amino-5-chloro-6-hydroxy-4-hexenoic acid

A. abrupta

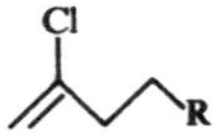

(2S)-2-Amino-5-chloro-5-hexenoic acid

A. miculifera

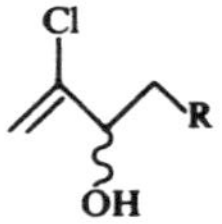

2-Amino-5-chloro-4-hydroxy-5-hexenoic acid

A. gymnopus

(2S)-2-Amino-3-cyclopropyl-propionic acid

A. virgineoides

41

ganisms nor any biological activity is known. Determination of the optical rotation suggested presence of the (2S) enantiomer.

Table 3 shows the distribution of unsaturated straight chain amino acids and cyclopropylalanine which will be noted in the next section.

1.4.11. Cyclopropylalanine (2-Amino-3-cyclopropylpropionic Acid) (32)

As a result of screening for biologically active substances of fungi, OHTA et al. (330) found that fruitbodies of Amanita virgineoides Bas contain 2-amino-3-cyclopropylpropionic acid which was not known before as a natural product. This amino acid inhibits spore germination in Pyricularia oryzae Cav., the cause of rice blast disease. The purification was carried out with Dowex 50 of formate-form and ammonium formate buffer. The concentration in the fruitbodies was extremely high for a single free amino acid. Thus, 198 mg of pure crystals were obtained from 243 g of fresh material. The results of the elementary analysis indicated the formula $C_6H_{11}NO_2$, while ^{1}H- and ^{13}C-NMR spectra indicated presence of a monosubstituted cyclopropane ring and a vicinal methylene adjacent to a methine. On catalytic hydrogenation over Adams catalyst at 60–65 °C, the cyclopropane ring was cleaved, giving leucine and isoleucine in the ratio 8:3. The (2S)-configuration was established by oxidation with L-amino acid oxidase of Habu-snake venom and CD measurement.

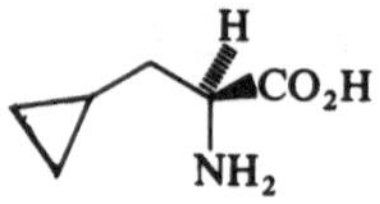

(32)

(2S)-2-Amino-3-cyclopropylpropionic acid

A stereospecific synthesis was carried out as follows (personal communication to Prof. S. NOZOE); N-benzyloxycarbonyl-L-allylglycine was refluxed with CH_2I_2 in the presence of copper powder and the product was separated on silica gel, followed by hydrogenolysis. All spectral comparisons of the natural and synthetic amino acids showed them to be identical. The synthetic racemate of the cyclopropylalanine had long been known as a potent antagonist to E. coli (284). Considering the high concentration of this amino acid in A. virgineoides, it is most likely that cyclopropylalanine plays an important role as an antibacterial and/or antifungal agent in the fungus. Seed germination might also be affected. It would be interesting and important to know whether this amino acid is produced in mycelia before the fruitbodies are formed.

Several amino acids containing a cyclopropane ring have been isolated from higher plants, e.g. hypoglycin A [2-amino-3-(methylene-cyclopropyl)propionic acid] (159, 89, 31) and its γ-glutamyl derivative (158) from unripe fruit of *Blighia sapida*, α-(methylenecyclopropyl)glycine and its γ-glutamyl derivative from the seeds of several *Acer* spp. (115), and β-(methylenecyclopropyl)-β-methylalanine from the seeds of *Aesculus californica* (117, 289). It is worth noting that some of them have hypoglycemic and/or teratogenic activity.

2-(1-Hydroxycyclopropyl)glycine acetic acid is a component of an antibiotic, cleomycin B_2 (228, 434).

A disulfiram-like substance, N^5-(1-hydroxycyclopropyl)glutamine in fungi, will be mentioned later.

1.5. Unsaturated Chloroamino Acids

Several chloroamino acids are known as antibiotics produced by *Streptomyces* spp. and other microorganisms (188). For instance, (2S,4S)-chloronorvaline was isolated from the culture broth of *Streptomyces griseosporeus* as a leucine analogue (313). This compound was active against *Pseudomonas aeruginosa*, weakly against *Serratia marcescens*, *Klebsiella pneumoniae* and *Bacillus subtilis*, but not against *Escherichia coli*. Chlorine-containing natural products in higher plants have been recently reviewed (91).

Since CHILTON and TSOU (54) reported 2-amino-5-chloro-4-hexenoic acid from *Amanita solitaria* (Fr.) Secr. *sensu* D. E. Stuntz as the first chlorine-containing amino acid with a straight carbon skeleton, several other chloroamino acids have been identified. The presence of chlorine atoms is easily verified by the characteristic $^{35}Cl/^{37}Cl$ doublets in the mass spectrum. The occurrence of free chloroamino acids seems so far to be restricted to the genus *Amanita*, sections *Phalloideae* and *Roanokenses*. Undoubtedly, interesting enzymological reactions are involved in their formation as well as in their degradation. An example will be described in the section dealing with 2-amino-4-chloro-4-pentenoic acid.

Some of the chlorine-containing amino acids are known to have an inhibitory or antagonistic effects on other microorganisms. Furthermore, a close relationship was reported between the occurrence of 2-amino-5-chloro-4-hexenoic acid and the extraordinarily high concentration of chlorine in fruitbodies of *Amanita smithiana* (54).

1.5.1. 2-Amino-4-chloro-4-pentenoic Acid (33)

This is a unique unsaturated norvaline with a chlorine atom known so far only from *A. pseudoporphyria* (*167*). Repeated fractionation on cellulose columns with the methyl ethyl ketone-containing solvent mentioned previously yielded 150 mg of crystals from 3 kg of mushrooms. The color reaction with ninhydrin was yellowish brown. After catalytic hydrogenation over Adams catalyst, norvaline hydrochloride was identified by its IR spectrum as the sole product. Elementary analysis, ^{1}H-NMR spectrum and oxidation with $KMnO_4$ suggested the structure. DL-2-Amino-4-chloro-4-pentenoic acid was synthesized by alkylation of acetylaminomalonate with 2,3-dichloropropene. Both natural and synthetic samples showed the same chromatographic behavior, mass spectral fragmentation pattern, and ^{1}H-NMR spectrum. The optical rotation of the natural substance indicated L-configuration according to the Clough-Lutz-Jirgensons rule.

(33)

(2*S*)-2-Amino-4-chloro-4-pentenoic acid

L-2-Amino-4-chloro-4-pentenoic acid was prepared by acylase as usual and its antibacterial activity was studied (*293*). The growth of a considerable number of bacterial strains, including species of *Achromobacter*, *Aerobacter*, *Alcaligenes*, *Arthrobacter*, *Bacillus*, *Escherichia*, *Micrococcus*, *Pseudomonas* and *Serratia*, was inhibited in minimal medium at concentration as low as 1.34×10^{-5} M, suggesting that this chloroamino acid is an antimetabolite against bacteria. Growth was recovered by addition of various protein amino acids whose reversal effects were dependent on the bacterial species.

These results suggest that this chloroamino acid is an antimetabolite.

MORIGUCHI et al. (*294*) reported bacterial intoxication by dehalogenation of 2-amino-4-chloro-4-pentenoic acid. They tested 81 strains and found that intact cells of *Arthrobacter oxydans*, *Bacillus sphaericus*, *B. subtilis*, *Proteus mirabilis* and *Serratia marcescens* possessed dehalogenation activity. Cell-free extracts, other than from *P. mirabilis*, showed no activity. Whether the enzymes are bound tightly to the membrane or are extremely unstable is not known. The enzyme from *P. mirabilis* was a membrane-bound protein. One of the detergents effective for solubilization of the enzyme was found to be octanoyl-*N*-methylglucamide. The addition of the chloroamino acid to the medium did not affect the enzymatic activity, indicating that the enzyme is constitutively produced. The highest total activity was obtained when the cells were cultured at 30 °C for 20 hr under shaking in a medium containing 0.5% glycerine and 0.5 % peptone as carbon and nitrogen sources, respectively. The release of chloride ion and ammonia formation were stoichiometric with the consumption of (2S)-2-amino-4-chloro-4-pentenoic acid. The enzyme was inert toward the D-isomer of this chloroamino acid. Neither oxygen consumption nor hydrogen peroxide formation was observed, suggesting that this enzyme is distinctly different from L-amino acid oxidase. The α,β-elimination reaction of L-3-chloroalanine was catalyzed by L-amino acid oxidase from rattlesnake. MORIGUCHI et al. proposed the α,γ-elimination mechanism shown in Scheme 2 for the dehalogenation of (2S)-2-amino-4-chloro-4-pentenoic acid since another product, 2-oxo-4-pentenoic acid was identified by precipitation as the 2,4-dinitrophenylhydrazone, followed by catalytic reduction, giving norvaline.

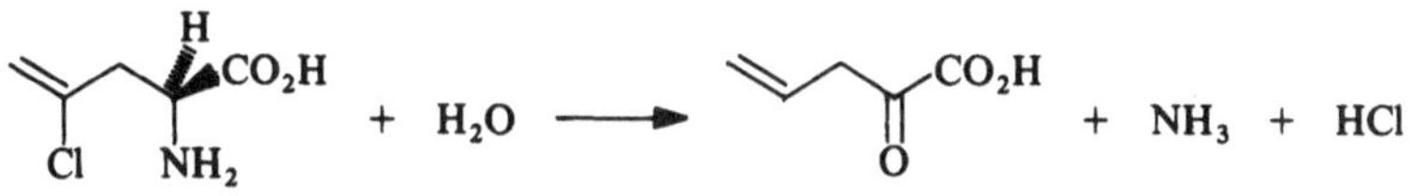

Scheme 2. Dehalogenation of 2-amino-4-chloro-4-pentenoic acid by *Proteus mirabilis* (from *294*)

As a mechanism for the antibacterial function of 2-amino-4-chloro-4-pentenoic acid, ESAKI et al. (*92*) proposed its irreversible covalent binding to L-methionine γ-lyase, using synthetic DL-[2-^{14}C]2-amino-4-chloro-4-pentenoic acid and the enzyme purified from the extract of *Pseudomonas putida* extract. The extent of inactivation of the enzyme was linearly dependent on the chloroamino acid/enzyme molar ratio; complete inactivation was observed when the ratio was about 5/1. The chloroamino acid was, therefore, very likely to be dehalogenated to form an active intermediate which binds to an amino acid residue at the active site and thus inactivates the enzyme. Spectral analysis as well as kinetic studies supported this possibility. From the enzyme complex with the chloroamino acid, acetopyruvate and pyridoxamine 5'-phosphate were demonstrated.

1.5.2. Chlorocrotylglycine (2-Amino-5-chloro-4-hexenoic acid) (**34**)

A few years after the report by CHILTON et al. on 2-amino-4,5-hexadienoic acid of *Amanita smithiana*, CHILTON and TSOU (*54*) described

(34)

(2S, Z)-2-Amino-5-chloro-4-hexenoic acid

the occurrence of 2-amino-5-chloro-4-hexenoic acid, a product which results from addition of HCl to the allenic acid, from the same species. This amino acid also gave the characteristic brown ninhydrin coloration. Paper and column chromatography were successful, so the enriched preparation as well as its N-acetyl and N-benzoyl derivatives were analyzed by ^{1}H-NMR and mass spectrometry. The methyl signal was deshielded by a double bond as well as by an electronegative group. The ^{1}H-NMR spectrum showed allylic coupling of the methyl protons ($J = 1.2$ Hz) with a vinylic hydrogen and homoallylic coupling ($J = 1.1$ Hz) with methylene hydrogens. This assignment was confirmed by spin decoupling. The mass spectrum exhibited the characteristic ^{35}Cl/^{37}Cl doublets. Chelate formation with Cu^{2+} and quantitative consumption of hydrogen on catalytic hydrogenation over Adams catalyst together with the results of instrumental analyses indicated the structure, trans-2-amino-5-chloro-4-hexenoic acid.

This was unequivocally confirmed by a comparison with a sample synthesized by alkylation of acetylaminomalonate. The 1,3-dichloro-2-butene used for the synthesis contained about 90 % of the trans-form. 2-Amino-4,5-hexadienoic acid could also be converted to the chloroamino acid in hot HCl, but the ratio of cis:trans isomer in the product was calculated as 1:9. The natural chloroamino acid did not contain the cis-isomer, but, under milder experimental conditions, no addition of HCl occurred.

Additionally, the chloroamino acid could be detected also in a fresh extract of the fruitbodies in the absence of added chloride. The possibility that the chloroamino acid in this fungus might be an artifact could therefore be excluded.

Fresh fruitbodies of A. smithiana contain ca. 2,000 ppm of chloride, an extraordinary high level when compared with the 50–500 ppm (f.w.) found in green land plants (54). Four per cent of this was in the form of the chloroamino acid. It would be interesting to investigate the chemical nature of the other chloride. In any case it is very likely that the fruitbodies of this species concentrate chloride from the soil. Also, mycelia before producing fruitbodies probably absorb chlorine ion or chloride selectively and concentrate it. Formation of the chloroamino acid might be an avoidance mechanism of chlorine ion or chloride, as in the case of the selenium-containing amino acids of *Astragalus* species (*390*).

Guinea pigs receiving 10–100 mg/kg DL-2-amino-5-chloro-4-hexenoic acid showed no marked physiological effects (*55*).

1.5.3. 2-Amino-5-chloro-6-hydroxy-4-hexenoic Acid (**35**)

This compound was the fourth unsaturated amino acid to be isolated from fatally toxic fruitbodies of *Amanita abrupta* Peck (*329*). As stated before, the other three are (2S)-2-amino-4-pentenoic acid, (2S)-2-amino-5-pentynoic acid and (2S)-2-amino-4,5-hexadienoic acid. During the separation of the above amino acids by preparative HPLC and cellulose chromatography, 19 mg of crystals were obtained from 320 g of fresh mushrooms. Ninhydrin coloration was first yellow, then turned brown and eventually violet. Peaks in the FD mass spectrum of the free amino acid at m/z 180 and m/z 182, and in those of the trimethylsilyl derivative at m/z 396 and m/z 398 implied the presence of a chlorine atom. Two further derivatives were prepared, methyl 5-chloro-2-trifluoroacetamido-6-trifluoroacetoxy-4-hexanoate and methyl 5-chloro-2-acetamido-6-acetoxy-4-hexanoate. High resolution mass as well as ^{1}H-NMR spectra were analyzed. The chloroamino acid was oxidized completely by the L-amino acid oxidase of Habu-snake venom, indicating the (2S)-configuration. The (Z)-configuration around the double bond was verified by the observed NOE between the hydrogen atoms attached to C-4 and C-6 in the ^{1}H-NMR spectrum of the diacetate. The structure is thus (2S,4Z)-2-amino-5-chloro-6-hydroxy-4-hexenoic acid.

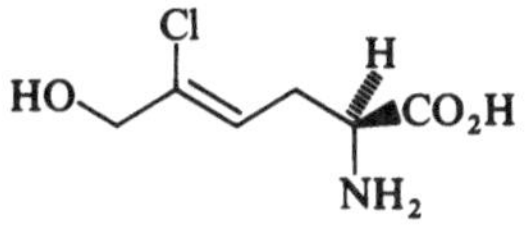

(**35**)

(2S, Z)-2-Amino-5-chloro-6-hydroxy-4-hexenoic acid

The content of unsaturated amino acids in *Amanita abrupta* was surprisingly high. From only one fruitbody of 16 g (f.w.) we once isolated 11.1 mg of 2-amino-4,5-hexadienoic acid and 16.5 mg of crude chloroamino acid crystals. In view of the fact that this species is deadly poisonous, the toxicity of this chloroamino acid should be examined in more detail.

1.5.4. 2-Amino-5-chloro-5-hexenoic Acid (36)

This is another chlorine-containing amino acid which occurs in low concentration in fruitbodies of *Amanita miculifera* (*181*). Eighteen mg of crude crystals were obtained from 580 g of fresh material. It gave the usual violet coloration with ninhydrin. Oxidation with acidic $KMnO_4$ produced glutamic acid, while catalytic hydrogenation gave norleucine. Optical rotation measurements indicated that it was an L-amino acid. This amino acid was prepared by refluxing synthetic 2-amino-5-hexynoic acid for 16 hr in 3 N HCl. The chloroamino acid was separated from starting material on a cellulose column, the yield being ca. 6%. Chemical as well as physical characteristics of synthetic and natural chloroamino acid were identical.

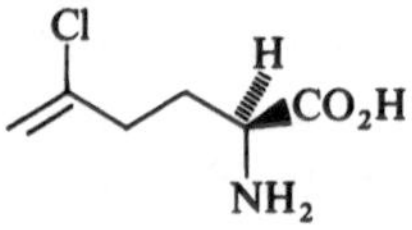

(36)

(2*S*)-2-Amino-5-chloro-5-hexenoic acid

1.5.5. 2-Amino-5-chloro-4-hydroxy-5-hexenoic Acid (37)

It has already been mentioned that we recently isolated and clarified the structure of an isomer of the above chlorohydrine amino acid, 2-amino-5-chloro-4-hydroxy-5-hexenoic acid. It was obtained in a good yield from fruitbodies of *Amanita gymnopus*. The FAB-MS showed a pair of intense peaks, m/z = 180 and 182 ($[M+H]^+$) in the ratio 3:1, and EI-MS m/z = 134 and 136 ($[M-COOH]^+$) in the same ratio. These results indicated the presence of one chlorine atom per molecule. The elementary composition, $C_6H_{10}O_3NCl$, was obtained from an accurate mass measurement of the ion of mass 180 in the FAB-mass spectrum (Found: 180.0444, Calcd: 180.0427). In the ^1H- and ^{13}C-NMR spectra, six non-exchangeable protons and six carbon atoms were observed and the

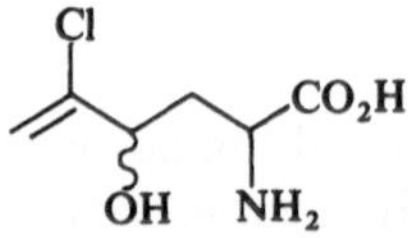

(37)

2-Amino-5-chloro-4-hydroxy-5-hexenoic acid

chemical shifts as well as the coupling constants were coincident with the ones expected from the above structure. The absolute configuration at C-4 has not yet been determined.

The distribution of unsaturated amino acids of the norvaline and norleucine type together with that of the chloroamino acids and of a cyclopropane amino acid is shown in Table 3. Most species containing these amino acids belong to the subgenus *Lepidella* (Table 4). No species in subgenus *Amanita* are so far known to produce them, but the scattered occurrence of ibotenic acid, muscimol, muscaflavin, and related compounds is characteristic of this subgenus. They have not been reported from any species of the subgenus *Lepidella*. Both subgenus also contain many species in which no specific non-protein amino acids have been detected. *Amanita phalloides* and *A. pseudoporphyria*, both belonging to the section *Phalloideae*, are quite different from each other in respect of nitrogen constituents. There is no evidence for the presence of *Amanita*-toxins in the latter species. In the present system, the section *Phalloideae* must be regarded as heterogenous in secondary nitrogen metabolism.

Table 4. *Subgenera and Sections of the Genus* Amanita *(from 391)*

Amanita Pers. ex Hooker
   Subgenus *Amanita*
      Sect. 1. *Caesareae* Sing. (1950)
          2. *Ovigerae* Sing.
          3. *Vaginatae* (Fr.) Quél.
          4. *Amanita*
   Subgenus *Lepidella* (Gilbert) Veselý (1933)
          5. Amidellae (Gilbert) Konr. & Maubl. (1948)
          6. *Phalloideae* (Fr.) Quél.
          7. *Mappae* Gilbert
          8. *Validae* (Fr.) Quél.
          9. *Roanokenses* Sing. ex Sing. (*Lepidella* aut.)

Most amino acids of *Lepidella* species are structurally related to each other and their concomitant occurrence in a single species suggests biosynthetic relationships. As stated before, the coexistence of 2-amino-4-pentenoic acid, 2-amino-4-pentynoic acid and 2-amino-4-chloro-4-pentenoic acid in *A. pseudoporphyria* suggests simply stepwise enzymatic dehydrogenation and halogenation. Fruitbodies of *A. abrupta* also contain the two pairs of amino acids, 2-amino-4-pentenoic acid and 2-amino-4-pentynoic acid, and 2-amino-4,5-hexadienoic acid and 2-amino-5-chloro-6-hydroxy-4-hexenoic acid. Other examples are the

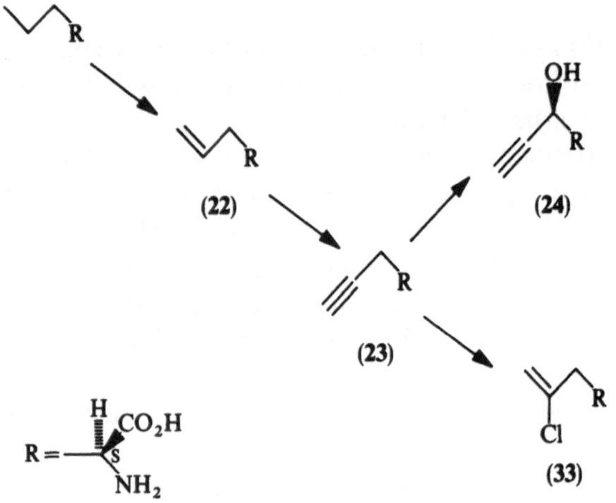

Scheme 3. Possible biosynthetic pathways to unsaturated norvalines

concomitant occurrence of 2-amino-4,5-hexadienoic acid and 2-amino-5-chloro-4-hexenoic acid in *A. smithiana*, and four amino acids, 2-amino-5-hexenoic acid, 2-amino-5-hexynoic acid, 2-amino-4,5-hexadienoic acid and 2-amino-5-chloro-4-hydroxy-4-hexenoic acid, in *A. gymnopus*. As shown in Scheme 3, biosynthetic routes leading to the above C_5 amino acids can be postulated based on possible chemical reactions of the amino acids and their cooccurrence in fungal species. Stepwise dehydrogenation at the end of the carbon skeleton of norvaline may lead to 2-amino-4-pentynoic acid. Hydroxylation occurs at C-3 in *Sclerotium*. Addition of hydrochloric acid would yield 2-amino-4-chloro-4-pentenoic acid in *Amanita pseudoporphyria*. Similarly, the existence of the related C_6 amino acids in *A. gymnopus* and other Amanitas suggests strongly that these species are able to synthesize the above amino acids by the biosynthetic routes shown in Scheme 4. The terminal acetylenic bond can be isomerized to the allene and further to the internal acetylene. Hydroxylation at C-3 or C-6 leads to the two hydroxy amino acids. Chlorination and hydroxylation could be possible in two ways; thus explaining the formation of four kinds of chloroamino acids. 2-Amino-5-chloro-4-hydroxy-5-hexenoic acid might also be formed from 2-amino-5-hexynoic acid. Reductive cyclization of the allenic amino acid may be the way cyclopropylalanine is biosynthesized.

There remain some stereochemical problems. For instance, 2-amino-6-hydroxy-4-hexynoic acid is present in *Amanita miculifera* as the optically pure (2R)-, while among the proposed precursors in this fungus the allenic amino acid is a racemate and the 2-amino-4-hexynoic acid is the (2S)-enantiomer in this fungus.

References, pp. 117–140

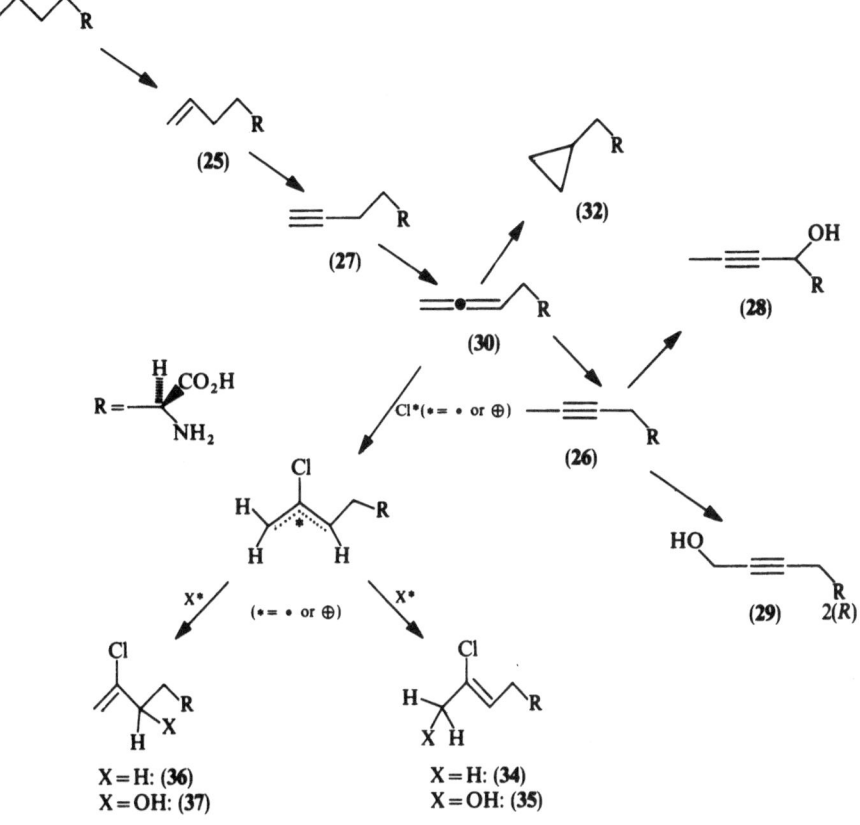

Scheme 4. Possible biosynthetic pathways to unsaturated norleucines

Furthermore, while the C_5–C_7 amino acids in *A. pseudoporphyria* are very likely to be formed by successive enzymatic elongation, it is not clear, at which step and how this occurs.

Schemes 3 and 4 do not imply any direct phylogenetic relationships or derivation of the various fungal species, which produce these amino acids.

2. Acidic Amino Acids

Among the large number of non-protein amino acids now known to occur in biological systems, acidic amino acids can be detected comparatively easily on two dimensional paper chromatography. Although they generally have small Rf-values in the solvent systems used conventionally, they are well separated from each other and are easily recognized. Most can be purified without much difficulty on a Dowex 1 column in acetate-form, being eluted with acetic acid of appropriate concentration.

Paper chromatographic screening in our laboratory for the free amino acids of mushrooms reveals that 2-aminoadipic acid is one of the ubiquitous constitutents of their amino acid pools.

Most of the acidic amino acids occurring in higher plants are derivatives of aspartic acid or glutamic acid. Acidic amino acids from mushrooms are no exception to this rule.

2.1. Hydroxyaspartic Acid

Free hydroxyaspartic acid is known only in species of *Medicago*, *Astragalus* (*203*), and *Trifolium* in the Leguminosae of higher plants (*463*). The stereochemistry is (2S,3R)-3-hydroxyaspartic acid (*erythro*-3-hydroxy-L-aspartic acid). As stated in the preceding chapter, (2R,3R)-3-hydroxyaspartic acid (**38**) is a component of three acidic phallotoxins, phallacin, phallacidin, and phallisacin (*453*).

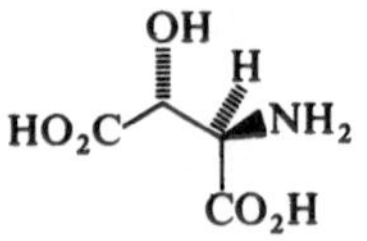

(38)

(2R, 3R)-3-Hydroxyaspartic acid

2.2. 4-Substituted Glutamic Acids

A large number of γ-substituted glutamic acids has been reported from various species of higher plants. Isolation of 4-methyleneglutamic acid and its amide, 4-methyleneglutamine, by DONE and FOWDEN (*79*) from *Arachis hypogaea* was essentially the first report on this group of amino acids in plants.

Thus, (2S,4R)- and (2S,4S)-4-hydroxyglutamic acids were present in *Phlox* (*439, 26*). 4-Methyl- and 4-methyleneglutamic acids were reported subsequently from *Phyllitis scolopendrium* (Pteridophyte) (*438*), liliaceous plants (*119*), *Polygala vulgaris* (*218*), *Tulipa gesneriana* (*34*), *Lathyrus maritimus* (*354*), etc. An unequivocal structural study of these was made by BLAKE and FOWDEN (*34*). Most specimens of 4-methylglutamic acid represented the (2S,4R) diastereomer (*erythro*-). The possible occurrence of the (2S,4S) diastereomer in some mushrooms will be discussed later.

Both (2S,4R)- and (2S,4S)-4-hydroxy-4-methylglutamic acid occur in *Ledenbergia roseo-aenea* (*214*) and in *Pandanus veitchii* (*26*). Further examples of the concomitant occurrence of two diastereomers are (2S,3S,4R)- and (2S,3R,4S)-3-hydroxy-4-methylglutamic acid in *Gymnocladus dioicus* (*68, 67*). Also (2S,3S)-3-hydroxy-4-methyleneglutamic acid was isolated from *Gleditsia caspica* (*72*). The configuration of the hydroxy and/or methyl group was clarified mainly by analyses of the ^{1}H-NMR spectra.

Threo-3-hydroxyglutamic acid occurs as a cell wall component in *Microbacterium lacticum* (*383*).

2.2.1. 4-Methylene-, 4-Ethylidene-, and 4-Propylideneglutamic Acid (39), (40), (41)

As stated above, 4-methyleneglutamic acid was first isolated together with its amide from *Arachis hypogaea* (*79*). Subsequently, it was also found together in *Amorpha fruticosa* (*426*), *Tetrapleura tetraptera* (*145*), and the liliaceous *Tulipa gesneriana* (*474, 34*), *Lilium maximowiczii* (*336*) and *Lilium candidum* (*365*). A pteridophyte, *Phyllitis scolopendrium* also

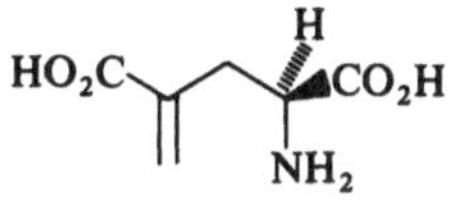

(39)

(2S)-4-Methyleneglutamic acid

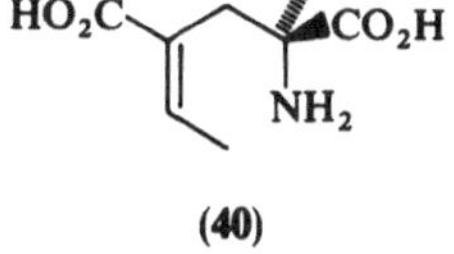

(40)

(2S, E)-4-Ethylideneglutamic acid

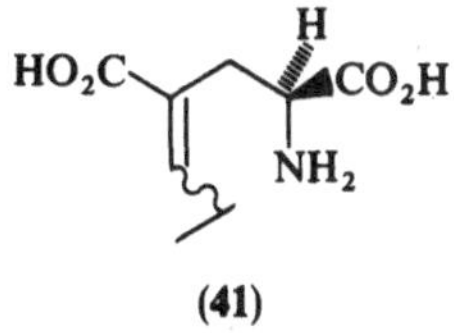

(41)

(2S)-4-Propylideneglutamic acid

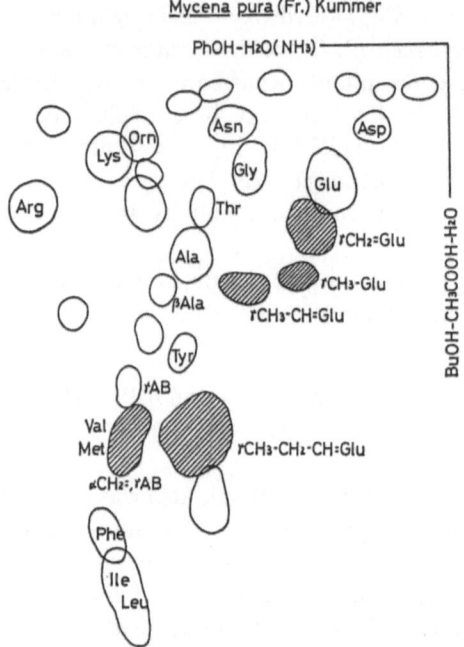

Fig. 1. Free amino acids in *Mycena pura*. γCH$_2$=Glu, γ-methyleneglutamic acid; γCH$_3$–Glu, γ-methylglutamic acid; γCH$_3$–CH=Glu, γ-ethylideneglutamic acid; γCH$_3$–CH$_2$–CH=Glu, γ-propylideneglutamic acid; αCH$_2$=γAB, α-methylene-γ-amino-butanoic acid (from *177*)

contains 4-methyleneglutamic acid (*34*). In some liliaceous plants, in particular, its concentration is much higher than any other free amino acids.

In our screening of non-protein amino acids of mushrooms, five unidentified spots were observed in a two dimensional paper chromato-gram of the amino acid-fraction prepared from the fruitbodies of *Mycena pura* (Fr.) Kummer (Fig. 1) (*168*). Four of these gave a brown coloration with ninhydrin reagent, suggesting strongly that they contained unsatur-ation. Additionally, from their locations on the chromatogram, the slowest moving amino acid seemed to be 4-methyleneglutamic acid and two others its higher homologues. After treatment of the ethanol extract of 8.2 kg of mushrooms with Amberlite IR-120 (H$^+$-form), the amino acids were fractionated on a Dowex 1 (acetate-form) column. Because 4-ethylidene- and 4-propylideneglutamic acid were displaced together with glutamic acid, they were further separated using a cellulose column. 4-

Methyleneglutamic acid was eluted from a Dowex 1 column after the above two higher homologous amino acids. The yields were 4.48 g of 4-methyleneglutamic acid, 70 mg of 4-ethylideneglutamic acid, and 740 mg of 4-propylideneglutamic acid. Results of elementary analyses, ^{1}H-NMR spectra, and oxidation products with $KMnO_4$ were all in accordance with the proposed structures.

4-Ethylideneglutamic acid was first reported from tulip (*Tulipa gesneriana*) by FOWDEN (*106*) and subsequently from two species of legumes, *Guilandina crista* (*322*) and *Tetrapleura tetraptera* (*145*). The first two plants are known to contain 4-methyleneglutamic acid. According to GMELIN and LARSEN, γ-ethylideneglutamic acid isolated from *T. tetraptera* is (*S, E*)-isomer (*145*) (**40**). The common occurrence of these homologous 4-substituted glutamic acids suggests sequential or closely related biosynthetic pathways. The configuration at C-2 of 4-methylene- and 4-ethylideneglutamic acid was assumed to be L- after comparison of their optical rotations with those of authentic samples or with such rotations mentioned in the literature. As for 4-propylideneglutamic acid, a positive shift of the optical rotation was observed in more acidic solutions. This amino acid is known so far only in the fruitbodies of *Mycena pura*.

The fourth substance was eluted from a Dowex 1 column in a neutral and a basic fraction (*176*). Fractionation and purification were carried out as usual by cellulose column chromatography and n-butanol-acetic acid-water. To remove a small amount of concomitant valine, the sample was again passed through a Dowex 50 (H^+-form) column and eluted with 1.5 N HCl. The yield was 1.56 g of optically inactive crystals. Results of elementary analysis, catalytic hydrogenation in the presence of Adams platinum oxide, and the ^{1}H-NMR spectrum were all satisfactory for the structure, 2-methylene-4-aminobutanoic acid (**42**). This substance was earlier reported from groundnut (*110, 104*) and is structurally a decarboxylation product of 4-methyleneglutamic acid. 4-Methyleneglutamic acid and (*2S,3S*)-3-hydroxyglutamic acid are known to be decarboxylated by the glutamate decarboxylase of *Escherichia coli* (*192*). The 4-methyleneglutamic acid isolated from *Mycena pura* was treated similarly with commercial *E. coli* glutamate decarboxylase yielding the expected product. 2-Methylene-4-aminobutanoic acid in *Mycena pura* is probably formed by the action of glutamate decarboxylase or one of the similar enzymes in this fungus.

$$HO_2C \diagup\!\!\diagdown\!\!\diagup NH_2$$

(42)

2-Methylene-4-aminobutanoic acid

2.2.2. 4-Methylglutamic Acid (43), (44)

The remaining unidentified spot which gave the usual ninhydrin coloration on paper was very likely 4-methylglutamic acid. 4-Methylglutamic acid was partially overlapped by glutamic acid and two other 4-substituted glutamic acids on a Dowex 1 column. Repeated column chromatography with cellulose and Dowex 50 (H$^+$-form with 1.5 N HCl finally yielded a pure sample. Two diastereomers of (2S)-4-methylglutamic acid had been separated from one another on a Dowex 1 column by BLAKE and FOWDEN (34). Using their method, we separated our sample on a long column (169 cm) with 0.2 M acetic acid and obtained 6.5 mg of the *threo* (43) and 9.9 mg of the *erythro* diastereomer (44) in the form of thick syrups. Neither diastereomer could be crystallized. Comparable amounts of each diastereomer, prepared from the authentic 4-methyleneglutamic acid were then applied to the same Dowex 1 column and eluted under as exactly the same conditions as possible. Almost the same elution patterns were obtained. Although further chemical characterization was not carried out, it is very likely that 4-methylglutamic acid in *Mycena pura* is a mixture of the *threo* and *erythro* diastereomers (177).

(43)

(2S, 4S)-4-Methylglutamic acid
(*threo*-form)

(44)

(2S, 4R)-4-Methylglutamic acid
(*erythro*-form)

The absolute configuration at C-2 was confirmed by oxidative deamination with L-amino acid oxidase prepared from Habu-snake venom as before. The L-configuration of 4-methylglutamic acid isolated from this species was further verified by decarboxylation using *E. coli* glutamate decarboxylase (164).

Fruitbodies of *Mycena pura* are frequently found in mountaineous regions and vary widely in color and size. In many older books on mushrooms they were described as edible, but a small amount of muscarine was recently detected (394, 397). Consequently, this species is now listed among the poisonous fungi. It is doubtful, however, that intoxication (186) was really caused by muscarine (40), since the latter is mainly present in the pharmacologically inactive *epi* form. According to our experience, fruitbodies of *Mycena pura* develop twice yearly in central Japan, i.e. in the July rainy season and again in October in early fall.

Fruitbodies used for the isolation of the above amino acids were all collected in October. Patterns of non-protein amino acids of material collected in July are quite different and only 4-methylglutamic acid was detected. Its configuration is unknown.

This species is a saprophyte and the mycelium is cultured comparatively easily. In order to investigate the biosynthetic relationships among the several 4-substituted glutamic acids and also to find their common precursor, mycelial cultures were studied. However, neither the 4-substituted glutamic acids nor their decarboxylated products were detected on two dimensional paper chromatograms. Whether 4-substituted glutamic acids are produced only in fruitbodies or are produced elsewhere and transported there at the time of sporulation is not known. Possibly our cultures are not sufficiently adapted to laboratory media to produce them.

In a closely related, similarly small and violet mushroom, *Baeospora* sp., an amino acid corresponding to 4-methylglutamic acid was detected by paper chromatography.

4-Methylglutamic acid was also detected in the fruitbodies of *Wynnea gigantea* Berk. et Curt. Although the sample purified was too scanty for further characterization, the spot on the chromatogram of this species was always "pea nut shell-shaped", suggesting again the presence of a mixture of the *threo* and *erythro* diastereomers.

Shortly after the discovery of several 4-substituted glutamic acids, studies on their biosynthesis were reported, using several higher plants, such as *Lilium regale* (*452*), *Arachis hypogaea* (*120*), and *Adiantum pedatum* (*271*). The results were, however, not always consistent and no conclusive information was obtained.

An attractive idea is that the carbon skeleton of naturally occurring 4-substituted glutamic acids might be formed from the mother compound 2-oxo-4-hydroxy-4-methylglutaric acid which can be formed by condensation of two molecules of pyruvic acid. If an amino group were transferred to C-2 of the product by transaminase, 4-hydroxy-4-methylglutamic acid would be formed. Starting with this amino acid, several other 4-substituted glutamic acids might be produced. In fact, a few positive results were obtained using labeled pyruvate and intact plants (*452*, *120*). From the results of these experiments, however, no conclusive information could be drawn. Marcus and Shanon (*276*) doubted the hypothesis, because they observed that aldolase in *Arachis hypogaea* catalyzed a reaction between 4-hydroxy-4-methylglutamic acid and pyruvic acid. Although the former was labeled, the reaction did not contribute to the synthesis. Peterson and Fowden (*341*) studied this problem in a novel way using the legume *Gleditsia triacanthos*. Seeds of this plant germinate rapidly, and germination is accompanied by production of considerable amounts of several 4-substituted glutamic acids. They showed that the concentration of 4-substituted glutamic acids in dry seeds was very small. The predominantly formed amino acids were (2S,4R)-4-methyl-, (2S)-4-methyleneglutamic acid and (2S,4S)-4-hydroxy-4-methylglutamic acid. First, DL-[1-^{14}C]glutamic acid and L-[methyl-^{14}C]methionine was added separately to 3-day-old seedlings of this plant to check the possibility that C_1 transfer from methionine to glutamic acid would be necessary. Essentially no radioactivity was incorporated into any of the 4-substituted glutamic acids. They then administered L-[1-^{14}C]leucine to the seedlings and observed that incorporation into 4-methylglutamic acid was six times higher than after the administration of DL-[^{14}C]-glutamic acid. Additionally, the predominant incorporation of radioactivity into 4-hydroxy-4-methylglutamic acid in this species was much lower and occurred much later. It seemed likely that (2S,4S)-4-hydroxy-4-methylglutamic acid was derived by direct hydroxylation of 4-methylglutamic acid. Direct involvement of leucine in the biosynthesis of 4-methylglutamic acid was further verified by ninhydrin-mediated decarboxylation of the product which had been formed from radioactive precursors. The loss of radioactivity measured after decarboxylation of 4-methylglutamic acid produced from [1-^{14}C]glutamic acid was only 10.5 %. However, the recovery of radioactive [1-^{14}C]leucine was 85.3 %. Furthermore, if 4-methylglutamic acid were formed from [^{14}C]glutamic acid, 16.7 % of the total activity would have been lost as $^{14}CO_2$ in this reaction.

Although some of the radioactivity of [1-^{14}C]pyruvate was incorporated, it was concluded from similar experiments that pyruvate was not a direct precursor of 4-hydroxy-4-methylglutamic acid, as was earlier assumed.

[2-^{14}C]*Threo-* and [2-^{14}C]*erythro*-4-methylglutamic acid was also administered separately to the seedlings to check the possibility that 4-methyleneglutamic acid might be formed from some of the diastereomeric 4-hydroxy-4-methylglutamic acid. The results, however, were negative. 4-Methyleneglutamic acid seemed not to participate directly in the conversion of 4-methylglutamic acid into 4-hydroxy-4-methylglutamic acid.

Recently 4-methyleneglutaminase was purified by two research groups independently from *Arachis hypogaea* (*353, 195*).

2.3. 2-Amino-4-methylpimelic Acid (45)

The genus *Lactarius* contains many morphologically diverse species. Patterns of non-protein amino acids are also very different from species to

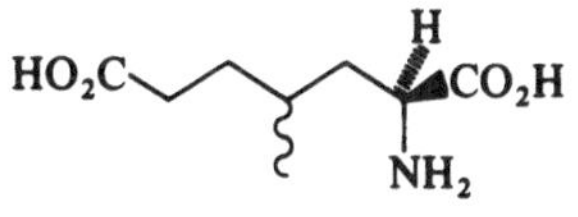

(45)

(2S)-2-Amino-4-methylpimelic acid

species. 4-Hydroxyisoleucine from *Lactarius camphoratus* has already been described.

L-2-Amino-4-methypimelic acid is a characteristic non-protein amino acid occurring in the fruitbodies of *Lactarius quietus*. Fr. and in a few other related *Lactarius* spp. (*165*). This amino acid was displaced from a Dowex 1 column (acetate-form) with 0.15 M acetic acid just before 2-aminoadipic acid. An example of the yield was 410 mg from 2.7 kg fresh mushrooms. The results of the elementary analysis agreed with the formula $C_8H_{15}NO_4$. The structure was easily clarified from the potentiometric titration and the determination of ^{1}H-NMR spectrum. DL-2-Amino-4-methylpimelic acid was synthesized by the condensation of diethyl acetylaminomalonate with (3-bromoisobutyl)malonate, which had been prepared from the chlorocompound and KB_r. The product was extracted with diethylether, the protecting groups were removed by refluxing with HCl, and then purified on Amberlite IRA-120 (H^+-form), followed by Dowex 1 (acetate-form).

2.4. 3-Nitraminoalanine (**47**) and 2-Amino-4-nitraminobutanoic Acid (**50**)

Amino acids containing nitro groups are quite rare except for components of some antibiotics, such as L-3-nitro-4-hydroxyphenylalanine in the ilamycins (*418*). 1-Amino-2-nitrocyclopentanecarboxylic acid is reported to be one of the fermentation products of *Aspergillus wenttii* (*44*); this compound causes unusual morphological changes as well as growth inhibition in plants (*41*). *Fomes robiniae* is reported to produce 1,4-dimethoxy-2-nitro-3,5,6-trichlorobenzene (*45*) (**46**).

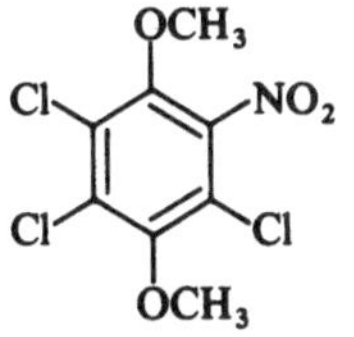

(46)

1,4-Dimethoxy-2-nitro-3,5,6-trichlorobenzene

CHILTON and HSU (51) reported isolation of 3-nitramino-L-alanine
(47) and its decarboxylation product, N-nitroethylenediamine (48), from
fruitbodies of *Agaricus silvaticus* Vitt.: Fr. Electrophoresis of the extract
at pH 2 and 6 suggested the existence of both acidic and basic ninhydrin-
positive substances. Both gave an initial bright yellow reaction which
turned to brown and finally violet. This strongly suggested that their
structures were unusual. Isolation was performed on Dowex 50 (H^+-
form) and Dowex 3 (acetate-form) columns followed by paper chromatography. Yields were 90 mg of acidic amino acid and 137 mg of the basic
substance from 384 g of fruitbodies.

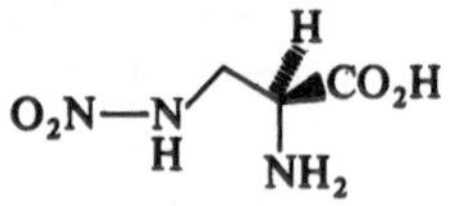

$$O_2N{-}N\underset{H}{\overset{H}{}}{\diagup}\overset{H}{\diagdown}\underset{NH_2}{\overset{}{|}}CO_2H$$

(47)

(2*S*)-3-Nitraminoalanine

$$O_2N{-}NH{-}CH_2{-}CH_2NH_2$$

(48)

N-Nitroethylenediamine

Elementary analysis of the acidic amino acid showed the formula
$C_3H_7N_3O_4$ which was confirmed by the mass spectrum. The ^{1}H-NMR
spectrum indicated the presence of a 3-substituted alanine. Thus, the
remaining group composed of HN_2O_2 must be attached to C-3 of
alanine. Formulae (47) or (49) are possible and stable. They may be
distinguished on the basis of their UV-spectra. Although the λ_{max} of 3-
nitraminoalanine and that of the *N*-nitrosohydroxyl compound both
occur at 230 nm, that of the former does not change over pH 1–13, while,
on the other hand, the absorption maximum of the latter shifts to 250 nm
on addition of dilute HCl. Furthermore, compound (49) is known as
alanosine (61, 307). Synthesis of this antibiotic has also been reported
(253). The electrophoretic as well as chromatographic behavior of the

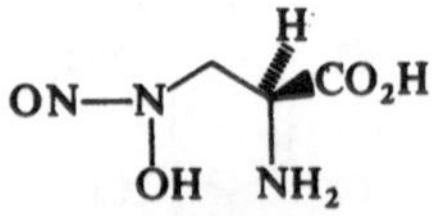

$$ON{-}N\underset{OH}{\overset{H}{}}{\diagup}\overset{H}{\diagdown}\underset{NH_2}{\overset{}{|}}CO_2H$$

(49)

Alanosine

basic compound in *A. silvaticus* suggested that it is the decarboxylation product of (47). This was verified by ^{1}H-NMR, UV, and IR spectrometry. For final confirmation, *N*-nitroethylenediamine was synthesized by nitrating *N*-carbethoxyethylenediamine according to the method of HALL and WRIGHT (*156*).

Shortly after the report by CHILTON and HSU, we published the results of analyses of amino acids in fruitbodies of the closely related. *A. subrutilescens* (Kauffm.) Hotson et Stuntz (*163*). An ethanol extract from about 10 kg of fresh fruitbodies was treated successively with columns of Amberlite IR-120B, Dowex 1, Dowex 50, and cellulose. We obtained 4.2 g of 3-nitraminoalanine, 2.7 g of 2-amino-4-nitraminobutanoic acid, 2.8 g of *N*-nitroethylendiamine, and 880 mg of the γ-glutamyl derivative of the last mentioned substance. The second and the last compounds were unknown previously. Spectral data of the two known compounds were in good agreement with those reported by CHILTON and HSU.

$$O_2N-\overset{H}{N} \diagdown\diagup\overset{H}{\diagup}CO_2H$$
$$NH_2$$

(50)

(2S)-2-Amino-4-nitraminobutanoic acid

From the values of the optical rotations 3-nitraminoalanine, $[\alpha]_D = -36°$ (c 1; H_2O), $-10°$ (c 0.5; 3 N HCl), and of 2-amino-4-nitraminobutanoic acid, $[\alpha]_D = +18°$ (c 1; H_2O), $+38°$ (c 0.5; 3 N HCl), the C-2 configurations of both amino acids seemed to be (2S). The UV spectrum of 2-amino-4-nitraminobutanoic acid were nearly the same as that of its lower homologue. The ^{1}H-NMR spectrum as well as the elementary analysis supported this structure. Hydrogenolysis with Adams catalyst yielded 2,4-diaminobutanoic acid and a small amount of 2-amino-n-butanoic acid. Under the same conditions 3-nitraminoalanine gave alanine, 2,3-diaminopropionic acid, and ammonia.

On treatment of 3-nitraminoalanine with commercial glutamate decarboxylase of *Escherichia coli* at pH 3.4, crystalline *N*-nitroethylene-diamine could be isolated in 80% yield. 2-Amino-4-nitraminobutanoic acid could not be decarboxylated. The molecule is probably too large to act as a substrate for glutamate decarboxylase.

3. Heterocyclic Amino Acids

A large number of heterocyclic imino acids are distributed in plants. With few exceptions, they are derivatives of proline or its higher

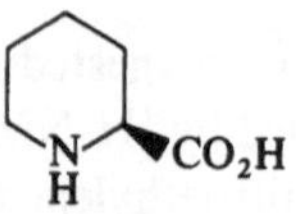

(51)

(2S)-Pipecolic acid

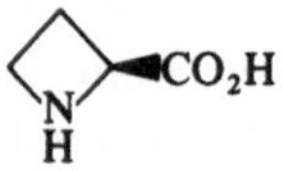

(52)

(2S)-Azetidine-2-carboxylic acid

homologue, pipecolic acid (**51**). However, azetidine-2-carboxylic acid
(**52**), a lower analogue of proline, and its derivatives are also known.

3.1. Azetidine-2-carboxylic Acid and Derivatives

3.1.1. Azetidine-2-carboxylic Acid (**52**)

This lower analogue of proline is accumulated in several species of
Liliaceae (*105*), Leguminosae (*442, 97*) and a few other families of plants
(*200*). This imino acid seems to be distributed more widely than expected.
As for fungi, IKEDA et al. (*197*) isolated azetidine-2-carboxylic acid from
fruitbodies of *Clavaria miyabeana* S. Ito, noting that it is a growth
inhibitor of *Phaseolus mungo* seedlings. They also studied the distribution
of this imino acid in the Clavariaceae and detected it in nine of eleven
species by thin-layer chromatography. Azetidine-2-carboxylic acid is one
of the many non-protein amino acids which inhibit the growth of
hypocotyls as well as radicles of lettuce (*464*).

It is known that azetidine-2-carboxylic acid competitively inhibits the
incorporation of proline into protein. The mechanism of the inhibition is
based on the competitive binding of this imino acid to prolyl-tRNA
synthetase. NORRIS and FOWDEN (*321*) reported that the prolyl-tRNA
synthetase of *Phaseolus aureus*, *Ranunculus bulbosa* and other non-
producer plants of this imino acid bound it as well as proline. By contrast,
the corresponding enzyme of *Convallaria majalis* and other producer
plants of azetidine-2-carboxylic acid discriminated well between the
imino acid and proline and did not bind the former. Similar results of
other studies along this line on the other non-protein amino acids have
been reviewed by FOWDEN and LEA (*111*).

References, pp. 117–140

Such a phenomenon might be encountered for many non-protein amino acids produced by plants and fungi. As stated before, many non-protein amino acids are structurally similar to protein amino acids and behave as competitive inhibitors in the metabolism of protein amino acids. Differences in the affinities of aminoacyl-t-RNA synthetase of producers and non-producers of non-protein amino acids could well explain, in many cases, why they are toxic to the latter plants and not to the former.

Azetidine-2-carboxylic acid is also one of those non-protein amino acids whose biosynthesis has been rather well studied in higher plants. LEETE (257) first reported that methionine was a relatively effective precursor in the formation of this imino acid, and this finding was soon confirmed by SU and LEVENBERG (400). However, it was later shown (258) that azetidine-2-carboxylic acid derived from [1-^{14}C,2-^{3}H]methionine had almost completely lost the tritium present in the precursor. SUNG and FOWDEN (404) reported that 2,4-diaminobutanoic acid was a good

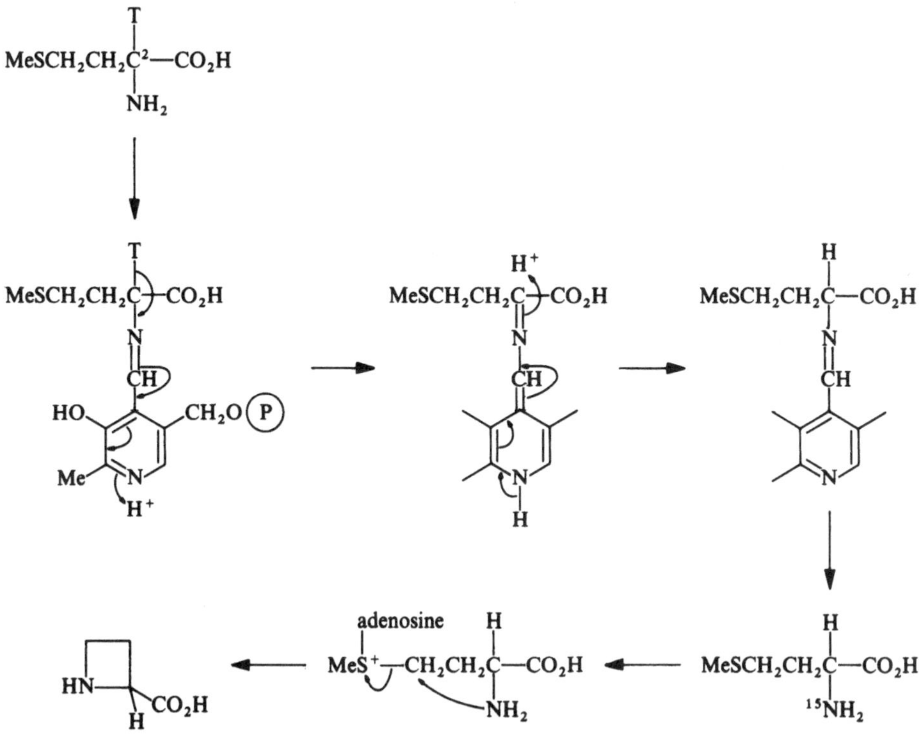

Scheme 5. Hypothesis for the loss of tritium from C-2 of methionine in the biosynthesis of azetidine-2-carboxylic acid (from 259)

precursor in seeds of the legume *Delonix regia*. A modified hypothesis is that the loss of tritium from C-2 of labeled methionine occurs by way of 1-aminocyclopropane-1-carboxylic acid, followed by ring enlargement and reduction of the resultant azetidinium ion. The cyclopropane amino acid is an established intermediate in the biosynthesis of ethylene in green plants. Recently LEETE *et al.* (*259*) reexamined the earlier hypothesis and proposed that the tritium loss occurs as shown in Scheme 5. When the Schiff base is formed from [2-³H]methionine and pyridoxal phosphate, then a tautomeric shift should result in the loss of the tritium. Hydrolysis yields tritium-free methionine. Racemization of α-amino acids is considered to occur in this manner (*103*).

The next well known 4-membered cyclic imino acid is nicotianamine. It was first isolated from tobacco leaves (*319*) and subsequently from *Fagus silvetica* and a few other plants (*246*). The initially assigned structure was revised to 2(*S*),3′(*S*),3″(*S*)-*N*-[*N*-(3-amino-3-carboxypropyl)-3-amino-3-carboxypropyl]azetidine-2-carboxylic acid. Nicotianamine has been isolated from many higher plants, but, there have been no reports from mushrooms.

3.2. Derivatives of Proline

Various proline and pipecolic acid derivatives are distributed in plants and have been well summarized by BELL (*23*). (2*S*,3*S*)-3-Hydroxyproline (*trans*-L-form) (*402*), (2*S*,4*R*)-4-methylproline (*trans*-L-form) (*193*), (2*S*,4*R*)-4-carboxy-L-proline (*trans*-L-form) (*202, 449*), etc. are known. Both (2*S*,3*S*)- and (2*S*,3*R*)-3-hydroxyproline are known in hydrolyzates of a mediterranean sponge (*205*) and also in the hydrolyzate of the antibiotic telomycin (*204, 205*). It is interesting that 4-hydroxymethylproline in *Eriobotrya japonica* (loquat) is a mixture of the (2*S*,4*R*)- and (2*R*,4*S*)-isomers and that these are accompanied also by 4-methylene-DL-proline (*154, 153*). The occurrence of 3,4-dihydroxy-L-proline in cyclic peptides of *Amanita*-species and in diatoms has already been mentioned.

3.2.1. 3-Aminoproline (53)

Fruitbodies of *Morchella esculenta* Pers. ex St. Adams, a morel, are considered among the best of edible mushrooms. In 1969, I reported (2*S*,3*R*)-3-aminoproline (*cis*-3-amino-L-proline, morchelline) (**53**) from *M. esculenta* and from two related species, *M. conica* Pers. and *M. crassipes* (Vent.) Pers. (*160*). Although its concentration depends upon the sample,

(53)

(2S, 3R)-3-Aminoproline

habitat, etc., this amino acid always occurs in the highest concentration in these mushrooms. After the "amino acid-fraction" is passed through a column of Dowex 50 (H^+-form) and the neutral and acidic amino acids are displaced with 1 M pyridine, this imino acid is eluted with 1 M ammonia. Crystals were obtained as the hydrochloride. The yields were 270 mg from 900 g of fresh *M. esculenta*, 490 mg from 980 g of *M. conica*, and 400 mg from 1500 g of *M. crassipes*.

The results of the elementary analysis agreed with formula $C_5H_{10}N_2O_2$. The substance formed a copper complex which interfered with the ninhydrin reaction (256). The optical rotation suggested an α-amino(imino) acid. On deamination, it gave both *cis*- and *trans*-3-hydroxyproline, probably because of Walden inversion. However, by comparing the coupling constant of H-2 with those in the literature, it was concluded that the structure is (2S,3R)-3-aminoproline.

Morchelline could also be isolated in good yield from the cultured mycelia of *M. esculenta* (296). It gives a sky-blue or greenish blue coloration with ninhydrin and is easily detected with two-dimensional paper chromatography. The occurrence of morchelline in other genera has not yet been reported.

Feeding of ^{14}C-proline to cultured mycelia of *M. esculenta* showed rapid incorporation of the radioactivity into morchelline (162). The mechanism for introducing the amino group is not yet clear. Trials to detect 2,3-diaminoglutamic acid or 3-hydroxyproline as possible precursors have so far been unsuccessful.

Along with the formation of ^{14}C-morchelline, a small amount of radioactivity was observed in a neutral compound. Its chromatographic behavior corresponded to that of γ-glutamyl-*cis*-3-aminoproline which was later isolated from the cultured mycelia of *M. esculenta* (297). *Cis*- and *trans*-3-amino-DL-proline were synthesized by GALLINA et al. (132).

Although no biological activity of 3-aminoproline has been reported, 3-amino-3-carboxypyrrolidine (cucurbitine) was separated from seeds of *Cucurbita moschata* as an inhibitor of the growth of immature *Schistosoma japonicum* (99). Another isomer, 1-amino-D-proline is a constituent of linatine which was isolated as a growth inhibitor from linseed meal

(*234*). This imino acid has been shown to be a vitamin B_6 antagonist (*234, 379*) and an inhibitor of methionine metabolism (*231*).

3.3. *Pipecolic Acid* (**51**) *and Derivatives*

Pipecolic acid, a six-membered proline analogue, is encountered as frequently in fungi as in plants. The accumulation of pipecolic acid is shown by its unique position and ninhydrin coloration (bluish violet) on the paper chromatograms. In contrast to azetidine-2-carboxylic acid, pipecolic acid is not known to inhibit competitively the incorporation of proline into proteins. This is probably one of the simple cases in which molecular sizes of analogues play an important role in competitive inhibition. Pipecolic acid has been established as an intermediate in lysine catabolism in plants and in legumes in particular, a group in which many species contain a high level of lysine. Several derivatives of pipecolic acid are known as natural products, e.g. (2S,4S)-4-hydroxypipecolic acid (*58, 306*), (2S,4R)-4-hydroxypipecolic acid (*58, 227, 362*), (2S,5R)-5-hydroxypipecolic acid (*466, 185*), (2S,5S)-5-*hydroxypipecolic acid* (*77, 166, 97, 35*) (2S,5S,6S)- and (2S,5R,6S)-5-hydroxy-6-methylpipecolic acid (*247, 227*), and baikiain (4,5-dehydropipecolic acid) (*232*).

In the last decade, many naturally occurring derivatives of pipecolic acid have been reported. Thus, all four stereoisomers of (2S)-4,5-dihydroxypipecolic acid have been added to the list; (4R,5S)- (*277, 362*), (4R,5R)-(*388, 35*), (4S,5S)-(*277, 362*), and (4S,5R)-form (*36*). Furthermore, the occurrence of (2S,4R)-carboxy-2-acetylamino-4-piperidine (*278*) and of the sulphate ester of (2S,4S)-4-acetylaminopipecolic acid (*98*) were reported. In particular, from a tropical legume genus *Calliandra* no fewer than nine pipecolic acid derivatives are known (*362*). Mono- and dihydroxypipecolic acids seem to be formed sequentially from pipecolic acid (*406*).

3.3.1. 5-Hydroxypipecolic Acid (**54**)

As for fungi, (2S,5R)-5-hydroxypipecolic acid (*trans*-L-isomer) was isolated in our laboratory from *Stereum ostrea* (Blume et Nees) Fr. along with L-pipecolic acid (*161*). There have been only a few reports on the isolation of pipecolic acid and its modified forms from mushrooms, but we frequently detect pipecolic acid in fungal materials. We found pipecolic acid in six of 73 species which had been collected at random (*178*). It can therefore be predicted that pipecolic acid and at least several other derivatives are present in many other unexplored fungal species.

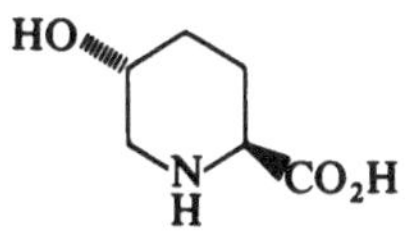

(54)

(2S,5R)-5-Hydroxypipecolic acid

3.3.2. 3-Hydroxybaikiain (55)

Baikiain (4,5-dehydropipecolic acid) **(56)** has long been known as a constituent of *Baikiaea plurijuga* (Leguminosae) (*232*). It has also been isolated from *Caesalpina tinctoria* and some related leguminous species (*442*). Baikiain inhibits the activity of glutamate as a neurotransmitter (*436*).

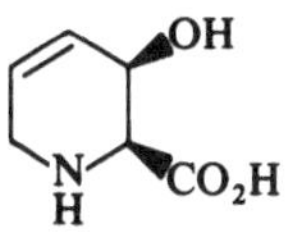

(55)   **(56)**

(2S, 3R)-3-Hydroxybaikiain   (2S)-Baikiain

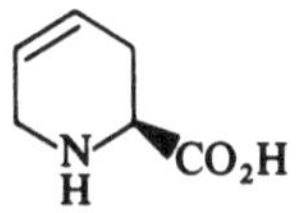

Recently KUSANO et al. (*248*) reported the isolation and structural elucidation of (2S,3R)-3-hydroxybaikiain **(55)**, together with (S)-baikiain and (S)-pipecolic acid, from a toxic mushroom, *Russula subnigricans* Hongo. Methanol extracts of the dried fruitbodies were partitioned between ethyl acetate and water and the water layer was applied to fractionation with Amberlite- IRC-50, -IR-120, and finally with cellulose. The yield was relatively high (3.5 g crystals from 250 g of dried fruit-bodies).

3-Hydroxybaikiain gives a yellow coloration with ninhydrin. The positive shift of $[\alpha]$ in acid indicates that it is an L-amino acid. The structure, (2S,3R)-3-hydroxybaikiain, was derived from the results of elementary analysis, MS, ^{1}H- and ^{13}C-NMR spectra. The coupling constant, $J = 3$ Hz, between the carbinyl hydrogen and the α-hydrogen suggested the *cis*-configuration. The structure of the hydrogenation product was assigned on (2S,3R)-3-hydroxypipecolic acid on the basis of the mass and ^{1}H-NMR spectra and it was further reduced with HI and red phosphorus to (2S)-pipecolic acid.

Although intoxication including fatalities after ingestion of fruit-bodies of *R. nigricans* was reported (*199*), (2S,3R)-3-hydroxybaikiain showed no toxicity in mice given 1 g/kg perorally. It is highly interesting that KUSANO *et al.* failed to detect 3-hydroxybaikiain by paper chromatography in the fruitbodies of the two very closely related species, *R. nigricans* and *R. adusta*. They are not considered to be toxic (*199*). Detection of 3-hydroxybaikiain may be useful in differentiating the fatally toxic *R. subnigricans*.

4. Amino Acids Containing Heterocycles

4.1. Pyrrolidine Amino Acids

4.1.1. 1-(3-Amino-3-carboxypropyl)-5-oxo-2-pyrrolidinecarboxylic Acid (57)

This is a derivative of pyroglutamic acid which was recently isolated and characterized from fruitbodies of *Lactarius piperatus* (Fr.) S.F. GRAY (*129*). The mixture of the acidic amino acids was treated with a Dowex 50 column which was equilibrated with an ammonium-formate buffer (pH 2.70) and the relevant fraction of the eluate was separated with Cellulofine GCL-25-m. The final purification was carried out on a cellulose column with ethanol-water. The yield was 225 mg of crude amino acid from 7 kg of the fruitbodies. The field desorption mass spectrum (FDMS) showed m/z 231 $[M+H]^+$. Careful examination of the ^{1}H-NMR spectrum of the amino acid and its *tert*-butoxycarbonyl dimethyl ester and other data led to the postulate that the structure was 1-(3-amino-3-carboxypropyl)-5-oxo-2-pyrrolidinecarboxylic acid. Starting with a derivative of L-glutamic acid and aspartic-β-semialdehyde (Scheme 6), the synthetic amino acid was obtained in good yield and was identical with the natural product in every respect. Because $[\alpha]_D$ of the natural amino acid, $-17.5°$, was very close to that of the synthetic sample, $-17.8°$, the absolute configuration

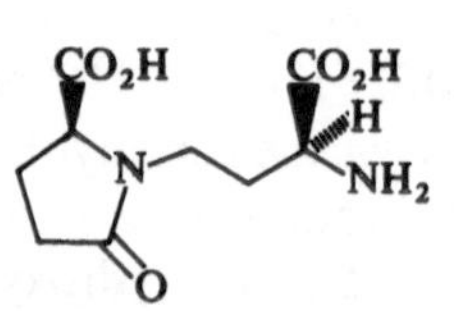

(57)

(2*S*, 3'*S*)-1-(3-Amino-3-carboxypropyl)-5-oxo-
pyrrolidinecarboxylic acid

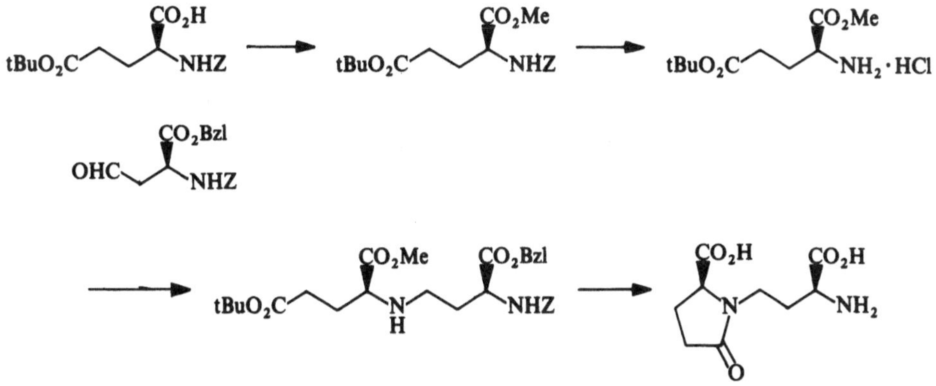

Scheme 6. Synthesis of 1-(3-amino-3-carboxypropyl)-5-oxo-2-pyrrolidinecarboxylic acid (from 129)

of the natural isomer was determined to be $(2S, 3'S)$. No biological activity has so far been reported.

4.2. Isoxazoline Amino Acids

Various derivatives of N-substituted isoxazolin-5-one are distributed in higher plants. The first report on the natural occurrence of this class of compounds, 2-alanyl-3-isoxazolin-5-one, appeared in 1969 by LAMBEIN and associates (250) as an UV-sensitive heterocyclic compound from pea seedlings. LAMBEIN's group extended their studies to other leguminous plants, including *Lathyrus odoratus*, and reported additional new iso-xazoline amino acids (251). The enzymatic introduction of side chains was also studied by MURAKOSHI et al. (305).

Isoxazoline amino acids were reviewed by LAMBEIN et al. (249). Much attention has been paid also to the non-protein amino acids of this type isolated from *Amanita* spp. They include substances causing poisoning by ingestion of the fruitbodies of *Amanita muscaria*, *A. pantherina*, and several other toxic mushrooms.

4.2.1. Tricholomic Acid (58)

In farmhouses in some northern districts of Japan, fruitbodies of *Tricholoma muscarium* Kawamura are sometimes used as a folk in-secticide against houseflies. They are heated with a small amount of water

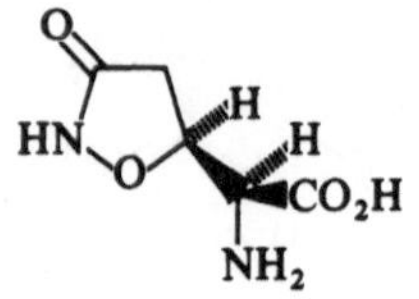

(58)

Tricholomic acid

in a tray. Physiological effects begin rapidly with leg paralysis in flies ingesting the material; the intoxication is lethal under these conditions. This mushroom was, however, considered to be edible. In 1964, TAKEMOTO and NAKAJIMA reported the isolation of the responsible constituent and named it tricholomic acid (*410*). The water extract was fractionated successively on Amberlite IR-120 and IR-4B, using the insecticidal activity as an indicator. They obtained 10 mg of crude crystals from 80 g of powdered dried mushrooms. The ninhydrin coloration was reported to be reddish-violet, the color later turning to the usual violet. In a subsequent report (*411*) the same authors showed that the substance underwent hydrogenation with the consumption of 1 mol/eqvt. of hydrogen and that the product yielded 3-hydroxyglutamic acid on reduction with hydriodic acid and red phosphorus. On hydrolysis with conc. HCl, this amino acid produced *erythro*-3-hydroxyglutamic acid, succinic acid, hydroxylamine and ammonia. From these results, together with the ^{1}H-NMR spectrum, the structure, α-amino-3-oxo-5-isoxazolidineacetic acid was assigned. Structurally, this is a cyclic form of 3-hydroxyglutamine.

Another noteworthy property of tricholomic acid is a pleasant flavor that is much stronger than that of monosodium glutamate. The flavor may result from the structural similarity of tricholomic and glutamic acids. Tricholomic acid was synthesized by IWASAKI *et al.* (*211*, *212*).

4.2.2. Ibotenic Acid (59) and Muscimol (60)

Fruitbodies of *Amanita pantherina* also contain a fly-killing substance(s) (*335*). TAKEMOTO and associates reported isolation of an amino acid similar to tricholomic acid from fruitbodies of *Amanita strobiliformis* (Paul.) Quél. (*414*). They named this substance ibotenic acid, basing the name on the Japanese name for the fungus ("ibo-tengu-take"). The yield was 12 mg of crude crystals from 800 g of fresh fruitbodies. Ibotenic acid gives a yellow color reaction with ninhydrin and the color turns violet with time. The structure, α-amino-2,3-dihydro-

(59)

Ibotenic acid

(60)

Muscimol

3-oxo-5-isoxazoleacetic acid, was elucidated mainly by studies of the decarboxylation product (413). Following the report on A. strobiliformis, the same authors reported the isolation of ibotenic acid from fruitbodies of A. muscaria (L.: Fr.) Hooker (fly agaric) and A. pantherina (DC.: Fr.) Secr. (412). The yields were 56 mg from 2.2 kg of the former and 640 mg from 3 kg of the latter species. As with tricholomic acid, ibotenic acid has a pleasant flavor similar to that of monosodium glutamate.

In the next year, two other research groups, EUGSTER et al. (95, 150) and BOWDEN et al. (37, 38) reported isolation of the same amino acid from fruitbodies of A. muscaria. The distribution of Tricholoma muscarium, the only known source of tricholomic acid, appears to be limited to Japan where it is found infrequently. On the other hand, A. muscaria is widely distributed and, if needed, considerable quantities of the fungus can be collected. Thus, EUGSTER once isolated 250 g of pure ibotenic acid from 770 kg (0.1%) of the 1962 harvest (95, 150). The chemistry and pharmacology of ibotenic acid have been studied in much more detail than those of tricholomic acid.

However, a strong reason for the intensive study of A. muscaria is that the toxicity of fruitbodies of this species was difficult to explain. That is, the concentrations of known constituents, such as muscarine, acetylcholine, muscaridine, etc., were too low for the reported symptoms.

In 1965 alone, EUGSTER and associates published many papers covering the isolation and structure of muscimol (303), praemuscimol (=ibotenic acid) (95, 150), the synthesis of ibotenic acid (130), and the structure of muscazone (124), etc.

Similarly to tricholomic acid, ibotenic acid is unstable in hot water, and it is easily racemized and decarboxylated, e.g. in 0.05 formic acid or even during two-dimensional paper chromatography (27), to give muscimol. The physiological effects of ibotenic acid on the central nervous system, called "pantherina syndrome", are considered mainly to be caused by this decarboxylated product (64). Muscimol is mostly an artifact created during the isolation procedure (94) and decarboxylation of ibotenic acid may occur in vivo as well. An improved isolation method to minimize the decarboxylation of ibotenic acid was reported by CHILTON and OTT (53). Both substances are present in much higher concentration in the flesh and gill than in the "peel" and stipe (*151*). The activity of muscimol is 5–10 times higher than that of ibotenic acid (*420*). Muscimol can be considered as a structural analogue of γ-aminobutanoic acid (a neurotransmitter) with restricted conformation (*39*). The relationship to synapic γ-aminobutanoic acid receptors has been reported (*76*). After various trivial names had been proposed for these two isoxazole amino acids, "ibotenic acid" and "muscimol" were adopted by the agreement of the authors (*96*).

As is well known, fruitbodies of *A. muscaria* contain a toxic quaternary ammonium base, muscarine. The absolute configuration of the natural form of muscarine has been determined (*420*); its physiological effects are due to the structural analogy with acetylcholine. The "muscarine-syndrome", characterized by strong perspiration, salivation and lacrimation, is not recognized in poisoning by *A. muscaria*. This is because the concentration of muscarine in this species is only ca. 0.0003%. The chemistry of the toxins in *A. muscaria* has been reviewed several times by EUGSTER (*93, 94*), one review appearing in this series in 1969 (*94*).

Pantherina syndrome, caused by *A. muscaria* and *A. pantherina*, was well documented by BENEDICT (*27*) and recently by BRESINSKY and BESL (*40*). Analytical methods for ibotenic acid and muscimol with an automated amino acid analyzer (*151*), gas chromatography (*357*), and HPLC (*272*) have been reported.

Ibotenic acid was reported also from three other species of *Amanita*, *A. gemmata* (*27, 53*), *A. cothurnata* (*53*) and *A. regalis* (Fr.) R. Mre. (*40*). A paper chromatographic survey in our laboratory revealed the presence of this amino acid also in *A. rubrovolvata* Imai. These species all belong to the subgenus *Amanita*, section *Amanita*, and the occurrence of ibotenic acid appears to be one of the specific properties of this section.

In regard to the ibotenic acid and muscimol contained in fruitbodies of *A. pantherina*, reported experimental results varied remarkably. Thus, EUGSTER failed to detect ibotenic acid and muscimol in one of his harvests (*96*). The concentration of these compounds seems to depend greatly on the season, locality and/or habitat, and other factors.

Some confusion occurred in former times regarding *Amanita strobiliformis* (Paul) Quél., studied by TAKEMOTO *et al.* This species is considered to be conspecific with *A. solitaria* (Fr.) Secr. However, in specimens of the latter species, BENEDICT *et al.* (*29*) could not detect any of the isoxazole narcotic-intoxicant compounds. Fruitbodies of "*A. strobiliformis*" (Ibo-tengu-take) used for the initial isolation of ibotenic acid might have been of *A. pantherina* itself. At present, Ibo-tengu-take is not yet recognized as a single species. We once had an opportunity to check the amino acid pattern of specimens called Ibo-tengu-take by experts in northern Japan.

No significant differences in free amino acid composition were observed between this sample and specimens of *A. pantherina* (DC.: Fr.) Secr. collected in a nearby location.

The decarboxylation product of 3-hydroxyglutamic acid, 4-amino-3-hydroxybutanoic acid, has also been reported to be present as the cyclic form, 4-hydroxypyrrolidine, in the same species (*282*). Occurrence in *A. muscaria* of this compound which is seldom encountered as a natural product does not appear to be accidental. They are probably closely related to the biosynthesis of tricholomic acid and ibotenic acid or, if not, to their biological degradation.

Among antibiotics, several closely related isoxazole amino acids are known, e.g., U-42, 126 (*280*) and U-43, 795 (*279*). These two amino acids have the same configuration at the two asymmetric carbon atoms, (L-*erythro*) as that of tricholomic acid, suggesting strongly ring closure of a common precursor.

It is highly interesting that the fruitbodies of *A. muscaria* produce a housefly attractant, 1,3-diolein (*310*). The biological significance of the concomitant occurrence of an attractant and an insecticide for the same insect remains unknown. However, Prof. R.J. BANDONI suggested to me (personal communication) that the mushroom might be luring flies with the attractant, killing them with the toxin and then digesting their bodies. That is, many fungi are capable of utilizing proteinaceous materials of insect carcasses. *A. muscaria* and similar mushrooms may have relatively elaborate mechanisms for insuring a supply of nitrogenous substances from this source, just as some mushrooms are capable of capturing and digesting nematodes.

4.2.3. Muscazone (61)

In the papers on isolation and characterization of muscimol, EUGSTER et al. (*95, 150*) also reported the presence of muscazone in *Amanita muscaria*. This amino acid has always been isolated as a racemate and does not undergo decarboxylation or exchange hydrogen readily. These facts suggest that muscazone is an artifact of isomerization of (2RS)-

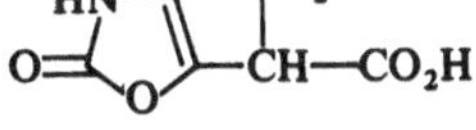

(61)

Muscazone

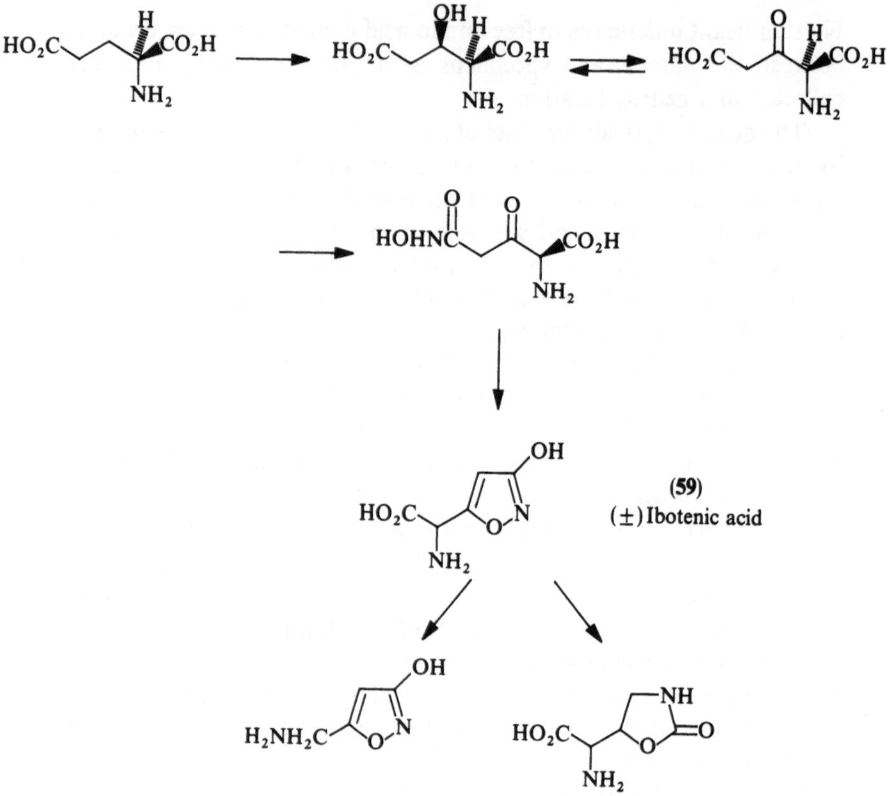

(59) (±) Ibotenic acid

(60) Muscimol

(61) (±) Muscazone

Scheme 7. Possible biosynthetic pathways of ibotenic acid, muscimol, and muscazone. Cited by TURNER and ALDRICH (*432*)

ibotenic acid (*50*). The photochemical conversion of ibotenic acid to muscazone has been reported (*152*).

Possible biosynthetic routes for ibotenic acid, muscimol, and muscazone were shown by EUGSTER, as summarized in Scheme 7.

4.3. Amino Acids Derived from Phenylalanine

4.3.1. Stizolobic Acid (**62**) and Stizolobinic Acid (**63**)

In 1959, HATTORI and KOMAMINE (*184*) isolated two new amino acids from exudate of velvet bean seedlings (*Stizolobium hassjoo*, Leguminosae);

(62)

Stizolobic acid

(63)

Stizolobinic acid

they named these stizolobic acid and stizolobinic acid. The former amino acid gave rise to an orange spot with ninhydrin, the latter gave a yellow to brown one. The structures of these α-pyrone amino acids were conclusively elucidated by synthetic studies (*385, 386*).

Seeds of *Stizolobium hassjoo* have long been known to contain unusually large amounts of (2*S*)-3,4-dihydroxyphenylalanine (L-DOPA) and it was assumed from the start of the study that stizolobic acid and stizolobinic acid are both derived from it. In fact, using epicotyls cut from the 4-day-old seedlings of *S. hassjoo*, SAITO et al. (*371, 372*) demonstrated incorporation of DL-[β-^{14}C]DOPA into stizolobic and stizolobinic acid in the dark. If an intramolecular rearrangement is assumed only in the aromatic ring of DOPA, the side chain of the DOPA should be introduced intact into these amino acids. In order to confirm this prediction, purified stizolobic acid biosynthesized from DL-[β-^{14}C]DOPA was subjected to alkaline degradation. The specific radioactivity of glutamic acid separated from the degradation products was as expected. From the results of feeding experiments with L-[G-^{14}C]phenylalanine, L-[G-^{14}C]tyrosine and DL-[3-^{14}C]DOPA, DOPA proved to be the most efficient precursor, tyrosine came next and phenylalanine was the poorest. Hydroxylation of phenylalanine to tyrosine had already been demonstrated in the same species by KOMAMINE (*238*). Two of extradiol cleavage modes of phenylalanine are possible, one of which leads to stizolobic acid and the other to stizolobinic acid (Scheme 8). When L-[3,5-^{3}H]- and L-[G-^{14}C]tyrosine were administered, the isolated stizolobinic acid had about 55% of the ratio ^{3}H/^{14}C of the fed precursor, but in stizolobic acid, only 4.4% of the precursor's activity was recovered (Scheme 8).

By cuticular administration to young leaves of *Mucuna deeringiana* with [G-^{14}C]tyrosine or [G-^{14}C]DOPA, ELLIS (*90*) also obtained incorporation of radioactivity into stizolobic acid. The genus name *Mucuna* is considered to be synonymous with *Stizolobium*. Using doubly labeled [2,3-side chain-^{3}H; G-^{14}C]tyrosine, he also observed the loss of 50% of the tritium in the biosynthesized stizolobic acid. Another experiment by

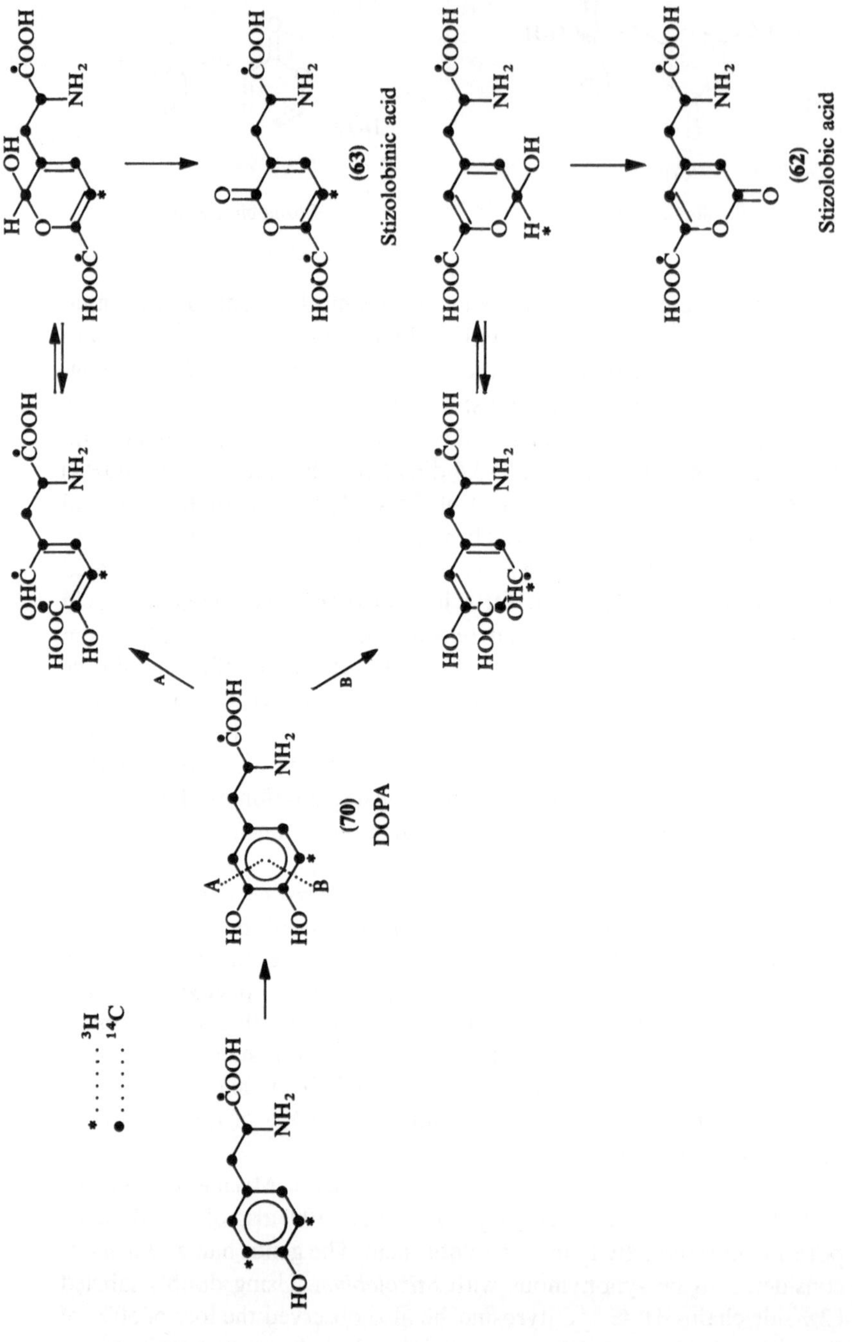

Scheme 8. Possible biosynthetic pathways leading to stizolobinic acid and stizolobic acid in *S. hassjoo* (from *371*)

ELLIS (*90*) demonstrated that the radioactivity of the carboxyl group of L-tyrosine was not randomized in the formation of stizolobic acid.

SAITO and KOMAMINE (*368*) subsequently obtained a cell-free system catalyzing the above sequential reaction by using an insoluble polyphenol adsorbent (Polyclar AT) and a reducing agent (araboascorbic acid) on an extract of *S. hasjoo* seedlings. This permitted separation of two enzymes, stizolobic synthase and stizolobinic synthase (*369*), from each other by fractionation of the cell-free extract with $(NH_4)_2SO_4$, Sephadex G-100, DEAE-cellulose, and hydroxyapatite. The molecular weights were both 45000, with Km-values for 3,4-dihydroxyphenylalanine of 1.67 mM and 1.39 mM, respectively. Both required $NADP^+$ under aerobic condition were activated by Zn^{2+}, and inhibited by sulfhydryl binding or zinc chelating agents. Without doubt, they are the enzymes, by which stizolobic acid and stizolobinic acid are formed in *Stizolobium*.

In 1974, CHILTON et al. (*52*) isolated both of the above amino acids from fruitbodies of *Amanita pantherina*. They used ion exchange and cellulose chromatography, and finally preparative paper chromatography and electrophoresis. The amino acids were detected in small amounts also in *A. muscaria*, *A. muscaria* var. *muscaria*, var. *formosa*, and *A. gemmata* (*53*). In the case of *A. gemmata*, the concentrations of ibotenic acid and muscimol, as well as stizolobic acid and stizolobinic acid, were reported to be variable, particularly depending on the locality (*53*).

SAITO et al. (*370*) administered DL-[3-^{14}C]DOPA to intact fruitbodies of *A. pantherina* through a paper strip inserted into a slit made longitudinally on their stipes. After 24 hours' incubation, 0.020% and 0.021% of the radioactivity of the immersion medium was found in stizolobic acid and stizolobinic acid, respectively. After 72 hours, the radioactivities of these amino acids increased 5–8 times compared with those after 24 hours' incubation. Incorporation of radioactivity into stizolobic acid was confirmed by purification and repeated crystallization with a cold authentic sample. Both of these amino acids are probably biosynthesized quickly from DOPA, thus resulting in little accumulation of DOPA in the mushrooms.

4.3.2. Muscaflavin (**64**)

The striking red color of the caps of *Amanita muscaria* has long drawn the attention of organic chemists. For nearly two decades, knowledge has accumulated and it is now known that the "peels" of the caps contain a mixture of various kinds of pigments. A great contribution to this subject was made by MUSSO and associates who have published several reviews on the pigments of *Amanita muscaria* (*308, 309*).

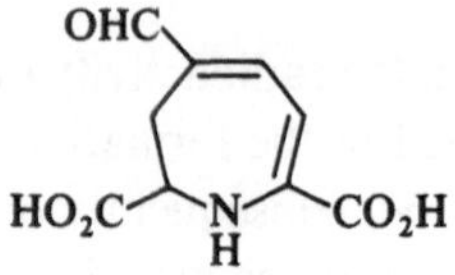

(64)

Muscaflavin

The first of these pigments to be characterized was muscaflavin; the remainder are condensation products of betalamic acid (65) (see below) and various kinds of amino acids including ibotenic acid and stizolobic acid (Table 5). Muscaflavin is an unstable yellow pigment ($\lambda_{max} = 425$ nm); the initially proposed structure (81, 82) was later revised to 64 (440) and confirmed by chemical synthesis (20). It proved to have a dihydroazepine structure. The resolution of the synthetic enantiomers of dimethyl ester has been achieved by chromatography on potato starch in 1 M potassium citrate buffer at pH 7.0 (20).

Further occurrences of muscaflavin in yellow and red species of *Hygrocybe* were reported by VON ARDENNE et al. (440). The yield was 0.2% of dry weight. Identification was made as stable dimethyl ester by means of m.p., chromatographic behavior, ^{1}H-NMR, mass, and IR spectrometry. Careful examination of the extracts of these *Hygrocybe* spp. has so far failed to detect derivatives of betalamic acid.

For the biosynthesis of muscaflavin, an extradiol cleavage of DOPA is also predicted without any difficulties. When the intermediate glutaconaldehyde derivative recyclizes by route b as shown in Scheme 9, muscaflavin is formed. Route a gives rise to betalamic acid which will be mentioned in the next section. In the biosynthetic routes to stizolobic and stizolobinic acids, the alanine moiety of DOPA do not participate in the ring closure. However, in the biosynthesis of muscaflavin and betalamic acid, alanine is directly involved. As stated above, no betalamic acid derivatives could be detected in any of the *Hygrocybe* spp. It can therefore be assumed that the genus *Hygrocybe* has the enzyme system only for the route b, and that the enzyme system in *Amanita* is able to catalyze both routes a and b.

4.3.3. Betalains (65)

The term "betalains" was introduced in 1968 by MABRY and DREIDING to denote inclusively the betacyanins and betaxanthins which are found specifically in most members of the order Centrospermae. It is of high interest that the betalains are not accompanied by any of the anthocya-

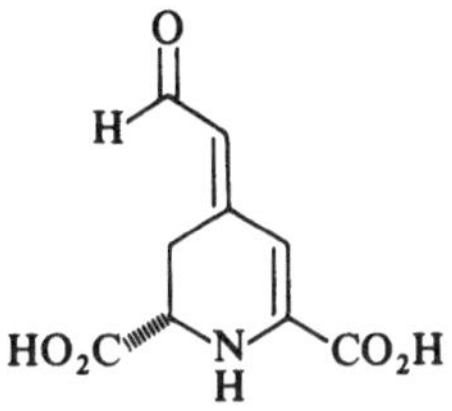

(65)

Betalamic acid

nins which are otherwise very widely encountered in nature (*275*). Many reviews have been published on betalains in plants (*273, 274, 343*).

Over 50 different betacyanins are now known and are pigments contributing to the red-violet coloration of certain plants. They mostly occur in living cells as glycosides of betanidin or isobetanidin. The structure of betanidin was elucidated by WYLER et al. (*469*). Isobetanidin is a C-15 epimer of betanidin. Thus, betanin, the red beet pigment, is formulated as 5-*O*-β-D-glucopyranosyl betanidin, and amaranthin and isoamaranthin in the leaves of *Amaranthus tricolor* are *O*-(β-D-glucopyranosyluronic acid)-5-*O*-β-D-glucopyranosides of betanidin and isobetanidin, respectively (*348, 345*). Free betalamic acid was also detected in numerous species which contain betalain pigments (*230*).

The betaxanthins are conjugates of betalamic acid (**65**) and various amino acids. The first well-studied betaxanthin was indicaxanthin from the fruit of a cactus, *Opuntia ficus-indica* (*344, 347, 290*) and several other betaxanthins are now known. Recently, humilixanthin, a new betaxanthin, the 5-hydroxynorvaline-immonium conjugate of betaxanthin, was reported from the fruits of *Rivina humilis* (Phytolaccaceae) (*398*).

On acid hydrolysis, all betaxanthins liberate an amino acid or amine, and similar to betacyanins, alkaline fusion yields 4-methylpyridine-2,6-dicarboxylic acid.

It is very interesting that the "peels" of the caps of *Amanita muscaria* contain several betaxanthins which are known to be distributed in Centrospermae. As early as 1930, KÖGL and ERXLEBEN (*235*) reported the structure of a pigment called "muscarufin" which was widely accepted. However, none of the pigments isolated in later studies of this species corresponded to "muscarufin".

From 19.6 kg of red "peels" obtained from 294 kg of *A. muscaria* collected in the Schwarzwald, MUSSO et al. (*82*) extracted the pigments with cold methanol. The colored substances were extremely sensitive to temperature and light, as well as to acid and alkali. After careful evaporation and lyophilization, successive fractionations were carried out on various Sephadex columns. In particular, column chromatography on DEAE-Sephadex with NaCl-gradient elution was successful. In order to avoid decomposition of the pigments, some steps were performed rapidly in the dark at 3–5° on a small amount of the pigment-mixture. The yellow sodium salt ($\lambda_{max} = 420$ nm) was named muscaflavin (see the preceding section), the seven orange-red salts ($\lambda_{max} = 475$ nm) musca-aurins I–VII, the red violet salt ($\lambda_{max} = 540$ nm) muscapurpurin, and the reddish brown salt ($\lambda_{max} = 495$ nm) muscarubin. On acid hydrolysis, each of them gave rise to betalamic acid and various amino acids (Table 5). Betalamic acid was identified conclusively in the form of its dimethyl ester, and the amino acids by liquid chromatography-mass

Table 5. *Muscaurins from* Amanita muscaria *and Their Occurrence in Plants* (from *308*)

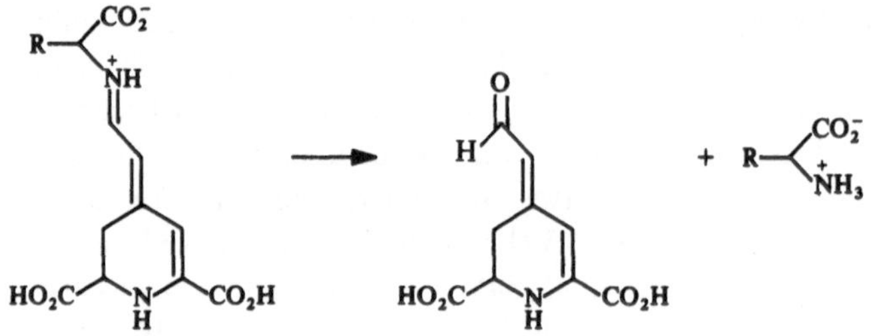

Muscaurins (Betaxanthins)	(65) Betalamic acid	Amino acids

Muscaurins	Amino acids	Occurrence in plants
I	Ibotenic acid	
II	Stizolobic acid	
III	Glutamic acid.	*Beta vulgaris*
	+	(Vulgaxanthin I)
	as minor components	
	α-Aminoadipic acid	
	Aspartic acid	
IV	Aspartic acid . . .	*Mirabilis jalapa*
	+	(Miraxanthin III)
	as minor components	
	Glutamic acid	
V	Glutamine . . .	*Beta vulgaris*
	Asparagine	(Vulgaxanthin II)
	Leucine	
	Valine	
VI	Proline	*Opuntia ficus-indica*
VII	Histidine	(Indicaxanthin)

References, pp. 117–140

spectrometry in the form of their *N*-triacetyl methyl esters. The muscaurins were synthesized and their structures were verified (*80*). The amino acids in muscapurpurin and muscarubin have not yet been unequivocally identified.

It is rather surprising that fruitbodies of *A. muscaria* contain pigments which normally occur in red beet, *Beta vulgaris*, and other plants. Glutamic acid and glutamine containing betalains are known as vulgaxanthin I (*349*) and vulgaxanthin II (*349*), respectively. Both pigments are known from *Beta vulgaris*. Similarly, the betalain containing aspartic acid is the same as miraxanthin III from *Mirabilis jalapa* (*346*) and the proline containing betalain is identical with indicaxanthin from the fruit of a cactus, *Opuntia ficus-indica* (*344, 347, 290*). All of these structures have been confirmed by chemical synthesis. When betanin (R : β-D-glucopyranosyl) was hydrolyzed with ammonia in the presence of a great excess of an amino acid, an equilibration reaction occurred, yielding the corresponding betalain pigment. A possible structure of muscapurpurin has been discussed (*308*).

In red beets and figs use of [14]C- and ^{3}H-precursors has shown that betalamic acid is formed from L-tyrosine *via* 3,4-dihydroxyphenylalanine (*288, 88, 102, 201*). The intermediate leading to the formation of stizolobic acid can also cyclize between the amino group and the enolized 2-oxogroup in position 6 (Scheme 9), giving betalamic acid. Extradiol cleavage of DOPA at another position leads to another intermediate which gives rise to stizolobinic acid. Muscaflavin would be biosynthesized by an alternative way of ring closure between amino group and C-7 (Scheme 9)*.

Pigment profiles of the fruitbodies of many samples of *A. muscaria* from various regions in Europe and from the United States, have been examined in MUSSO's laboratory. All specimens were found to contain the same pigments, although the relative concentrations varied in some cases (*83*). In all other Amanitas tested (i.e. *A. citrina, A. pantherina, A. rubescens, A. spissa,* and *A. fulva*) the pigments were absent (*309*). Exceptionally, two yellow-orange pigments were found in *A. flavoconia* (*83*). The similarly orange-capped and edible *Amanita caesarea* seems to contain all of the pigments of betalamic acid-amino acid conjugates present in *A. muscaria* with the exception of muscaurin I, an ibotenic acid conjugate (*309*). As one might expect in an edible species, ibotenic acid was not present in detectable amounts.

---

* Quite recently P.-A. DIROD and J.-P. ZRYD reported "Biogenesis of Betalains: Purification and Partial Characterization of DOPA 4,5-dioxygenase from *Amanita muscaria*" [Phytochem. **30**, 169 (1991)].

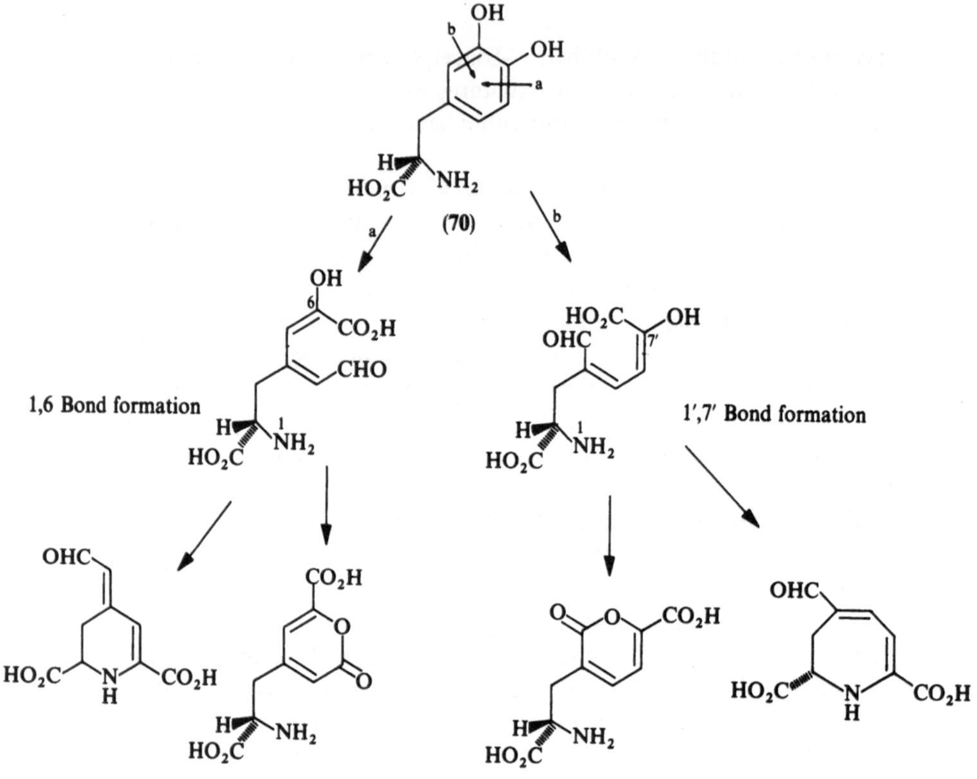

(65) Betalamic acid (62) Stizolobic acid (63) Stizolobinic acid (64) Muscaflavin

Scheme 9. Biosynthesis of betalamic acid, muscaflavin, and related compounds (from *308*)

4.3.4. Acromelic Acid A (**66**) and Acromelic Acid B (**67**)

Fruitbodies of *Clitocybe acromelalga* Ichimura exhibit an unusual toxicity. Accidental ingestion causes violent pains and a marked reddish edema in the hand and foot after several days and the pain continues for several weeks. The symptoms are similar to acromelagia and erythromelagia (*242*). SHIRAHAMA and associates have studied extensively the toxic substances occurring in this mushroom. Using an assay mainly the lethal effect in mice, they have isolated and characterized five new compounds. Two of them were named acromelic acid A and acromelic acid B, both being obtained in minute amounts as potent neuroexcitatory amino acids (*244*). These amino acids showed yellow colorations with ninhydrin and strong blue fluorescence under UV light. Structures were originally postulated only on the basis of the ^{1}H-NMR, UV, and CD

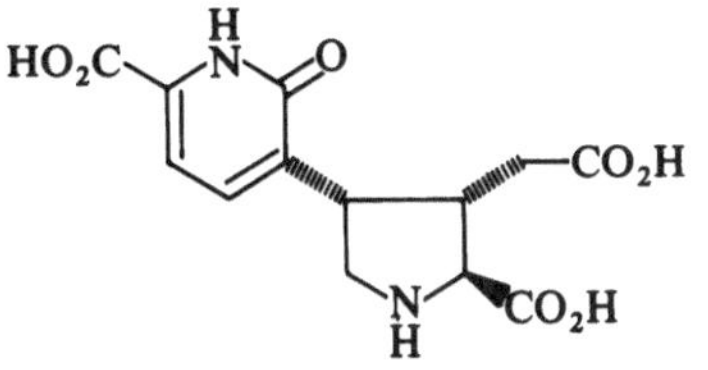

(66)   (67)

Acromelic acid A   Acromelic acid B

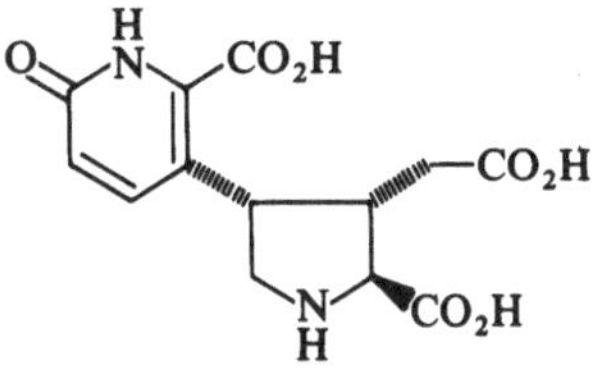

spectra. Biogenesis of kainic acid, domoic acid, and stizolobinic acid also gave important clues to the chemical structures. Shortly afterwards, the proposed structures of acromelic acid A (66) (239, 409, 240, 19) and B (67) (157, 240) were verified by synthesis. Similar to other naturally occurring kainoids, such as kainic acid and domoic acid, acromelic acids A and B cause a marked depolarization of the mammalian central nervous system, particularly of the spinal cord and invertebrate muscle fiber (389, 206).

Acromelic acids A and B may also considered to be derived from phenylalanine as shown in Scheme 10 (240). Extradiol cleavage followed by cyclization with ammonia would form pyridine carboxylic acid. If this condenses with glutamic acid, acromelic acid A is formed. Similarly extradiol cleavage at the alternative side leads to acromelic acid B. Although acromelic acids A and B are highly toxic to the central nervous system, their relationship to the unique symptoms which arise after ingestion of the fruitbodies of C. acromelalga is still not well understood.

4.4. Oxygen Heterocyclic Amino Acids

L-3-(2-Furoyl)alanine was isolated from buckwheat (Fagopyrum esculenta Moench) seeds (196, 225). Its ninhydrin coloration was orange-yellow. Synthetic β-2-furylalanine (57) and β-3-furylalanine (265) are both microbial growth inhibitors, and are considered to be analogues of phenylalanine.

4.4.1. 3-(3-Carboxyfuran-4-yl)alanine (68)

This is the only furan amino acid known to-date from mushrooms. In 1974, DOYLE and LEVENBERG (85) reported this amino acid from fruitbodies of Phyllotopsis nidulans (Pers.: Fr.) Sing. Separation and purification were carried out by treatment of a methanol extract with Amberlite IRA-400 (Cl⁻-form) followed by preparative paper chromatography.

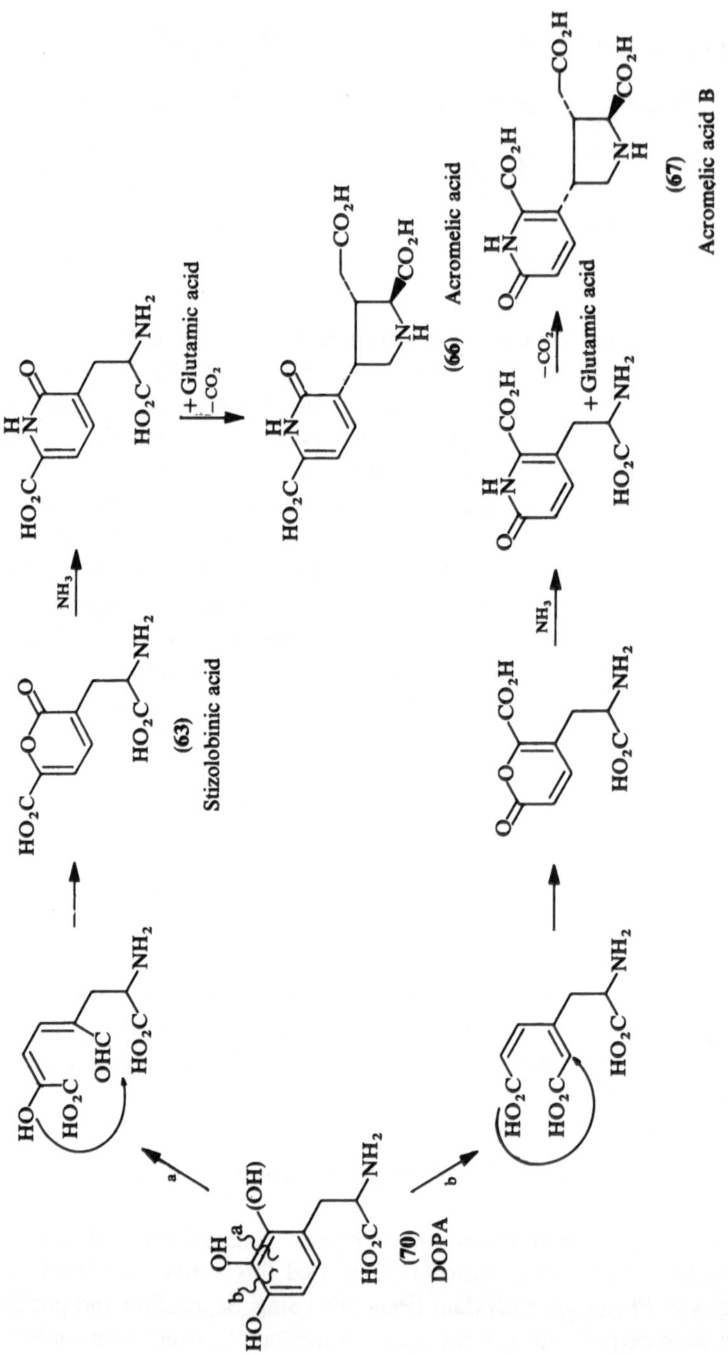

Scheme 10. Possible biosynthetic routes of acromelic acid A and acromelic acid B (from 240)

(68)

3-(3-Carboxyfuran-4-yl)alanine

Comparison of the UV spectrum, $\lambda_{max} = 241$ ($\varepsilon = 3500$), with spectra of known compounds indicated the presence of a furan ring. The ^{1}H-NMR spectrum showed that this amino acid was a furanylalanine containing an extra carboxyl group. Furthermore, the positions of the substitutions were at C-3 and C-4, a deduction confirmed by comparing the chemical shifts of the furan ring protons with those of the model compounds.

Results of the elementary analysis were also in good agreement with the proposed chemical formula. The optical rotation was not measured because of the scanty sample available. However, the benzoyl derivative of a small amount was oxidized with 0.1 M $KMnO_4$ and, after debenzoylation, L-aspartic acid was demonstrated quantitatively using malate dehydrogenase-glutamate oxalacetate transaminase procedure (*342*).

In the following year, HATANAKA and NIIMURA (*172*) reported isolation of the same amino acid from *Tricholomopsis rutilans* and from several other species in the same genus. These also contain several unsaturated norleucine-type amino acids. The concentrated amino acid fraction was applied to a column of Dowex 1 (acetate-form) and the acidic amino acids were displaced with 0.5 N acetic acid. The furan amino acid was eluted in pure form after aspartic acid unaccompanied by other amino acids. The yield was 695 mg from 3 kg of fruitbodies. The optical rotations $[\alpha]_D = -48°$ (c 1; H_2O) and $-28°$ (c 0.5; 3 N HCl) indicated that this amino acid belonged to the L-series. In the ^{1}H-NMR spectrum the signals of two furan protons at 7.45 (s) and 8.05 (d, $J = 1.2$ Hz) could be assigned to those of the positions at C-5 and C-2, respectively. The latter signal was split by a small coupling constant.

Fruitbodies of *Phyllotopsis nidulans* and *Tricholomposis* spp. thus contain the same rare amino acid. So far, from the former fungus, unsaturated norleucine-type amino acids have not been reported.

4.4.2. Lycoperdic Acid [3-(5-Carboxy-2-oxotetrahydrofuran-5-yl)-2-alanine] (69)

In 1979 RHUGENDA-BANGA *et al.* (*358*) reported an interesting heterocyclic acid from the fruitbodies of *Lycoperdon perlatum*. The ninhydrin

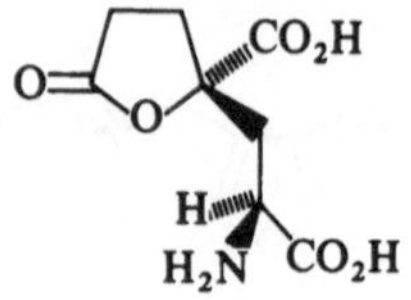

(69)

(αS, 2S)-Lycoperdic acid

reaction was strongly yellow and after 24 hr changed to violet. An ethanol extract of 750 g of fresh mushrooms was treated with a cation exchanger, Lewatit. The neutral and acidic amino acids were eluted by 1 N pyridine and were further fractionated on Dowex 1 (acetate-form) as usual. The new amino acid was displaced after aspartic acid and its behavior on high voltage electrophoresis also suggested a strong acidic nature. The yield was 260 mg. The pure sample showed a positive Cotton effect while the optical rotations were $[\alpha]_D = +14.9°$ (c 0.47; H_2O) and $+36.5°$ (c 1.37; 1 N HCl). If the effect of C-5 were negligible in this case, the configuration at C-2 should probably be S. Elementary analysis, potentiometric titration, and α-amino group analysis according to the Van Slyke method indicated presence of a monoaminodicarboxylic acid of formula $C_8H_{11}NO_6$. Characteristically, the IR spectrum of lycoperdic acid showed an absorption band at 1770 cm^{-1} which suggested that it was a γ-lactone. The entire structure including the absolute configuration of the two asymmetric carbon atoms was elucidated by a thorough analysis of the mass, ^{1}H- and ^{13}C-NMR spectra.

As expected, the lactone ring of lycoperdic acid was opened by hydrolysis with 3 N NaOH. The ninhydrin reaction of the open form was initially green-olive changing soon to violet. In the product the absorption band of the lactone had disappeared.

Although the open type might have been formed during treatment of the extract with cation exchangers, the authors considered both lactone and open chain amino acids to be normal metabolites of this fungus. They detected, together with a large amount of the lactone on a paper chromatogram, a small spot corresponding to the open chain acid without treatment with cation exchanger. The crystal and molecular structure of this amino acid were determined by X-ray analysis and proved to be 3-[(5S)-5-carboxy-2-oxotetrahydrofuran-5-yl]-(2S)-alanine (252).

5. Aromatic Amino Acids

Relatively few aromatic amino acids are so far known to be specific to higher fungi. It is expected, however, that fungal taxa might be found in the future which, like the Curciferae and Resedaceae (255), can synthesize a variety of aromatic amino acids.

5.1. Derivatives of Phenylalanine

5.1.1. 3,4-Dihydroxyphenylalanine (70)

As stated previously, 3,4-dihydroxyphenylalanine (DOPA) is a precursor of various kinds of betalain pigments in plants and fungi as well as of stizolobic acid and stizolobinic acid. According to our experience (178), DOPA is rather rare in fungi. STEGLICH and ESSER reported the existence of DOPA in fruitbodies of Strobilomyces floccopus (Vahl: Fr.) Karst. (395). Similar to some higher plants, such as Vicia faba and Stizolobium hassjoo, the color of this mushroom changes to reddish to grey-black when it is wounded or bruised. This color change is considered to be caused by oxidation of DOPA. Furthermore, STEGLICH and PREUSS (396) isolated DOPA from the yellow to scarlet fruitbodies of Hygrocybe conica (Scop.: Fr.) Kummer. Fruitbodies of this species contained DOPA in 3.2% of dry weight together with minor amount of muscaflavin. The "grey mushroom" Hygrocybe ovina (Bull.: Fr.) Kühn. contains DOPA and it is also responsible for the color change to black and red of the fruitbodies when they are bruised.

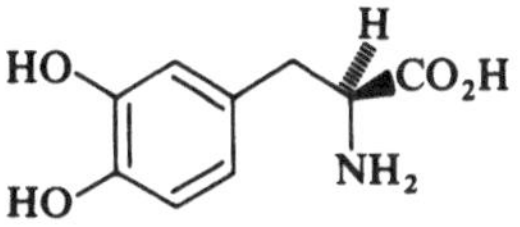

(70)

(S)-3,4-Dihydroxyphenylalanine

5.1.2. 4-Hydroxy-3-methoxyphenylalanine (71)

This amino acid was reported from the fruitbodies of Cortinarius brunneus Fr.: Pers. by DARDENNE et al. (74). The ninhydrin color reaction was violet brown. Isolation was carried out on columns of Lawatit S 1080 and Dowex 1 × 2, successively, and finally with preparative paper chromatography. The yield was 200 mg from 2 kg of fruitbodies. Elementary analysis, Rosen test, IR, mass, ^{1}H- and ^{13}C-NMR spectra, etc. revealed the structure. From the positive Cotton effect, the absolute configuration was considered to be (2S).

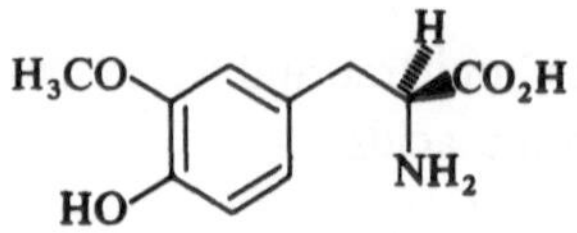

(71)

(S)-4-Hydroxy-3-methoxyphenylalanine

5.2. Derivatives of Tryptophan

All phallotoxins, amatoxins, and virotoxins contain each one molecule of a modified tryptophan. Bridges between C-2 of the tryptophan molecule and cysteine in these cyclopeptides are considered to be formed after ring closure of the peptides is completed. In the case of the virotoxins, the sulfur atoms in the bridges are first methylated and the bridges are broken, followed by oxidation of the sulfur atom.

As a free amino acid, 5-hydroxytryptophan (72) is known from the fruitbodies of *Panaeolus campanulatus* (433).

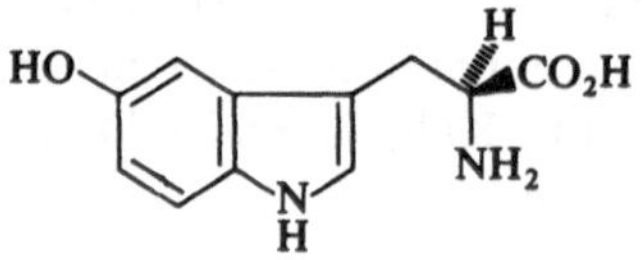

(72)   **(73)**

5-Hydroxytryptophan   Bufotenine

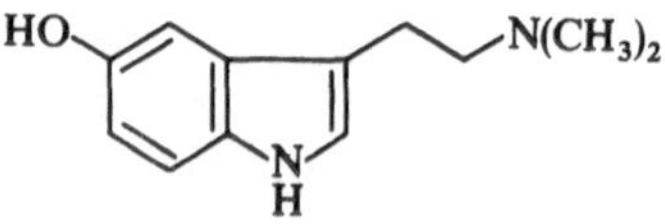

A few Amanitas are known to contain 5-hydroxytryptophan, 5-hydroxytryptamine, and bufotenine (73). They are *A. citrina*, *A. porphyria* and SINGER uses this character to delimit two Stirps within the section *Mappe*, Stirps *Citrina* and Stirps *Brunnescens* (392). 5-Hydroxytryptophan (25, 355) as well as L-DOPA (25, 356) are presumed to be defense substances of some legume seeds against insect predation.

Among tryptophan derivatives of mushrooms, hallucinogenic compounds such as psilocybin (74), psilocin (75), baeocystin (76), and norbaeocystin (77) are well-known (191, 260, 261). Although they are not amino acids, they are mentioned here briefly as nitrogen-containing products of mushrooms with unique physiological activities.

The history of research on the hallucinogenic mushrooms, started from an ethnobiological expedition to Mexico by WASSON (191, 267, 40). Psilocybin and psilocin were isolated first from *Psilocybe mexicana* Heim and later shown to be present not only in other species of *Psilocybe* but

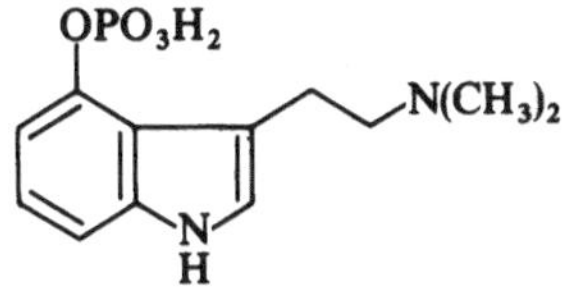

(74)

Psilocybin

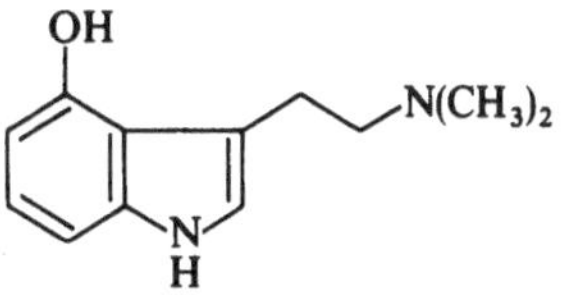

(75)

Psilocin

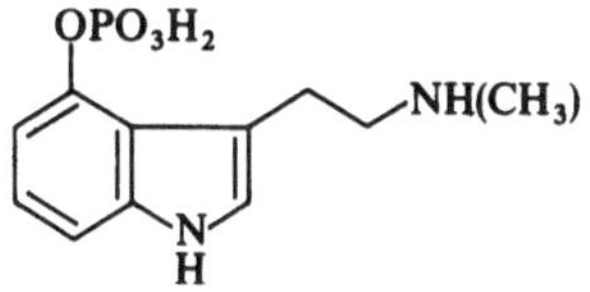

(76)

Baeocystin

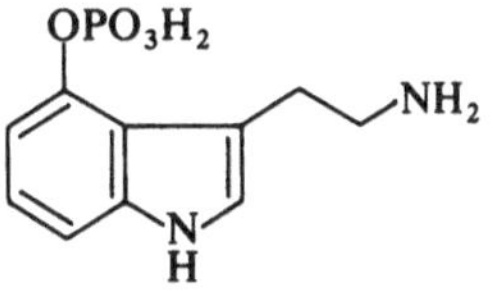

(77)

Norbaeocystin

also in mushrooms of other genera. Baeocystin (**76**) and norbaeocystin (**77**) were isolated and characterized also as hallucinogenic tryptophan derivatives from a saprophytic culture of *Psilocybe baeocystis* (*260, 261*).

4-Dimethylallyltryptophan (**78**) was isolated from ergot cultures as an efficient precursor of both clavine- and lysergic acid-type ergot alkaloids (*3*).

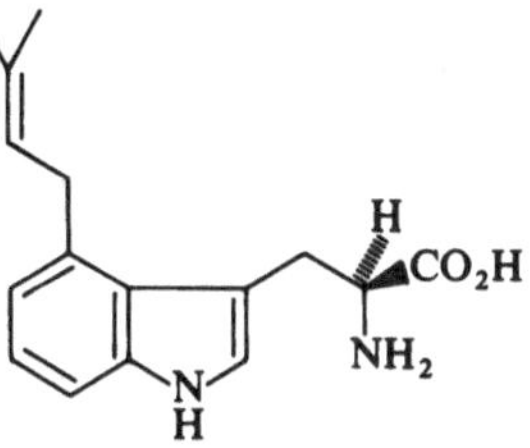

(78)

4-Dimethylallyltryptophan

A large number of species belonging to the various genera contain one or more compounds from among the above hallucinogenic tryptophan derivatives, namely, *Agrocybe, Conocybe, Copelandia, Gymnopilus, Inocybe, Panaeolina, Panaeolus, Pluteus, Pholiota,* and *Psathyrella* [Reference cited in (*40*), and (*28, 361, 267, 183, 447, 381, 133, 328*). OHENOJA *et al.* reported that psilocybin was detected in over 0.8% of the dry weight of *Psilocybe semilanceata* (Fr.) Kumm. and in 0.4% dry weight of *Conocybe cyanopus* (Akt.) Kühn. (*328*). KOIKE *et al.* (*236*) reported concentrations of psilocybin of up to 0.55% in dry samples of *Psilocybe argentipes* K. Yokoyama.

The hallucinogenic functions of psilocybin and related compounds are similar to those of LSD and are considered to be antagonistic to serotonin. Biosynthetic studies were carried out, using the cultured mycelia of *Psilocybe cubensis* (*4*).

Many tryptophan derivatives are known as metabolites of *Claviceps purpurea* and other *Claviceps* species, *Aspergillus* spp., and *Penicillium* spp. They have been reviewed concisely by TURNER and ALDRIDGE (*432*).

A structure of clavicepitic acid which was isolated from submerged cultures of *Claviceps* sp. SD-58 was proposed originally on the basis of instrumental analysis and biosynthetic studies (*360*). However, extensive studies of its mass and ^{1}H-NMR spectra subsequently led to the revision of the structure as in (**79**) (*233*). Using cell-free systems from *Claviceps* sp. SD-58 and *C. purpurea* PRL 1980, the conversion of 4-dimethylallyltryptophan to (**79**) was also studied (*18*).

γ-Glutamyl aromatic compounds will be described in a separate chapter.

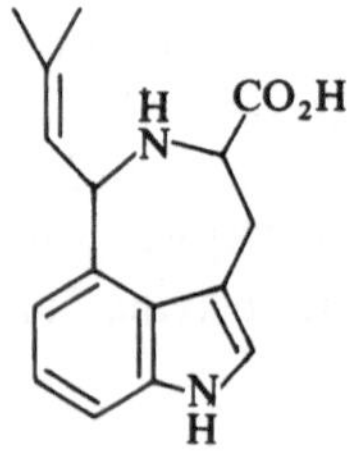

(**79**)

Clavicepitic acid

6. Miscellaneous Amino Acids

Commercially cultivated mushrooms can be easily obtained and, consequently, more intensive studies have generally been carried out on these species than on fruitbodies collected from natural habitats. The following enumeration deals mostly with amino acids and their derivatives that were already known from other living organisms and have been relatively recently demonstrated to occur in mushrooms.

Saccharopine (**80**) was reported from *Lentinus edodes* (Berk.) Sing. (Shii-take) (*11*) and *Agaricus bisporus* (*331*). From the former, nicotianine (**81**) and cystathionine (**82**) (*10*) have also been isolated. Cystathionine and saccharopine were reported to occur in *Flammulina veltipes* (*326*). Cystathionine was isolated from *Boletus erythropus* (*215*). N^δ-Acetyl-L-ornithine (**83**) was isolated and identified from the fruitbodies of *Pleurotus ostreatus* (*333*). Although this amino acid occurs rather frequently in

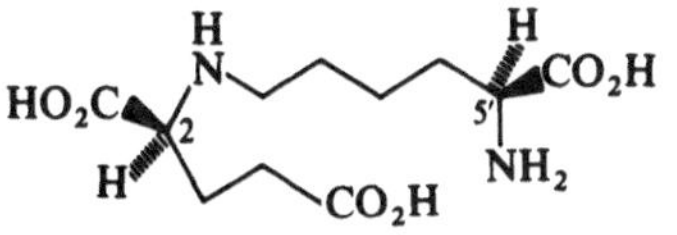

(80)

(2S, 5'S)-Saccharopine

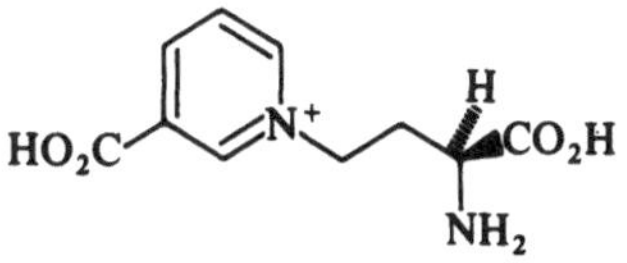

(81)

Nicotianine

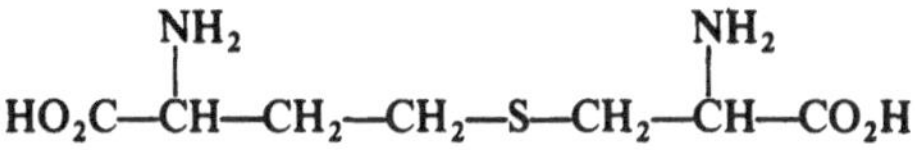

$$HO_2C-\overset{\overset{\displaystyle NH_2}{|}}{CH}-CH_2-CH_2-S-CH_2-\overset{\overset{\displaystyle NH_2}{|}}{CH}-CO_2H$$

(82)

Cystathionine

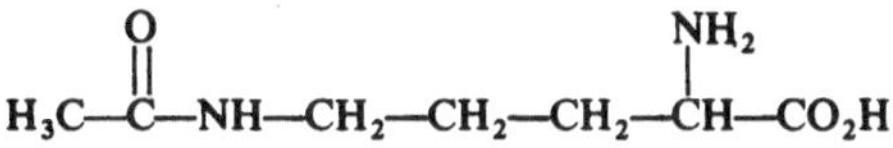

$$H_3C-\overset{\overset{\displaystyle O}{||}}{C}-NH-CH_2-CH_2-CH_2-\overset{\overset{\displaystyle NH_2}{|}}{CH}-CO_2H$$

(83)

N^δ-Acetylornithine

$$H_2N-CH_2-CH_2-S-CH_2-\overset{\overset{\displaystyle NH_2}{|}}{CH}-CO_2H$$

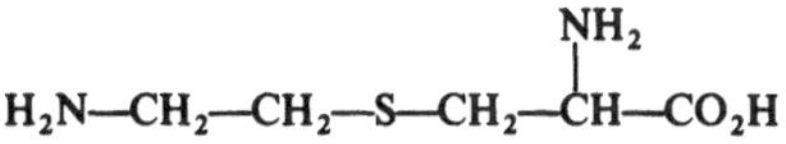

(84)

(2S)-S-2-Aminoethylcysteine

green plants, this was probably the first report of its occurrence in mushrooms. S-2-Aminoethyl-L-cysteine (**84**) was isolated by MATSUMOTO (*281*) from basidiocarps of *Rozites caperata* as an antibacterial amino acid. This is one of the lysine analogues and various biological activities are already known (*281*).

As mentioned previously, acromelic acids A and B were isolated from poisonous fruitbodies of *Clitocybe acromelalga*. From the same mushrooms three further compounds have been reported. Clitidine (**85**), a new pyridine nucleoside, is a weak toxin and 190 mg was obtained from 1 kg of fruitbodies. The structure, 3-carboxy-4-imino-1-(β-D-ribofuranosyl)-1,4-dihydropyridine, was confirmed by chemical synthesis (*241, 242*). Clithioneine (**86**), an amino acid betaine, has an imidazole nucleus and is closely related to ergothioneine (*243, 245*). The ninhydrin reaction is dark yellow and the yield was 38 mg from 16.2 kg fruitbodies. No biological activity is so far detected. Recently, 4-aminopyridine-2,3-dicarboxylic

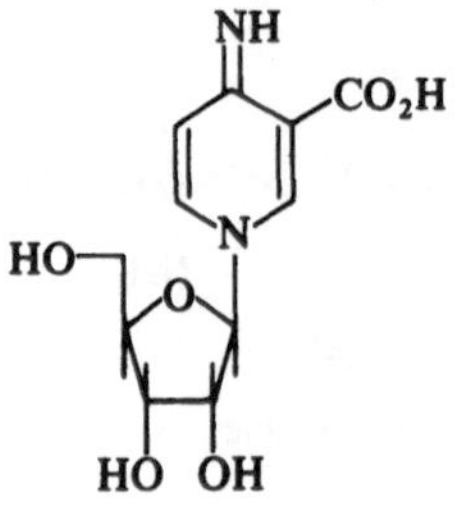

(85)

Clitidine

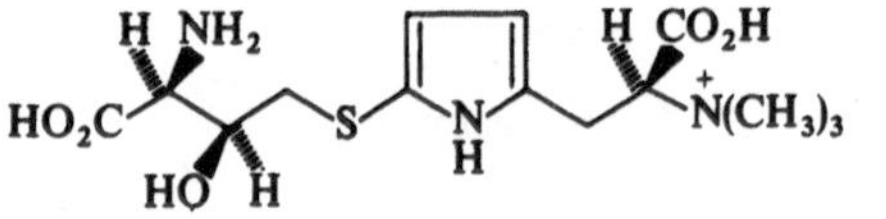

(86)

Clitioneine

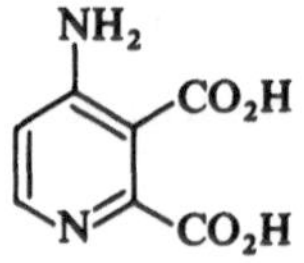

(87)

4-Aminopyridine-2,3-dicarboxylic acid

acid (**87**), a non-lethal compound was reported from fruitbodies of the above species (*189*).

The genus *Clitocybe* seems to be greatly diverse with respect to its chemical constituents. Thus, muscarine has been detected in a number of *Clitocybe*-species (*40*). As stated before, fruitbodies of *Clitocybe clavipes* have antabuse-like property and those of *C. infundibuliformis* are considered to be good edible mushrooms.

S-(Amino-2'-ethyl)cysteine and S-(acetamido-2'-ethyl)cysteine were isolated by WARIN *et al.* (*441*) from the edible mushrooms, *Rozites caperata*. The structures were verified by chemical synthesis. Later MATSUMOTO (*281*) also reported the former amino acid from the same species, together with 2-amino-3-butenoic acid, both as antibacterial amino acids.

FUGMANN and STEGLICH (*125*) reported several unique constituents from fruitbodies of *Lyophyllum connatum* (Schum.: Fr.) Sing. Connatin (**88**) is a derivative of citrulline, i.e. N^δ-hydroxy-N^ω,N^ω-dimethyl-

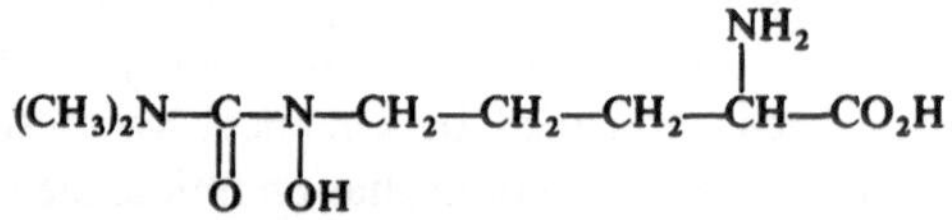

$$(CH_3)_2N-\underset{\underset{O}{\|}}{C}-\underset{\underset{OH}{|}}{N}-CH_2-CH_2-CH_2-\underset{\overset{NH_2}{|}}{CH}-CO_2H$$

(88)

Connatin

citrulline. Hydrolysis with 5% H_2SO_4 gives (S)-3-amino-1-hydroxy-2-piperidone, while use of 6 N HCl produces (S)-N^δ-hydroxyornithine hydrochloride. This is the first report of the occurrence of an N-hydroxyamino acid.

Information on basic amino acids in mushrooms is relatively scarce; however, 2-amino-4-(5-carboxythiazol-2-yl)butanoic acid (89) was isolated from fruitbodies of *Xerocomus subtomentosus* (217).

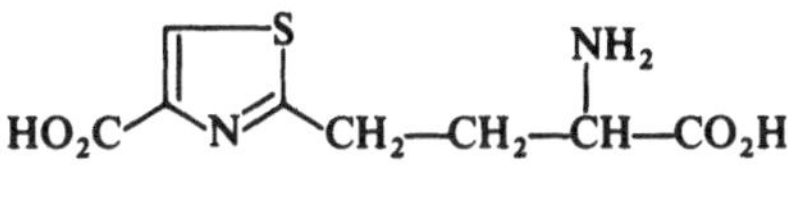

(89)

2-Amino-4-(5-carboxythiazol-2-yl)butanoic acid

7. γ-Glutamyl Compounds

According to KASAI and LARSEN's review (226) on γ-glutamyl derivatives from plants and mushrooms, it is very likely that more than 80 kinds of γ-glutamylpeptides are now known from plants and mushrooms. These are compounds in which protein amino acids, non-protein amino acids, or even some amines are bound to the γ-glutamyl moiety. Although the majority of the natural non-protein amino acids and amines are rather stable, some attain chemical stability only when bound to the γ-glutamyl moiety, e.g., some agaritine-related compounds and coprine. γ-Glutamyl compounds may be related to a γ-glutamyl cycle discussed by MEISTER and TATE (285). The fact that so many γ-glutamylpeptides are known suggests a common biological significance, i.e., the active transport of amino acids. In this connection, it is probably worth noting that, for instance, γ-glutamyl-D-alanine appears early in seedling development of *Pisum sativum*. This fact may give some indication of the physiological significance of this D-amino acid.

In the last ten years, non-protein amino acids and their γ-glutamyl compounds have been intensively studied in the two laboratories of SASAOKA and SUGAHARA in Japan.

7.1. γ-Glutamyl Compounds of Protein Amino Acids

γ-Glutamylpeptides of 19 of the 20 protein amino acids have been isolated and characterized; that of proline is not yet known from any mushroom or higher plant species. KASAI and LARSEN mentioned that this may simply reflect the difficulties involved in the isolation of these compounds. In fact, acidic amino acids and their derivatives can be purified and crystallized more easily than neutral or basic ones. The major sources are Fagaceae, Liliaceae, Alliaceae, and, in particular, seeds of leguminous plants. γ-Glutamylpeptides of several protein amino acids, leucine, phenylalanine and tyrosine, have been isolated from many plant species. In the following discussion, only a few recently isolated and characterized γ-glutamylpeptides of protein amino acids from mushrooms will be briefly mentioned.

7.1.1. γ-Glutamylglycine

The most commonly cultivated mushrooms, *Agaricus bisporus* and *Lentinus edodes* can easily be used as raw material for large scale extractions and many studies have been published on interesting compounds from these species. *Agaricus bisporus* contains γ-glutamylglycine, the simplest γ-glutamyl amino acid, as demonstrated by OKA and associates (*331*). They treated an ethanol extract successively with Amberlite IR-120 and Dowex 1, and one of the acidic amino acid fractions was purified by paper chromatography to obtain the pure peptide. It was easily hydrolyzed by 3 N HCl to form equimolecular amounts of glutamic acid and glycine. The configuration of the glutamic acid was also determined. By dinitrophenylation, glutamic acid was shown to be the N-terminal. Paper chromatographic behavior was the same as that of a synthetic sample.

7.1.2. γ-Glutamylmethionine

γ-Glutamylmethionine is known only from the excellent edible mushroom, *Boletus edulis* Bull.: Fr. (*425*). However, there must be many other instances in which these substances were detected, as "expected", but not studied closely. Little attention is often paid to natural products already known from other species, especially if no interesting biological functions are known for such substances.

7.1.3. γ-Glutamylcystine and Related Compounds

AOYAGI *et al.* reported isolation and identification of γ-glutamylcystine (*13*), N,N'-bis-γ-glutamylcystine (**90**) (*12, 13*) and N,N'-bis-γ-glutamylcystinylglycine (**91**) (*13*) from fruitbodies of *Lentinus edodes*. Fractionation and purification were performed using Sephadex and

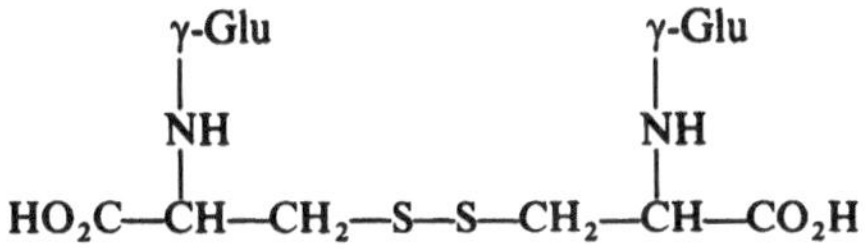

$$\gamma\text{-Glu} \qquad\qquad \gamma\text{-Glu}$$
$$| \qquad\qquad\qquad |$$
$$NH \qquad\qquad\qquad NH$$
$$| \qquad\qquad\qquad\qquad |$$
$$HO_2C\text{—}CH\text{—}CH_2\text{—}S\text{—}S\text{—}CH_2\text{—}CH\text{—}CO_2H$$

(**90**)

N,N'-bis-γ-Glutamylcystine

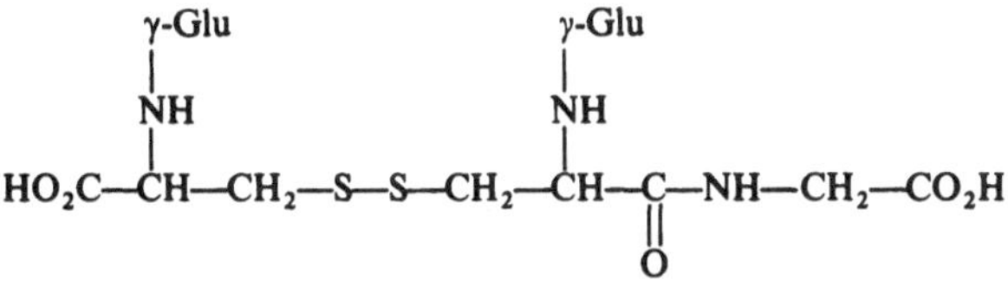

$$\gamma\text{-Glu} \qquad\qquad\qquad \gamma\text{-Glu}$$
$$| \qquad\qquad\qquad\qquad |$$
$$NH \qquad\qquad\qquad\qquad NH$$
$$| \qquad\qquad\qquad\qquad\qquad |$$
$$HO_2C\text{—}CH\text{—}CH_2\text{—}S\text{—}S\text{—}CH_2\text{—}CH\text{—}\underset{\displaystyle O}{\overset{\displaystyle \|}{C}}\text{—}NH\text{—}CH_2\text{—}CO_2H$$

(**91**)

N,N'-bis-γ-Glutamylcystinylglycine

Dowex 50 and finally paper chromatography. The N,N'-bis-γ-glutamylcystinylglycine was obtained in the form of colorless crystals. The structure was clarified by elementary analysis, determination of the hydrolysis products, N- and C-terminal analysis of peptides, pyridoxal test (*222*), etc. The reaction mixture resulting from reductive cleavage with Raney Ni was analyzed on an amino acid analyzer and γ-glutamylalanine and γ-glutamylalanylglycine were identified by comparison with authentic specimens.

In 1987, Ogawa et al. (*326*) reported a new glutathione analogue, N-(N-γ-L-glutamyl-3-sulfo-L-alanyl)glycine (**92**), from the cultured mushroom, *Flamulina veltipes*. Purification was carried out by preparative paper chromatography. Hydrolysis with 6 M HCl at 100° yielded equimolecular amounts of glutamic acid, 3-sulfoalanine and glycine, while under milder conditions (1 M HCl, 100°) liberation of glutamic acid was observed. N-Terminal determination using 2,4-dinitrofluorobenzene was also satisfactory. The absolute configurations of 3-sulfoalanine and glutamic acid were shown to be L. Final conclusive evidence was provided

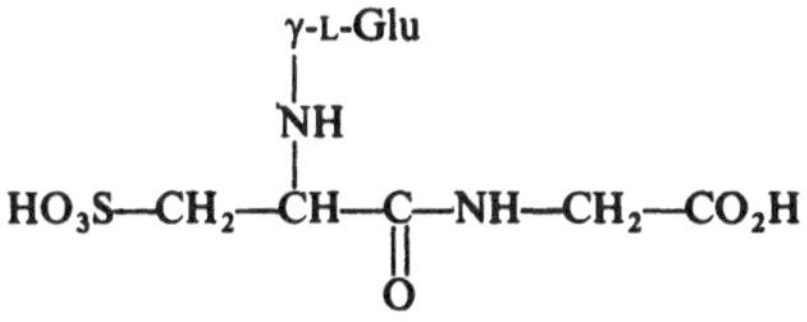

$$\gamma\text{-L-Glu}$$
$$|$$
$$NH$$
$$|$$
$$HO_3S\text{—}CH_2\text{—}CH\text{—}\underset{\displaystyle O}{\overset{\displaystyle \|}{C}}\text{—}NH\text{—}CH_2\text{—}CO_2H$$

(**92**)

N-(N-γ-Glutamyl-3-sulfo-L-alanyl)glycine

by comparison with an authentic sample prepared by performic acid oxidation of glutathione (46).

7.1.4. γ-Glutamylglutamic Acid

This compound is rather widely distributed in green plants (226) and was once isolated also from *Lentinus edodes* (16).

7.1.5. γ-Glutamylhistidine and γ-Glutamyllysine

Until 1980, three amino acids, proline, lysine and histidine were not known in the form of γ-glutamylpeptides. Two of them, γ-glutamyl α-lysine and -histidine, are now known from *Lentinus edodes* (17).

7.2. γ-Glutamyl Compounds of Non-Protein Amino Acids

7.2.1. γ-Glutamyl-β-alanine (93)

Isolation, chemical structure, and synthesis of this compound was reported independently in two papers; in one case (254), seeds of *Lunaria annua* L. (Cruciferae) were used, and in the second case (300), iris bulbs (Iridaceae) were the source. As stated before, β-alanine is one of the ubiquitous amino acids in plants and fungi and it might be difficult to find a species which lacks this compound. Its γ-glutamylpeptide is probably also distributed quite widely. In fact, we often observe in our chromatographic screening a spot which corresponds to γ-glutamyl-β-alanine.

$$HO_2C—CH_2—CH_2—NH—\gamma\text{-L-Glu}$$

(93)

γ-L-Glutamyl-β-alanine

An amino acid fraction from *Phaeolepiota aurea* (Fr.) Maire (170) was applied to a column of anion exchanger (DIAION SA #100, acetate-form) and fractionated with 0.1 N acetic acid. Displacement of γ-glutamyl-β-alanine occurred after basic and neutral amino acids and before glutamic acid, and without any accompanying amino acids. The yield was 530 mg from 1.4 kg of fresh mushrooms (0.04%). Acid hydrolysis of 380 μ moles with 3 N HCl yielded 374 μmoles of β-alanine and 370 μmoles of glutamic acid. These were determined colorimetrically with ninhydrin after separation on a small Dowex 1 column.

7.2.2. γ-Glutamyl-2-amino-4-hexynoic Acid (**94**) and γ-Glutamyl-2-amino-3-hydroxy-4-hexynoic Acid (**95**)

Shortly after the isolation and characterization of three acetylenic amino acids from fruitbodies of *Tricholomopsis rutilans* (*174, 314*), we reported γ-glutamylpeptides of two of the three (*316*). They were separated successively by displacement from Dowex 1 column in acetate-form with 2 M acetic acid. Results of the elementary analysis of the purified samples and mild acid hydrolysis products with 1 N H$_2$SO$_4$ were satisfactory. Determination of the molar ratios of the hydrolysis products and their optical rotations gave good results for each dipeptide. It is worth noting that the dipeptide of the (2S,3R)-isomer (*threo-L-*) of the acetylenic amino acid could not be detected. The free acetylenic amino acids present in *T. rutilans* gave yellow to brown products with ninhydrin on filter paper, while their γ-glutamylpeptides afforded the usual violet coloration similar to alanine, leucine, and others.

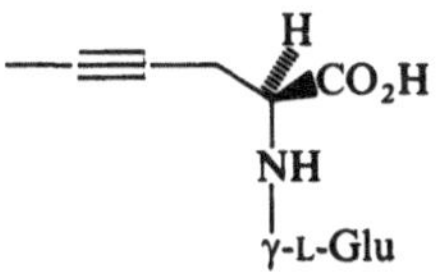

(**94**) (**95**)

γ-L-Glutamyl-(2S)-2-amino-4-hexynoic acid   γ-Glutamyl-(2S,3S)-2-amino-3-hydroxy-4-hexynoic acid

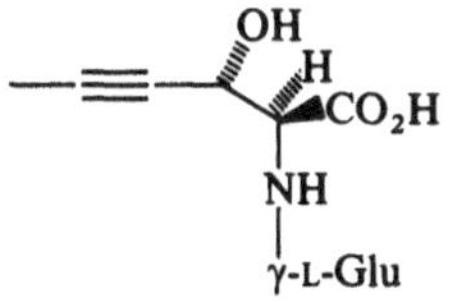

7.2.3. γ-Glutamyl-α-ornithine (**96**)

KASAI and LARSEN concluded that this compound is not easy to isolate (*226*). AOYAGI et al. (*17*) reported γ-glutamyl-α-ornithine in *Lentinus edodes*. Although it was not obtained in crystalline form, the structure could be clarified by acid hydrolysis, N-terminal determination, pyridoxal test (*222*), formation of the copper complex, etc.

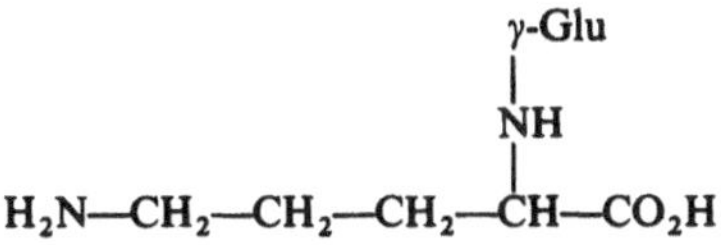

(**96**)

γ-Glutamyl-α-ornithine

7.2.4. γ-Glutamyl-3-aminoproline (97)

This compound was isolated from cultured mycelia of *Morchella esculenta* (*295*). As stated before free (2S,3R)-3-aminoproline is present in the cultured mycelia as well as in the fruitbodies of this species. In connection with this finding, specific γ-glutamyltranspeptidases in this fungus were partially purified and some of the enzymological properties were studied (*297*). Although the possibility of artificial modification during the purification procedure cannot be excluded, three isozymes were found and were completely separated from one another by DEAE-cellulose chromatography. The molecular masses were 102,000, 155,000, and 219,000, respectively, and all of them catalyzed both hydrolysis and transpeptidation. Isozymes of γ-glutamyltranspeptidase were reported recently also from *Proteus mirabilis*. Many amino acids and their γ-glutamylpeptides were tested; (2S,3R)-3-aminoproline and its glutamyl-peptide proved to be the best substrate as a γ-glutamyl acceptor and donor, respectively. Monovalent cations and maleate were reported to activate γ-glutamyltranspeptidase of animal tissues. However, these cations did not affect the enzyme of *M. esculenta*. This enzyme was inhibited, just like animal enzymes, by a mixture of L-serine and borate, a typical inhibitor of γ-glutamyltranspeptidases. The enzyme from *Morchella* seems to have characteristics slightly different from that in animals.

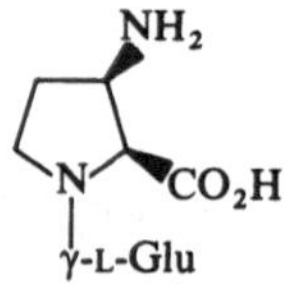

(97)

γ-L-Glutamyl-(2S, 3R)-3-aminoproline

7.2.5. γ-Glutamylnicotianine (98)

This dipeptide has also been isolated and characterized from *Lentinus edodes* (*16*); this represents the first report of this substance as a natural

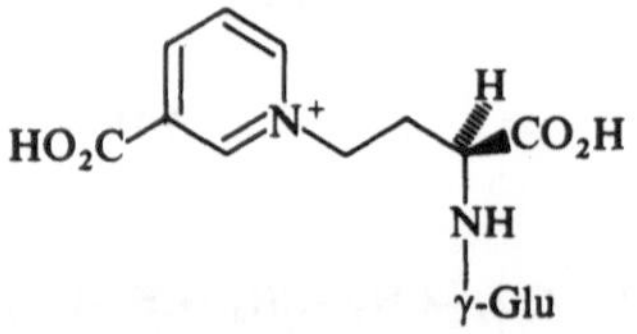

(98)

γ-Glutamylnicotianine

product. The final confirmation of the structure was made by comparison with an authentic sample prepared by reaction of γ-glutamyltranspeptidase (porcine kidney) with a mixture of nicotianine and reduced glutathione.

7.3. γ-Glutamylpeptides of Amines

7.3.1. N-(γ-Glutamyl)ethanolamine (99)

As stated before, ethanolamine is one of the most frequently encountered ninhydrin-positive substances in the amino acid-fractions of higher plants and fungi. Its γ-glutamyl derivative was first isolated and characterized by LARSEN (254) from the seeds of Lunaria annua L. OKA et al. (332) reported its occurrence in the mushroom, Agaricus bisporus, and the identity was confirmed by comparison with a synthetic preparation (266).

$$HO—CH_2—CH_2—NH—\gamma\text{-Glu}$$

(99)

N-(γ-Glutamyl)ethanolamine

7.3.2. Theanine (γ-Glutamylethylamine) (100)

This substance may be regarded also as N-ethyl-γ-glutamine or glutamic acid γ-ethyl amide. Its trivial name indicates that it is one of the characteristic chemical constituents of tea (Camellia sinensis) leaves. The first isolation was reported in 1950 by SAKATO et al. (373) from Japanese green tea. Theanine was soon synthesized by the same authors (374) and by FURUYAMA et al. (128).

Theanine is the first γ-glutamyl derivative isolated from a species of mushroom. In 1960, CASIMIR and coworkers reported its isolation from fruitbodies of Xerocomus badius (48). They found, in an ethanol extract of this fungus, a large unknown spot formed by treatment with ninhydrin on a two dimensional paper chromatogram. Fractionation was carried out on a cation exchanger and the relevant fraction was purified by paper chromatography. They obtained 430 mg from 2.14 kg of fresh mushrooms. The compound was readily hydrolyzed with 1 N HCl and water-saturated Ba(OH)$_2$, yielding glutamic acid and ethylamine. These were

$$H_3C—CH_2—NH—\gamma\text{-L-Glu}$$

(100)

Theanine

separated by electrophoresis in the presence of phosphate buffer, pH 5.6. The molar ratio was also determined and two moles of ammonia were liberated by the Van Slyke method. These results suggested the probability of a γ-peptide linkage. The postulated structure was confirmed by synthesis according to LICHTENSTEIN (*266*) from ethylamine and 5-pyrrolidone-2-carboxylic acid. CASIMIR *et al.* also discussed the possible biosynthetic route to theanine, e.g., decarboxylation of γ-glutamylalanine, participation of γ-glutamyltranspeptidase, or amide formation similar to that of glutamine from L-glutamic acid and ATP.

Intensive studies on the biosynthesis of theanine were carried out by SASAOKA and his associates. When [1-^{14}C]glutamate or [1-^{14}C]ethylamine was incubated with tea seedling homogenate in the presence of ATP and creatine phosphokinase system, the radioactivity was incorporated almost exclusively into theanine (*375*). In the hydrolysate of theanine formed in this experiment, radioactivity was detected only in glutamic acid or ethylamine, depending on the radioactive substrate used (*376*). Interestingly, the synthesis of theanine could be also carried out by using powdered acetone extracts of pea seeds (*374*) or by using extracts of pigeon liver (*377*) on mixtures of glutamic acid and ethylamine in the presence of ATP. However, the enzyme from the tea seedlings was not inhibited at all by high concentrations of ammonia. In contrast, those of the other two organisms were highly sensitive to ammonia and the mode of the inhibition was competitive with respect to ethylamine (*378*). Later, TAKEO (*415*) presented experimental evidence that ethylamine in tea seedlings originates from L-alanine. It was considered that the conversion of alanine to ethylamine was inhibited by excess ethylamine in the culture solution. Part of the alanine was also metabolized to glutamic acid and, consequently, each of these was incorporated into theanine. The same author also reported some properties of L-alanine decarboxylase prepared from cotyledons and roots of tea seedlings (*416*).

7.3.3. γ-Glutamyl-2-amino-3-hexanone (**101**)

WELTER *et al.* (*448*) reported γ-L-glutamyl-2-amino-3-hexanone from the fruitbodies of *Russula ochroleuca* as α-aminoacetone associated with glutamic acid. An ethanol extract from 2.6 kg of fruitbodies was passed through a column of Lewatit S 1080 (H$^+$-form) and the amino acids were displaced with 1 M pyridine. Fractionation was carried out on a Lewatit M 5080 column (acetate-form) and 0.5 M acetic acid as the eluting agent. This peptide was displaced before aromatic amino acids, the yield being 118 mg. On hydrolysis it gave glutamic acid and an unstable substance.

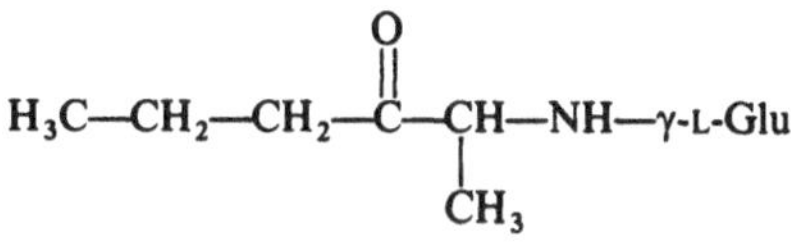

$$\text{H}_3\text{C}-\text{CH}_2-\text{CH}_2-\overset{\overset{\text{O}}{\|}}{\text{C}}-\underset{\underset{\text{CH}_3}{|}}{\text{CH}}-\text{NH}-\gamma\text{-L-Glu}$$

(101)

γ-L-Glutamyl-2-amino-3-hexanone

The structure was elucidated by elementary analysis, IR-, ^{1}H- and ^{13}C-NMR-spectra. A positive Cotton effect at 280 nm showed that the glutamic acid belonged to the L-series; however, the configuration at another carbon atom (C-6) could not be determined.

7.3.4. γ-Glutamyl-N-nitroethylenediamine (102).

Several nitramino compounds from mushrooms have already been described. During their isolation from *Agaricus subrutilescens* (Kauffm.) Hotson *et* Stuntz, we observed another ninhydrin-positive constituent which yielded glutamic acid on mild hydrolysis. The amino acid mixture of this mushroom was fractionated as usual on a Dowex 1 column in acetate-form and 0.1 M acetic acid; the nitramino compound was displaced together with phenylalanine and tyrosine. Further fractionation was then carried out on a cellulose column with n-butanol-acetic acid-water. The yield was 880 mg from 10 kg of mushrooms (*163*). The UV spectrum suggested that it was also a nitramino compound. Hydrolysis with commercial γ-glutamyltranspeptidase resulted in formation of glutamic acid and γ-nitroethylenediamine, which were separated, crystallized and identified by thin layer cellulose chromatography and IR spectra. Although N-nitroethylenediamine yielded a yellow coloration with ninhydrin, its γ-glutamylpeptide gave the usual violet reaction.

$$\text{O}_2\text{N}-\text{NH}-\text{CH}_2-\text{CH}_2-\text{NH}-\gamma\text{-L-Glu}$$

(102)

γ-L-Glutamyl-N-nitroethylenediamine

7.3.5. Coprine [N^5-(1-hydroxycyclopropyl)-L-glutamine] (103)

Young fruitbodies of inky caps (*Coprinus atramentarius* Bull.) are edible, but, when consumed together with alcohol, they can cause an antabuse-like syndrome. In most cases it occurs a few minutes after the

HO⟍ ⟋NH—γ-L-Glu

(103)

Coprine

drink and may continue for about 3 days. The main symptoms are a warm feeling in the face, neck, and chest, prickly irritation in the arms and legs, pulse increase, headache, breathing difficulties, dizziness, sweating, a decrease in blood pressure etc. (*40*). Such distinguishing properties of *Coprinus atramentarius* had long suggested the presence of an antabuse-like substance in this mushroom. In 1975, isolation and structural studies of this substance called coprine, N^5-(1-hydroxycyclopropyl)-L-glutamine, were reported independently by two research groups, LINDBERG *et al.* (*268, 269*) in Sweden, and HATFIELD and SCHAUMBERG (*182*) in the United States. In a large scale isolation procedure, the former group obtained 7.3 g of pure coprine from 65.2 kg of frozen fresh mushrooms. The ethanol extract was concentrated, the hexane soluble part removed, and the water layer was dialyzed against water. The dialyzable materials were then fractionated on a strongly acidic cation exchange column (Amberlite CG-120). Further fractionation was carried out using a weak anion exchanger (Amberlite CG-4B) column in acetate-form which resulted in isolation of pure coprine. After recrystallization, the elementary analysis agreed with the formula $C_8H_{14}N_2O_4$. As a bioassay during the procedure, rats were fed with an amount of test sample equivalent to 1 g of total extractable solids per kg of rat; after 6–10 hr, ethanol (4.5 g/kg, 43% by volume in water) was given. After the first lachrymation, maximal oedema appeared after about 28 hr and lasted for a few days (*269*).

Acidic hydrolysis afforded L-glutamic acid as one of the major products while weak alkaline hydrolysis yielded L-pyroglutamic acid and propioamide. 1-Aminocyclopropanol is unstable, but its hydrochloride in acidic solution is stable. On catalytic hydrogenation, coprine was converted to N^5-isopropyl-L-glutamine and a small amount of acetone and glutamine. The ^{1}H-NMR spectrum also suggested the presence of glutamine and cyclopropanone moieties by a comparison of the chemical shifts and their splitting patterns with those of model compounds. The suggested structure was supported by the ^{13}C-NMR spectrum and conclusively by chemical synthesis. Coprine and several analogous compounds were synthesized by *N*-acylation of 1-aminocyclopropanol generated from 1-hydroxycyclopropylammonium salts. Many of the 1-aminocyclopropanol derivatives proved to have the same inhibitory effects as coprine.

Soon after the isolation of coprine the inhibitory mechanism was studied using of mice and acetaldehyde dehydrogenase of mouse liver and yeast (*182, 424*). Coprine is an inhibitor of acetaldehyde dehydrogenase and has the same effect on ethanol metabolism as many other inhibitors of this enzyme, including the drug antabuse, bis(diethylthiocarbonyl)-disulfide. However, the inhibitory effects of coprine are observed only *in vivo*, not *in vitro* (*182, 424*). On hydrolysis, coprine decomposes into glutamic acid and unstable cyclopropanone hemiaminal. The latter further releases ammonia, yielding cyclopropanone hydrate. These two cyclopropanone derivatives both inhibit acetaldehyde dehydrogenase *in vitro* and *in vivo*. Furthermore, the inhibitory effect by cyclopropanone hydrate was demonstrated to be completely reversible and the inhibitor bound to the enzyme was released unchanged. Cyclopropanone hydrate inhibits a number of other SH-enzymes and a mode of action was proposed (*465*) (Scheme 11).

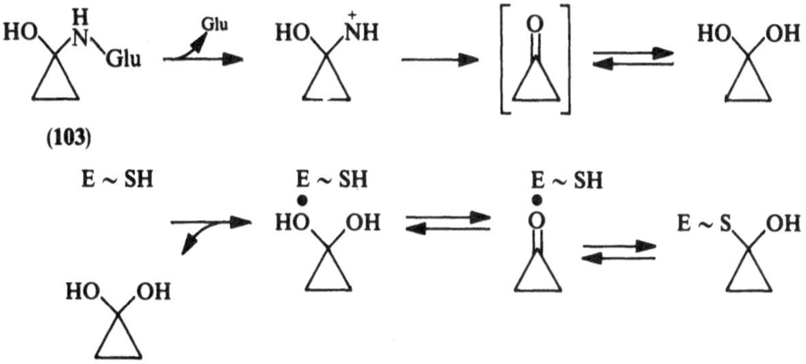

Scheme 11. Binding of coprine to acetaldehyde dehydrogenase (from *465*)

Although the toxic symptoms are not well documented, the fruit-bodies of *Clitocybe clavipes* (Pers.: Fr.) Kummer are also known to cause sickness when eaten together with alcoholic drinks. These mushrooms are also considered edible. We tried to detect coprine in them using two dimensional paper chromatography, but have so far been unsuccessful.

7.4. Agaritine and Related Compounds

Since 1960, many papers have been published dealing with isolation, structural studies and enzymic reactions of several aromatic γ-glutamyl compounds. The fungal materials used were mostly commercially available *Agaricus bisporus* or *A. hortensis*. The scientific name *Agaricus hortensis* used often in Europe is considered to be a synonym of *A. bisporus.*

7.4.1. γ-Glutaminyl-4-hydroxybenzene (104)

In 1960, JADOT *et al.* (*216*) reported separation and characterization of γ-glutaminyl-4-hydroxybenzene from *A. hortensis.* From an ethanol extract of 2 kg of mushrooms, aromatic amino acids were absorbed on charcoal, then eluted with 25% acetic acid containing 5% phenol. Fractionation was carried out successively with Amberlite IR-120 (H$^+$-form) and Dowex 1 (acetate-form). The yield was 140 mg. The pure substance had an UV spectrum characteristic of an aromatic compound. Hydrolysis with strong acid or alkali afforded glutamic acid and 4-hydroxyaniline which were separated on a preparative scale and identified by elementary analysis, determination of molecular weight, and titration. The synthesis from *p*-aminophenol and pyrrolid-2-one-5-carboxylic acid gave a substance which proved to be identical with the naturally occurring material in all respects. According to OKA *et al.* (*334*), γ-glutaminyl-4-hydroxybenzene is not restricted to the mushroom gill, but is found in every part of the fruitbodies. On the basis of the fresh weight, however, the content was highest in the gills.

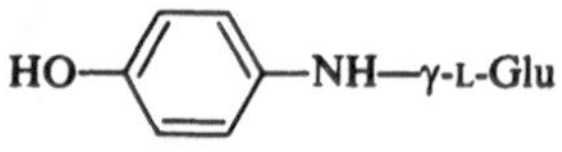

HO—⟨benzene ring⟩—NH—γ-L-Glu

(104)

γ-L-Glutaminyl-4-hydroxybenzene

The biosynthesis of γ-glutaminyl-4-hydroxybenzene has been studied intensively by SASAOKA and associates (*380*). When [G-^{14}C]shikimic acid was administered to fruitbodies of *A. bisporus,* radioactivity was incorporated into the above γ-glutamyl derivative as the main radioactive metabolite. The substance was identified unequivocally by means of spectroscopic analysis, co-chromatography with an authentic sample and gas chromatography-mass spectrometry. The radioactivity incorporated proved to be localized exclusively in the 4-hydroxyaniline moiety. Subsequently, TSUJI *et al.* (*427*) of the same research group showed that radioactive shikimic acid was also incorporated into N^5-(3,4-dihydroxyphenyl)glutamine and again only in the aniline moiety of the molecule. This substance had been isolated earlier from *Agaricus campestris* characterized and named agaridoxin (see below). In order to locate the position of amination in shikimic acid, 4-hydroxyaniline formed from ^{3}H- and ^{14}C-labeled shikimic acid was variously modified and decomposed (*428*). From the exact determination of the specific radioactivities of the products, it was clearly demonstrated that the amination occurs at C-4 of shikimic acid, analogously to the formation of *p*-aminobenzoate and *p*-aminophenylalanine (*73, 122*).

Several years later, a new FAD-dependent monoxygenase, 4-aminobenzoate hydroxylase, was purified to a high degree by $(NH_4)_2SO_4$ fractionation and various chromatographic procedures from the same mushrooms (*429*). This enzyme catalyzes the decarboxylative hydroxylation of 4-aminobenzoate and converts it to 4-hydroxyaniline in the presence of NAD(P)H and O_2. Experiments using $^{18}O_2$ showed that one oxygen atom is incorporated into 4-hydroxyaniline. The Km values for 4-aminobenzoate, NADH, and O_2 were 20.4, 13.6, and 200 µM, respectively. The value for NADPH was 133 M. This enzyme was insensitive to the iron- and copper-chelators, sodium arsenate, and KCN, but sensitive to heavy metal ions and *p*-chloromercuribenzoate. Isotopic studies with [4A-^{3}H]- and [4B-^{3}H]NAD(P)H demonstrated that the enzyme selectively transfers the hydrogen of the 4A position of NAD(P)H. This result was confirmed also by NMR spectrometry, using [4A-^{2}H]- and [4B-^{2}H]-NAD(P)H (*430*).

Experimental evidence was presented by STÜSSI and RAST (*399*) for the possible participation in melanogenesis of the spore wall of *A. bisporus*.

7.4.2. Agaritine [β-*N*-(γ-L-glutamyl)-4-hydroxymethyl-phenylhydrazine] (105)

In 1961 LEVENBERG reported a new γ-glutamyl derivative in fruit-bodies of *Agaricus bisporus* and named it agaritine (*262*). Separation and purification were carried out with Dowex 1 in the acetate-form, followed by repeated paper chromatography. Agaritine exhibited a characteristic UV spectrum, $\lambda_{max} = 237.5$ nm and 280 nm, $\varepsilon = 11,400$ and 1200, respectively. Elementary analysis and chemical and enzymatic analyses led to identification of the various functional groups or moieties, i.e., the α-amino α-carboxyl and primary hydroxyl groups, the glutamic acid residue and an aryl hydrazine, were carried out. Cleavage of agaritine

HOCH$_2$—⟨benzene ring⟩—NH—NH—γ-L-Glu

(105)

Agaritine

with dilute HCl gave glutamic acid; a second hydrolysis product could be oxidized with selenous acid to produce an aryl diazonium salt. On mild acid hydrolysis, the latter compound yielded 4-hydroxybenzyl alcohol. A hydrolytic enzyme prepared from the same mushroom, effected catalytic cleavage of agaritine to 4-hydroxymethylphenylhydrazine and L-glutamic acid.

This finding led the authors to study intensively the γ-glutamyl-transpeptidase of this species (141, 142). When L-glutamine and hydroxylamine were incubated with the partially purified enzyme, γ-glutamylhydroxamic acid was formed. Hydrazine was also an acceptor of the γ-glutamyl group and acted in the above process as a strong competitive inhibitor. Transfer of the γ-glutamyl group to the usual protein amino acids was not detected. Hydrolysis and transfer seemed to be carried out by the same enzyme. Several distinct features were observed which differentiate this from γ-glutamyltranpeptidases of other organisms. Cell-free extracts of various plants, animals, and micro-organisms were examined for the hydrolytic activity of agaritine or γ-glutamylphenylhydrazine. The results were negative in every case. Under the same conditions, agaritine was rapidly hydrolyzed by an extract from *A. bisporus*. Additionally, the partially purified enzymes of different kinds of mushrooms, e.g. species of *Boletus, Colosyphia, Coprinus, Cortinarius. Discina, Helvela, Pleurotus, Xeromphalina*, were tested, but the results were all negative. Agaritine was not detected in any of these test samples. More interestingly, even among the species of *Agaricus*, only in *A. bisporus, A. pattersonii, A. perrarus* and *A. xanthodermus*, which contain-ed agaritine, was the specific activity of γ-glutamyltranspeptidase demon-strated. In every species, both activities of hydrolysis and γ-glutamyl transfer were detected. These results suggest that the γ-glutamyltranspep-tidase in the agaritine-containing mushrooms plays a significant role in stabilizing the metabolism of agaritine and related compounds.

Immediately after the first paper by LEVENBERG, DANIELS *et al.* (66) reported an improved isolation procedure and synthesized agaritine and related compounds by condensation of γ-azide of *N*-carbobenzoxy-L-glutamic acid with the appropriate hydrazine.

In a later report (263), LEVENBERG also described the distribution of agaritine and related compounds. Fifteen species of *Agaricus* were tested. Ten of which contained agaritine in quantities comparable to those in *A. argentatus, A. campestris, A. crocodilinus, A. edulis, A. hortensis, A. micromelathus, A. pattersonii, A. perrarus*, and *A. xanthodermus*. In five other species, which belong to the *sylvaticus* subgroup, *A. benesi, A. sterlingii, A. subrutilescens, A. sylvaticus* and *A. sylvicola*, agaritine and related compounds were not detected.

From 1970 on, WEAVER *et al.* published a series of chemical and biochemical studies of this compound in relation to gill color changes during maturation of commercial *A. bisporus* (*443, 445, 446, 444*). It is well known that the gill color of *Agaricus* species including *A. bisporus* changes with time from pale pink to pinkish brown and finally dark brown to nearly black. WEAVER *et al.* showed that spore formation in *Agaricus bisporus* was accompanied by a marked reduction in mitochondrial respiration. Aerobic incubation of homogenates of the gill led to increase in the concentration of an inhibitor of respiration. Furthermore, it was observed that the increase in inhibitor concentration was proportional to the concentration of a red pigment in the gill homogenate. It suggested strongly the existence of precursor of the inhibitor(s) of respiration and a related enzyme system which catalyzes aerobically the formation of the inhibitor(s). Isolation and purification were carried out with the aid of a purified enzyme preparation. The precursor of the inhibitor was separated on a column of Sephadex G-25 in a single step, the yield being about 1 mg per 1 g of tissue. UV, and ^{1}H-NMR spectra and analysis of hydrolysis products led the authors to the structure, N^5-(4-hydroxyphenyl)glutamine; this was also confirmed by chemical synthesis. The coupling product of *p*-benzyloxyaniline with *N*-carbobenzoxy-L-glutamic acid-α-benzylester was catalytically hydrogenated with palladium on carbon. The product was identical with that isolated by JADOT *et al.* in 1960 (*216*).

The red inhibitor itself was isolated from gill tissue of *A. bisporus* with a column of Sephadex G-25. The same substance was prepared also by incubation of either natural or synthetic N^5-(4-hydroxyphenyl)glutamine with the enzyme. UV and IR spectrophotometric determination, analysis of the hydrolysate, and the stoichiometry of O_2 consumption in the enzymic reaction indicated the quinone nature of the product. This enzyme was sulfhydryl dependent and was thus found to inhibit pyruvic, 2-oxoglutaric and succinic dehydrogenases at concentrations low enough to account for the marked decrease of the respiration in mature gills of this species. Chicken liver xanthine dehydrogenase, dihydroorotic dehydrogenase, and triosephosphate dehydrogenase were also inhibited by this compound.

The enzymatic conversion of γ-glutaminyl-4-hydroxybenzene to the inhibitor was of the oxygenase type and the purification and some properties of the enzyme were described (*444*). The enzyme was purified with $(NH_4)_2SO_4$ from acetone powder with DEAE-cellulose, and finally with Sephadex G-200. The resulting enzyme preparation had high tyrosinase activity, suggesting that the precursor was oxidized. Analysis of the IR spectrum of the acetylated inhibitor showed that it was a quinone. The simplest possibility for the structure was γ-glutaminyl-3,4-benzoquinone (**106**). Although direct elucidation of the structure of the inhibitor was rendered difficult by its extreme instability, this inference is very plausible.

MINATO (*291*) isolated anthglutin (**107**) from *Penicillium oxalicum*. It has a closely related structure to agaritine and an inhibitor of γ-glutamyltranspeptidase.

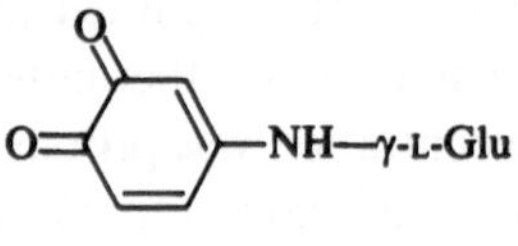

(**106**)

γ-Glutaminyl-3,4-benzoquinone

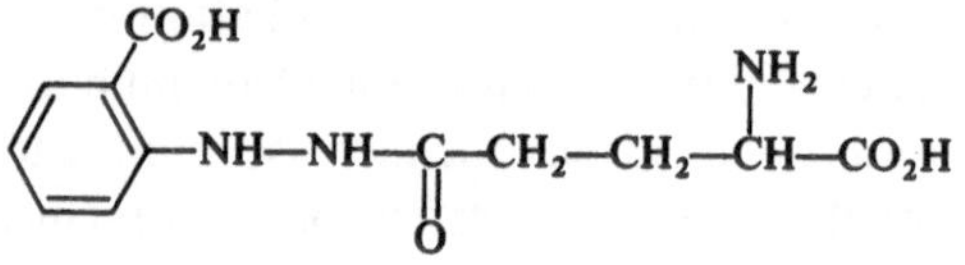

(**107**)

Anthglutin

7.4.3. Xanthodermine [(γ-Glutamyl-*N*'-(4-hydroxyphenyl)hydrazine)] (**108**)

In course of an investigation of antibiotics produced by various species of Basidiomycotina in the laboratory of STEGLICH, HILBIG et al. (*187*) reported an antibiotic, named xanthodermine, from the mushroom *Agaricus xanthoderma*. Extraction was carried out with SO_2-saturated methanol followed by chromatographic procedures with Sephadex LH 20 and LiChroprep RP 8 undertaken under argon. The structure was clarified by analysis of ^{1}H-NMR and mass spectra, and finally verified by chemical synthesis. It is a colorless substance, but when $K_3[Fe(CN)_6]/NaHCO_3$ or NaOH is added to a water solution, its color changes to intense yellow. This color reaction is based on the formation of an acylazo compound as shown in Scheme 12.

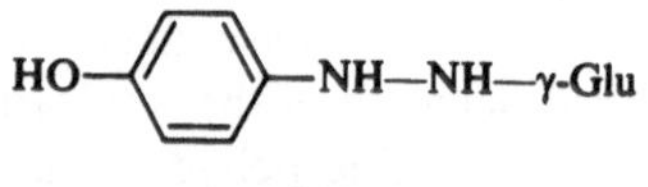

(**108**)

Xanthodermine

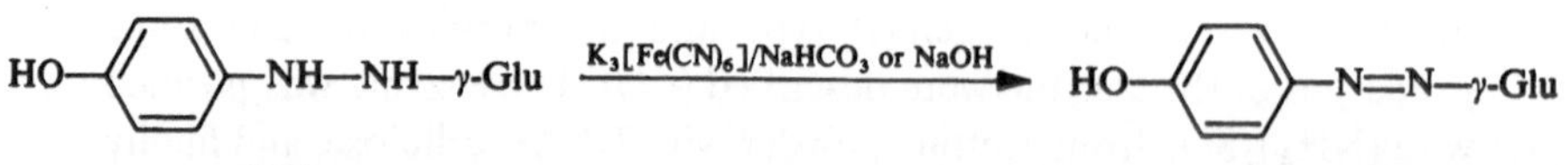

(**108**)

Scheme 12. Formation of an acylazo compound from xanthodermine (from *187*)

A water extract of this mushroom showed the same color reaction. Xanthodermine inhibited the growth of *Bacillus* spp. (*187*).

7.4.4. Agaridoxin ([N^5-(3,4-dihydroxyphenyl)glutamine], γ-glutaminyl-3,4-dihydroxybenzene) (109)

The putative intermediate in the enzymatic conversion of N^5-(4-hydroxyphenyl)glutamine to the 3,4-dioxo form, was mentioned before. N^5-(3,4-Dihydroxyphenyl)-L-glutamine was isolated from the closely related *Agaricus campestris* by SZENT-GYORGYI and coworkers (*407*). This substance is strongly autoxidizable; it was named agaridoxin. The methanol extract was treated with lead acetate and the supernatant was subjected to Sephadex G 10 chromatography. The assumed disposition of the hydroxy groups was confirmed by comparing the polarographic and UV spectrographic behavior with those of some model compounds. The glutamic acid was separated from the acid hydrolysate, and on the basis of its ORD curve, it was shown to have the L-configuration. Agaridoxin was synthesized by condensation of 3,4-(isopropylidenedioxy)anilide and *N*-phthaloylglutamic anhydride and the product was converted into (γ-L-glutamyl)-3,4-(isopropylidenedioxy)anilide by hydrazine. The protecting group was removed by boron trichloride, yielding agaridoxin.

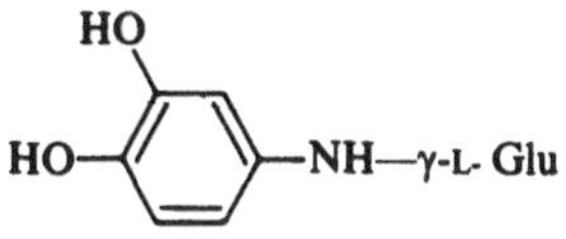

(109)

Agaridoxin

7.5. γ-Glutamyl Compounds Producing Odoriferous Substances

Mushrooms afford a great variety of odor or aroma substances such as alcohols, aldehydes, ketones, esters, or lactones. The most characteristic compounds, polythiepanes, are produced by species of several genera. In their evolution γ-glutamyl derivatives are known to be involved.

7.5.1. Lentinic Acid (111)

Lentinus edodes has been systematically cultivated in Japan as a food for more than 100 years. They are now also cultivated in several other

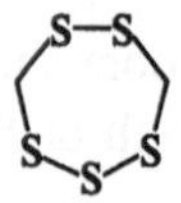

(110)

Lenthionine

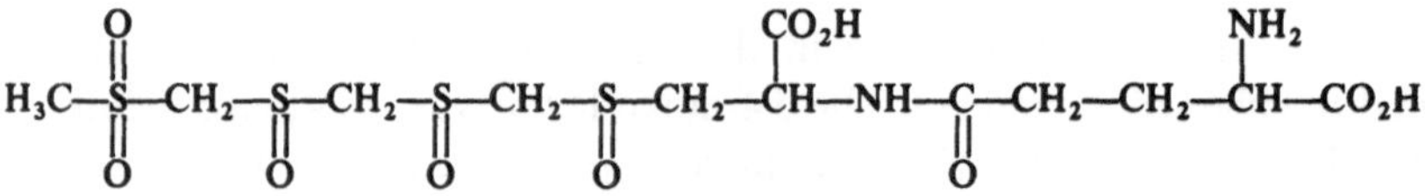

(111)

Lentinic acid

Asian countries as well as in other parts of the world. They have a very characteristic odor which is more apparent in the dried mushrooms.

In 1966, MORITA and KOBAYASHI (*298*) first reported isolation, structure, and synthesis of the odoriferous principle, 1,2,3,5,6-pentathiepane (**110**); they named this lenthionine. The structure determination was based mainly on mass spectrometry followed by an X-ray analysis (*318*). Subsequently the same group reported isolation of the analogous 1,2,4,6-tetrathiepane and 1,2,3,4,5,6-hexathiepane as minor constituents from the same mushrooms. All three of the odoriferous substances have antibacterial and antifungal activities (*299*).

Several years later, YASUMOTO and his associates (*471*) made an important observation. When dried mushrooms were extracted with water at pH 2, the chromatographic pattern of the free amino acids was similar to that of an ethanolic extract. However, water extraction at pH 7–8 yielded an extract with increased aroma and glutamic acid concentration and a decrease in the concentration of a substance assumed to be the precursor of the odoriferous substance. Considering the possibility that the odoriferous substance was enzymatically evolved from the "precursor", they began a study of the latter. For the isolation of the precursor, successive ion exchange chromatography on Amberlite IRA-400, IR-120, IRC-50, CG-120, was carried out. They obtained 1.5 g of pure precursor from 10 kg of fresh mushrooms and called it lentinic acid. Elementary analysis, various qualitative tests and hydrolysis with HCl showed that lentinic acid was a dipeptide composed of glutamic acid and a labile S-substituted cysteine. After desulfurization with Raney nickel and subsequent acid hydrolysis, glutamic acid and alanine were obtained in equimolecular amounts as well as several other substances.

Although the chemical structure of lentinic acid was not fully elucidated at that time, YASUMOTO et al. (472) detected an enzymic system converting lentinic acid into lenthionine. It consisted two enzymes; γ-glutamyltranspeptidase removed the γ-glutamyl moiety from lentinic acid and pyridoxal phosphate dependent cysteine sulfoxide lyase decomposed the desglutamyllentinic acid. In addition to lenthionine and glutamic acid, pyruvate, ammonia, and formaldehyde were produced. γ-Glutamyltranspeptidase of *Lentinus edodes* was obtained in a 200-fold concentrated state, although it was in a particulate or membrane fraction. In the cultured mycelium, however, this enzyme could be obtained in a soluble fraction and the activity increased remarkably during the formation of the fruitbodies. It catalyzed both hydrolysis of γ-glutamyl compounds and transfer of the γ-glutamyl moiety to another molecule. Lentinic acid was the most active substrate among the γ-glutamyl derivatives tested. An equimolecular mixture of serine and borate, a specific inhibitor of γ-glutamyltranpeptidase, inhibited this reaction. Activity of this enzyme in the fruitbodies was much higher than activity in the mycelia (210) and it was found to have several unique properties (209). Among these were the activation by various monovalent anions, e.g. Cl^-, Br^-, ClO_3^-, NO_3^-, I^-, F^-, SCN^-, BrO_3, and N_3^-, but not by alkali or earth cations known to activate the γ-glutamyltranspeptidase of animal sources. Nor was it activated by citric acid, a substance effective in activation of the enzyme in kidney beans (473). However, it should be noted that γ-glutamyltranspeptidase of this mushroom was studied in its bound form and an exact comparison of its properties with those of the soluble form might have been difficult. Later, the same research group reported purification, properties, and substrate specificity of C–S lyase, another enzyme participating in aroma evolution in *L. edodes*. They compared this enzyme with the corresponding one known in *Allium* species (208). Like other C–S lyases, *Lentinus* enzyme required pyridoxal phosphate as a cofactor. The enzyme occurs as two isozymes, one property of which is a great affinity for desglutamyllentinic acid. It is interesting that the C–S lyase in *L. edodes* attacks only desglutamyllentinic acid, not lentinic acid itself.

YASUMOTO et al. noted that aroma formation started with removal of glutamic acid by γ-glutamyltranspeptidase. The resulting product is then acted upon by C–S lyase; it decomposes, yielding thiosulphinate, pyruvate, and ammonia. Further degradation, disproportionation, and ring closure proceeds nonenzymatically, producing lenthionine and the analogous aroma compounds. Scheme 13 summarizes the aroma formation in *L. edodes*.

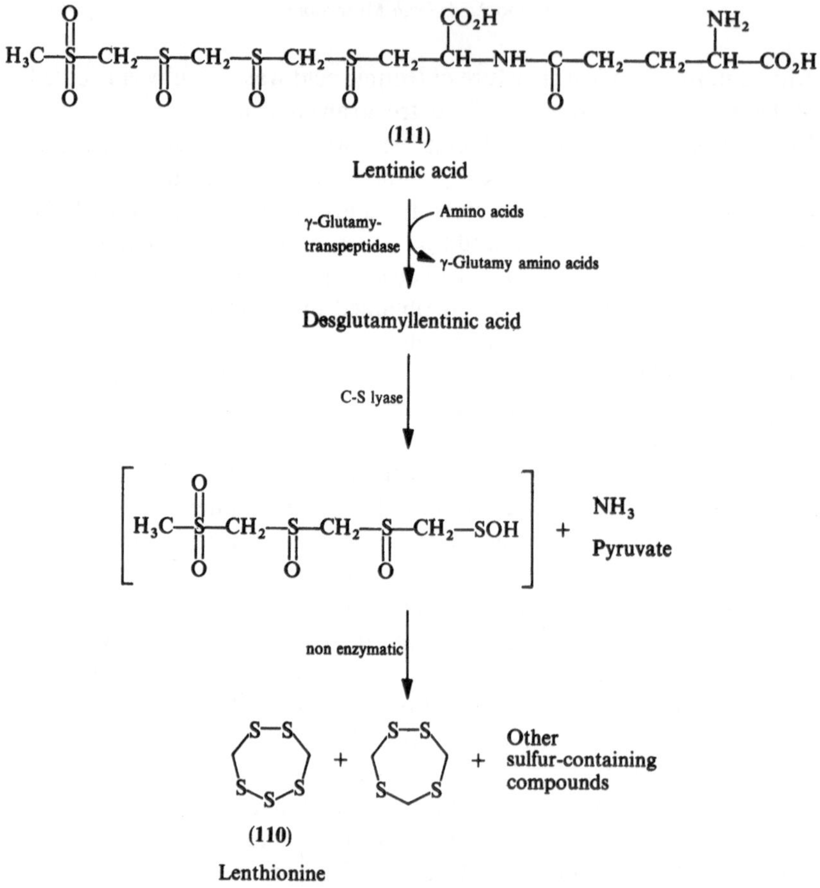

(111)

Lentinic acid

γ-Glutamy-
transpeptidase — Amino acids

— γ-Glutamy amino acids

Desglutamyllentinic acid

C-S lyase

+ NH₃

Pyruvate

non enzymatic

+ Other
sulfur-containing
compounds

(110)

Lenthionine

Scheme 13. Formation of lenthionine and its analogues from lentinic acid (from *147*)

7.5.2. γ-Glutamylmarasmine (**112**)

In 1976, GMELIN *et al.* (*146*) reported the isolation of another new
γ-glutamyl compound from fruitbodies of *Marasmius alliaceus* Jacq. and
M. scorodonius (Fr.) Sing. As suggested by the botanical epithets, these
develop garlic-like odors. The odors become more pronounced when the
fruitbodies are crushed, and there is concomitant formation of pyruvate
and ammonia. The glutamyl derivative gave a characteristic brick red
coloration with ninhydrin and decolorized K_2PtI_6 reagent on paper. The
latter test is often useful for detection of S-containing amino acids.
Isolation and purification were performed by repeated application to
Amberlite IR-120 (H^+-form), Lewatit MP 5080 (acetate-form), and
DEAE A25 Sephadex (acetate-form). The yields were strikingly high: 1%

$$H_3C-S-CH_2-\underset{\underset{O}{\|}}{S}-CH_2-\underset{\overset{|}{CO_2H}}{CH}-NH-\underset{\underset{O}{\|}}{C}-CH_2-CH_2-\underset{\overset{|}{NH_2}}{CH}-CO_2H$$

(112)

γ-Glutamylmarasmine

of dry weight. Mild hydrolysis gave glutamic acid and reduction with Raney nickel yielded γ-glutamyl-L-alanine. GMELIN *et al.* named the precursor of the odoriferous substance γ-glutamylmarasmine. Its complete structure, γ-L-glutamyl-3-(methylthiomethylsulphinyl)-L-alanine was established by analysis of ^{1}H-NMR and ^{13}C-NMR spectra. They studied also γ-glutamyltranspeptidase and C–S lyase in these mushrooms, both of which were necessary for aroma formation. The C–S lyase was very similar in its substrate specificity to that of seeds of *Albizzia lophanta* studied earlier by the same research group (*144*). It is also activated by added pyridoxal phosphate. The C–S lyase did not act on γ-glutamylmarasmine, but only on marasmine, as had already been shown for the C–S lyase of *Lentinus edodes* and also of the above legume (*384*). Six isozymes of this enzyme were detected by disc electrophoresis and this pattern was also similar to that of the legume. γ-Glutamylmarasmine was detected also in the cultured mycelium of *M. scorodonius* which evolved the same garlic-like odor as the fruitbodies. In the mycelium extract γ-glutamyltranspeptidase and C–S lyase were also detected.

GMELIN *et al.* concluded that glutamic acid was first removed from γ-glutamylmarasmine and that the marasmine formed is split by C–S lyase into pyruvate, ammonia, and presumably CH_3-S-CH_2-SOH. The latter passes through a few unstable and sharp tasting intermediates and finally produces several kinds of aroma substances.

GMELIN's group extended their research to two other mushrooms with garlic- or cabbage-like odors. *Micromphale perforans* (Hofm.: Fr.) Sing. and *Collybia hariolorum* (DC.: Fr.) Quél. (*190*). Prior to the extraction, the materials were dried and again a sulfur-containing γ-glutamyl compound was isolated in 1% yield.

From the results of their experiments, they concluded that this compound was unexpectedly identical with the lentinic acid of YASUMOTO *et al.* This was further verified by direct comparison. Because the structure of a part of lentinic acid had not been fully clarified at that time, they studied the structure intensively obtaining ^{1}H-NMR spectra which permitted identification of all protons. By treatment of the peptide with hexafluoracetone, they obtained the corresponding oxazolidinone-5. In addition to the signals of the γ-glutamyl and cysteinsulfoxide moieties, those of a group of three isolated AB-systems were observed which could be assigned to diastereotope protons. The signals of the fourth methylene group near the end sulfonyl could also be identified. The FD-mass spectrum of lentinic acid was also consistent with the formulation, but the configurations of the three chiral sulfoxide groups are not known. Lentinic acid, which was isolated from dried *Lentinus edodes* of commerce, contained ca. 30% of one of the diastereomers, whereas, the isolates from *M. perforans* and *C. hariolorum* did not. The structure of lentinic acid is 2-(γ-glutamylamino)-4,6,8,10,10-pentaoxo-4,6,8,10-tetrathiaundecanoic acid (*190*).

A few years later, an epimerized lentinic acid, epilentinic acid, was reported by GMELIN et al. (*147*) from *Micromphale foetidum* (Sow.: Fr.) Sing., *M. cauvetii* (Mre. u. Khn. ex Hora) and *Collybia impudica* (Fr.) Sing. The yields were ca. 2% of the dried samples. In the ^{1}H- and ^{13}C-NMR spectra the signals of the methyl and methylene group of epilentinic acid were shifted slightly from those of lentinic acid. The configuration of epilentinic acid, like that of lentinic acid, could not be determined by spectroscopic methods. It is interesting, that the isolates of different kinds of mushrooms consisted of these two epimers in various ratios. The ratios, calculated from the ^{13}C-NMR spectra, optical rotations, and melting points, showed good correlation with one another. For instance, the mp of pure epilentinic acid was 221° and that of pure lentinic acid 186°. Mp's of the mixtures had intermediate values depending on the ratios. A "lentinic acid" sample isolated from dry *Lentinus edodes* contained ca. 30% of epilentinic acid. Lentinic acid, epilentinic acid, and possibly other diastereomers might be produced in various ratios in different kinds of mushrooms. Drying under heat may greatly affect the ratios. In all of these mushrooms, γ-glutamyltranspeptidase and C–S lyase were present as well.

Distribution of polysulphides is rather restricted, but they have been reported in *Petiveria alliacea* (*2*), *Chondria californica* (red alga) (*468*), *Allium cepa* (*223*), *Parkia speciosa* (*148*), etc. The seeds of the last named leguminous plants are a favorite food in Indonesia because of their unique flavor. They are used also as a folk medicine, as they have antibacterial effects. Neither the mechanism of formation of such cyclic polysulphides in these organisms nor their precursors are known. They are very likely to be closely related to those of *Lentinus edodes*.

IV. Conclusions

1. Compared with higher plants, only a small portion of fungal species have been studied for their non-protein amino acids. A great

possibility thus exists that further chemically or biologically interesting amino acids of new types or unknown biochemical reactions involved in their formation and degradation will be found in future studies.

2. Most non-protein amino acids have been reported from the fruitbodies of Agaricales. Fungi of other taxa remain quite unexplored in this respect.

3. To discover the biological significance of newly isolated amino acids, it is often necessary to modify conventional assay methods or to develop new techniques, depending on possible specific functions.

4. The capability of fungal species to produce a specific non-protein amino acid can be considered as a characteristic of the taxon. In fact, detection of specific non-protein amino acids can be used as a tool in characterization of a species. Furthermore, chemotaxonomic studies of non-protein amino acids and the enzymes involved in their metabolism, if considered carefully together with morphological characters, contribute increasingly to our knowledge of taxonomy and phylogeny of fungi.

5. Chemical ecology is also an attractive prospect in mycology. Biologically active non-protein amino acids produced in growing mycelia or fruitbodies must affect other living systems in the same or nearby habitats. Experimental evidence of this should accumulate in the future.

Acknowledgments

I wish to express my cordial thanks to the late Professor H. GRISEBACH for generously giving me the opportunity to contribute to this series. Acknowledgments are also due to Professors R.J. BANDONI and G.H.N. TOWERS who read the manuscript and gave me many invaluable suggestions. I sincerely thank Professors T. HONGO, J. SUGIYAMA, S. TOMODA, Drs. I. BANNO, K. ANDO, T. AOKI, Messrs. H. NEDA and Y. TSUKAYA, for their kind help in various ways during the preparation of the manuscript.

Mushroom Index

Agrocybe
Amanita
Amanita abrupta Peck
Amanita caesarea
Amanita citrina
Amanita cothurnata
Amanita flavoconica
Amanita fulva
Amanita gemmata
Amanita gymnopus Corner & Bas
Amanita micurifera Bas & Hatanaka
Amanita muscaria (L.: Fr.) Hooker
Amanita muscaria var. *formosa*
Amanita muscaria var. *muscaria*
Amanita neoovoidea Hongo
Amanita ocreata
Amanita pantherina (DC.: Fr.) Secr.
Amanita phalloides (Vaill.) Secr.
Amanita porphyria
Amanita pseudoporphyria Hongo
Amanita regalis (Fr.) R. Mre.
Amanita rubescens
Amanita rubrovolvata Imai
Amanita smithiana
Amanita solitaria (Fr.) Secr. sensu D.E.
    Stuntz
Amanita spissa
Amanita strobiliformis (Paul.) Quél.
Amanita vaginata var. *fulva*
Amanita verna (Bull.) Pers.
Amanita virgineoides Bas
Amanita virosa Lam. ex Secr.
Baeospora sp.
Bankera fuligineoalba
Boletus
Boletus edulis Bull.: Fr.
Boletus erythropus
Boletus satanus Lenz
Boletus sp. (section *Ixocomus* group Nudi)
Boltitiaceae
Brunnescens
Citrina
Clavariaceae
Clavaria miyabeana S. Ito
Claviceps
Claviceps purpurea
Claviceps sp.
Clitocybe
Clitocybe acromelalga Ichimura
Clitocybe clavipes (Pers.: Fr.) Kummer

Clitocybe infundibuliformis
Collybia hariolorum (DC.: Fr.) Quél.
Collybia impudica (Fr.) Sing.
Colosyphia
Conocybe
Conocybe cyanopus (Atk.) Kühn.
Copelandia
Coprinus
Coprinus atramentarius Bull.
Cortinariaceae
Cortinarius
Cortinarius brunneus Fr.: Pers.
Cortinarius claricolor var. *tenuipes* Hongo
Cortinarius orelanus
Cortinarius orellanoides
Cortinarius speciossimus
Cortinarius violaceus
Discina
Flamulina veltipes
Fomes robiniae
Galerina
Gymnopilus
Helvela
Hygrocybe
Hygrocybe conica (Scop.: Fr.) Kummer
Hygrocybe ovina (Bull.: Fr.) Kühn.
Hygrocybe spp.
Inocybe
Lactarius camphoratus (Fr.) Fr.
Lactarius cimicarius (Secr.) Gill.
Lactarius helvus
Lactarius quietus Fr.
Lactarius piperatus (Fr.) S. F. Gray
Lactarius serifluus (Fr.) Fr.
Lactarius subsonarius Hongo
Lactarius spp.
Lentinus edodes (Berk.) Sing.
Lepidella
Lepiota
Leucocortinarius bulbiger (Alb. et Schwein)
    Singer
Lycoperdon perlatum
Lyophyllum connatum (Schum.: Fr.) Sing.
Lyophyllum ulmarium
Mappe
Marasmius alliaceus Jacq.
Marasmius scorodonius (Fr.) Sing.
Micromphale cauvetti (Mre. u. Khu. ex
    Hora)
Micromphale foetidum (Sow.: Fr.) Sing.

Micromphale perforans (Hoff.: Fr.) Sing.
Morchella conica Pers.
Morchella crassipes (Vent.) Pers.
Morchella esculenta Pers. ex St. Adams
Mycena pura (Fr.) Kummer
Oudemansiella platyphylla (Pers.: Fr.) Mos.
Panaeolina
Panaeolus
Panaeolus campanulatus
Phaeolepiota aurea (Fr.) Maire
Phalloideae
Pholiota
Phyllotopsis nidulans (Pers.: Fr.) Sing.
Platyphyllae
Pleurocybella porrigens
Pleurotus
Pleurotus ostreatus
Psathyrella
Psilocybe
Psilocybe argentipes K. Yokoyama
Psilocybe baeocystis
Psilocybe cubensis
Psilocybe mexicana Heim

Psilocybe semilanceata (Fr.) Kumm.
Rhodophyllus crassipes
Rhodophyllus nidorosus (Fr.) Quél.
Roanokenses
Rozites caperata
Russula cyanoxantha (Schw.) Fr.
Russula nigricans
Russula ochroleuca
Russula subnigricans Hongo
Sclerotium
Sclerotium rolfsii Sacc.
Stereum ostrea (Blume et Nees) Fr.
Strobilomyces floccopus (Vahl.: Fr.) Karst.
Tricholomopsis
Tricholomopsis decora
Tricholomopsis platyphylla (Pers.: Fr.) Sing.
Tricholomopsis rutilans (Schaef.: Fr.) Sing.
Tricholoma muscarium Kawamura
Wynnea gigantea Berk. et Curt.
Xerocomus badius
Xerocomus subtomentosus
Xeromphalina

References

1. ABELES, R.H., and C.T. WALSH: Acetylenic Enzymic Inactivator. Inactivation of γ-Cystathionase, *in vivtro* and *in vivo*, by Propargylglycine. J. Amer. Chem. Soc. **95**, 6124 (1973).

2. ADESOGAN, E.K.: Trithiolaniacin, a Novel Trithiolan from *Petiveria alliacea*. Chem. Commun. **1974**, 906.

3. AGURELL, S., and J.-E. LINDGREN: Natural Occurrence of 4-Dimethylallyl-tryptophan—an Ergot Alkaloid Precursor. Tetrahedron Letters **1968**, 5129.

4. AGURELL, S., and J.L.G. NILSSON: Biosynthesis of Psilocybin, Part II. Incorporation of Labelled Tryptamine Derivatives. Acta Chem. Scand. **22**, 1210 (1968).

5. ALBERICI DE CANAL, M., G. RODRÍGUEZ DE LORES ARNAIZ, and E. DE ROBERTIS: Glutamic Acid Decarboxylase Inhibition and Ultrastructural Changes by the Convulsant Drug Allylglycine. Biochem. Pharmacol. **18**, 137 (1969).

6. ALCOCK, N.W., D.H.G. GROUT, M.V.M. GREGORIO, E. LEE, G. PIKE, and C.J. SAMUEL: Stereochemistry of the 4-Hydroxyisoleucine from *Trigonella foenum-graecum*. Phytochem. **28**, 1835 (1989).

7. ANDERSON, J.W., and L. FOWDEN: Properties and Substrate Specificities of the Phenylalanyl-transfer-ribonucleic Acid Synthetases of *Aesculus* Species. Biochem. J. **119**, 677 (1970).

8. AOKI, T., S.I. HATANAKA, J. FURUKAWA, S. OKUDA, B. WATHELET, G. DARDENNE, and J. CASIMIR: unpublished.

9. AOYAGI, Y., N. NAKAMURA, and T. SUGAHARA: L-2-Amino-7-hydroxyoctanoic Acid: an Amino Acid from *Russula cyanoxantha*. Phytochem. **27**, 3305 (1988).

10. AOYAGI, Y., H. SASAKI, and T. SUGAHARA: Isolation and Identification of Nicotianine and Cystathionine from *Lentinus edodes.* Agric. Biol. Chem. **41**, 213 (1977).

11. — — —: Isolation and Identification of Saccharopine from *Lentinus edodes.* Agric. Biol. Chem. **42**, 1941 (1978).

12. — — —: On Peptides in Dried Shiitake *Lentinus edodes.* Nippon Nogeikagaku Kaishi **54**, 253 (1980).

13. AOYAGI, Y., H. SASAKI, T. SUGAHARA, T. HASEGAWA, and T. SUZUKI: Sulfur Containing Peptides of *Lentinus edodes.* Agric. Biol. Chem. **44**, 2667 (1980).

14. AOYAGI, Y., and T. SUGAHARA: 2(*S*)-Aminohex-5-ynoic acid, an Antimetabolite from *Cortinarius claricolor* var. *tenuipes.* Phytochem. **24**, 1835 (1985).

15. — —: β-Hydroxy-L-valine from *Pleurocybella porrigens.* Phytochem. **27**, 3306 (1988).

16. AOYAGI, Y., T. SUGAHARA, T. HASEGAWA, and T. SUZUKI: Constituents of a Cationic Peptide-rich Fraction of *Lentinus edodes.* Agric. Biol. Chem. **46**, 987 (1982).

17. — — — —: γ-Glutamyl Derivatives of Basic Amino Acids in *Lentinus edodes.* Agric. Biol. Chem. **46**, 1939 (1982).

18. BAJWA, R.S., R.-D. KOHLER, M.S. SAINI, M. CHENG, and J.A. ANDERSON: Formation of Clavicepitic Acid in Cell-free Systems of *Claviceps* sp. Phytochem. **14**, 735 (1975).

19. BALDWIN, J.E., and C.-S. LI: Enantiospecific Synthesis of Acromelic Acid A *via* a Cobalt-mediated Cyclisation Reaction. Chem. Commun. **1988**, 261.

20. BARTH, H., G. BURGER, H. DÖPP, M. KOBAYASHI, and H. MUSSO: Fliegenpilzfarbstoffe VII. Konstitution und Synthese des Muscaflavins. Liebigs Ann. Chem. **1981**, 2164.

21. BAS, C.: Morphology and Subdivision of *Amanita* and a Monograph of its Section *Lepidella.* Persoonia **5**, 285 (1969).

22. BAS, C., and S.I. HATANAKA: An Undescribed Species of *Amanita* Section *Lepidella* from Japan. Persoonia **12**, 321 (1984).

23. BELL, E.A.: Non-protein Amino Acids in Plants. In: Encyclopedia of Plant Physiology, New Series Vol. 8 (E.A. BELL and B.V. CHARLWOOD, eds.), pp. 403–432, Berlin Heidelberg New York: Springer. 1980.

24. BELL, E.A.: The Non-protein Amino Acids Occurring in Plants. In: Progress in Phytochemistry, Vol. 7 (L. REINHOLD, J.B. HARBORNE, and T. SWAIN, eds.), pp. 171–196. Oxford: Pergamon. 1981.

25. BELL, E.A., and D.H. JANZEN: Medical and Ecological Considerations of L-DOPA and 5-HTP in Seeds. Nature **229**, 139 (1971).

26. BELL, E.A., L.K. MEIER, and H. SØRESEN: Hydroxylated Glutamic Acids in *Phlox. Lepidium* and *Rheum* species. Phytochem. **20**, 2213 (1981).

27. BENEDICT, R.G.: Mushroom Toxins Other than *Amanita.* In: Microbial Toxins, Vol. **VIII** (S. KADIS, A. CIEGLER and S.J. AJL, eds.), pp. 281–320. New York: Academic Press. 1972.

28. BENEDICT, R.G., L.R. BRADY, A.H. SMITH, and V.E. TYLER, JR: Occurrence of Psilocybin and Psilocin in Certain *Conocybe* and *Psilocybe* Species. Lloydia **25**, 156 (1962).

29. BENEDICT, R.G., V.E. TYLER, JR., and L.R. BRADY: Chemotaxonomic Significance of Isoxazole Derivatives in *Amanita* Species. Lloydia **29**, 333 (1966).

30. BEUTLER, J.A., and A.H. DER MARDEROSIAN: Chemical Variation in *Amanita.* J. Nat. Prod. **44**, 422 (1981).

31. BLACK, D.K., and S.R. LANDOR: New Synthesis of Hypoglycin A. Tetrahedron Letters **1963**, 1065.

32. — —: Allenes. Part XVIII. The synthesis of Allenic Amino-acids from Allenic Bromides and Diethyl Formylaminomalonate. J. Chem. Soc. (London) **1968**, 283.

33. BLACK, S., and N.G. WRIGHT: Aspartic β-Semialdehyde Dehydrogenase and Aspartic β-Semialdehyde. J. Biol. Chem. **213**, 39 (1955).

34. BLAKE, J., and L. FOWDEN: γ-Methyleneglutamic Acid and Related Compounds from Plants. Biochem. J. **92**, 136 (1964).

35. BLEECKER, A.B., and J.T. ROMEO: 2,4-*trans*-4,5-*trans*-4,5-Dihydroxypipecolic Acid and *cis*-5-Hydroxypipecolic acid from Leaves of *Calliandra angustiflora* and Sap of *C. confusa*. Phytochem. **20**, 1845 (1981).

36. — —: 2,4-*cis*-4,5-*cis*-4,5-Dihydroxypipecolic acid—a Naturally Occurring Imino Acid from *Calliandra pittieri*. Phytochem. **22**, 1025 (1983).

37. BOWDEN, K., and A.C. DRYSDALE: A Novel Constituent of *Amanita muscaria*. Tetrahedron Letters **1965**, 727.

38. BOWDEN, K., A.C. DRYSDALE, and G.A. MONGEY: Constituents of *Amanita muscaria*. Nature **206**, 1359 (1965).

39. BREHM, L., H. HJEDS, and P. KROGSGAARD-LARSEN: The Structure of Muscimol, a GABÁ Analogue of Restricted Conformation. Acta Chem. Scand. **26**, 1298 (1972).

40. BRESINSKY, A., and H. BESL: Giftpilze, mit einer Einführung in die Pilzbestimmung, Ein Handbuch für Apotheker, Ärzte und Biologen. Stuttgart: Wissenschaftliche Verlagsgesellschaft. 1985.

41. BRIAN, P.W., G.W. ELSON, H.G. HEMMING, and M.E. RADLEY: An Inhibitor of Plant Growth—Produced by *Aspergillus wentii* Wehmer. Nature **207**, 998 (1965).

42. BUKU, A., H. FAULSTICH, TH. WIELAND, and J. DABROWSKI: 2,3-*trans*-3,4-*trans*-3,4-Dihydroxy-L-proline, an Amino-acid in Toxic Peptides of *Amanita virosa* Mushrooms. Proc. Nat. Acad. Sci. (USA) **77**, 2370 (1980).

43. BURNETT, G., P. MARCOTTE, and C. WALSH: Mechanism-based Inactivation of Pig Heart L-Alanine Transaminase by L-Propargylglycine. J. Biol. Chem. **255**, 3487 (1980).

44. BURROWS, B.F., and W.B. TURNER: 1-Amino-2-nitrocyclopentanecarboxylic Acid. J. Chem. Soc. (London) **1966**, 255.

45. BUTRUILLE, D., and X.A. DOMINGUEZ: Un nouveau produit naturel: Dimethoxy-1,4 nitro-2 trichloro-3,5,6 benzene. Tetrahedron Letters **1972**, 211.

46. CALAM, D.H., and S.G. WALEY: Some Derivatives of Glutathione. Biochem. J. **85**, 417 (1962).

47. CAMPOS, L., M.'MARLIER, G. DARDENNE, and J. CASIMIR: γ-Glutamylpeptides from *Philadelphus coronarius*. Phytochem. **22**, 2507 (1983).

48. CASIMIR, J., J. JADOT, and M. RENARD: Séparation et charactérisation de la N-éthyl-γ-glutamine à partir de *Xerocomus badius*. Biochim. Biophys. Acta **39**, 462 (1960).

49. CASIMIR, J., and A.I. VIRTANEN: Isolation of a New Neutral Amino Acid from *Lactarius helvus*. Acta Chem. Scand. **13**, 2139 (1959).

50. CHILTON, W.S.: Secondary Amino Acids of Mushrooms. In: Chemistry and Biochemistry of Amino Acids, Peptides, and Proteins Vol. **6** (B. WEINSTEIN, ed.), pp. 185–244. New York: Marcel Dekker. 1982.

51. CHILTON, W.S., and C.P. HSU: N-Nitroamines of *Agaricus silvaticus*. Phytochem. **14**, 2291 (1975).

52. CHILTON, W.S., C.P. HSU, and W. ZDYBAK: Stizolobic and Stizolobinic Acids in *Amanita pantherina*. Phytochem. **13**, 1179 (1974).

53. CHILTON, W.S., and J. OTT: Toxic Metabolites of *Amanita pantherina*, *A. cothurnata*, *A. muscaria* and other *Amanita* species. Lloydia **39**, 150 (1976).

54. CHILTON, W.S., and G. TSOU: A Chloro Amino Acid from *Amanita solitaria*. Phytochem. **11**, 2853 (1982).

55. CHILTON, W.S., G. TSOU, L. DE CATO, JR., and M.H. MALONE: The Unsaturated Norleucines of *Amanita solitaria*. Chemical and Pharmacological Studies. Lloydia **36**, 169 (1975).

56. CHILTON, W.S., G. TSOU, L. KIRK, and R.G. BENEDICT: A Naturally-Occurring Allenic Amino Acid. Tetrahedron Letters **1968**, 6283.

57. CLARK, D.A., and K. DITTMER: The Inhibition of Microbial Growth by β-2-Furylalanine. J. Biol. Chem. **173**, 313 (1948).

58. CLARK-LEWIS, J.W., and P.I. MORTIMER: The 4-Hydroxypipecolic Acid from *Acacia* Species, and its Stereoisomers. J. Chem. Soc. (London) **1961**, 189.

59. COETZER, J., R. BARTHO, M. LAING, and C. WEEKS: The Zwitterion 2-Amino-3-hydroxypent-4-ynoic Acid. Acta Crystallogr. **B33**, 3257 (1977).

60. COLONGE, J., and G. POILANE: Utilization du chloro-4 butyne-2 ol-1 en synthese organique. Bull. soc. chim. France **1955**, 502.

61. CORONELLI, C., C.R. PASQUALUCCI, G. TAMONI, and G.G. GALLO: Isolation and Structure of Alanosine, a New Antibiotic. Il Farmaco—Ed. Sci. **21**, 269 (1966).

62. COUFFIGNAL, R., M. GAUDEMAR, and P. PERRIOT: Sur la preparation des bromures propargylique du type R–C≡C–C$_2$–Br (note de Laboratoire). Bull. soc. chim. France **1967**, 3909.

63. COULTER, A.W., J.B. LOMBARDINI, and P. TALALAY: Structural and Conformational Analogues of L-Methionine as Inhibitors of the Enzymatic Synthesis of S-Adenosyl-L-methionine I. Saturated and Unsaturated Aliphatic Amino Acids. Mol. Pharmacol. **10**, 293 (1974).

64. CURTIS, D.R., D. LODGE, and H. MCLENNAN: The Excitation and Depression of Spinal Neurons by Ibotenic Acid. J. Physiol. **291**, 19 (1979).

65. DAKIN, H.D.: Hydroxyleucines. J. Biol. Chem. **154**, 549 (1944).

66. DANIELS, E.G., R.B. KELLY, and J.W. HINMAN: Agaritine: an Improved Isolation Procedure and Confirmation of Structure by Synthesis. J. Amer. Chem. Soc. **83**, 3333 (1961).

67. DARDENNE, G.A., E.A. BELL, J.R. NULU, and C. CONE: Absolute Configuration of β-Hydroxy-γ-methylglutamic Acids from *Gymnocladus dioicus*. Phytochem. **11**, 791 (1972).

68. DARDENNE, G., J. CASIMIR, E.A. BELL, and J.R. NULU: Two Stereoisomers of β-Hydroxy-γ-methylglutamic Acid from Seeds of *Gymnocladus dioicus*. Phytochem. **11**, 787 (1972).

69. DARDENNE, G., J. CASIMIR, S.I. HATANAKA, and T. AOKI: Note sur l'isolement d'une 4-hydroxyisoleucine à partir du champignon *Lactarius camphoratus*. Bull. Rech. Agron. Gembloux **21**, 125 (1986).

70. DARDENNE, G., J. CASIMIR, and J. JADOT: Séparation et charactérisation du L(−)5-méthyl-2-amino-4-hexenoïque à partir de *Leucocortinarius bulbiger*. Phytochem. **7**, 1401 (1968).

71. DARDENNE, G., J. CASIMIR, M. MARLIER, and P.O. LARSEN: Acid 2(R)-amino-3-butenoique (vinylglycine) dans les carpophores de *Rhodophyllus nidorosus*. Phytochem. **13**, 1897 (1974).

72. DARDENNE, G., J. CASIMIR, and H. SØRENSEN: 2(S),3(S)-Hydroxy-4-methyleneglutamic Acid from *Gleditsia capsica*. Phytochem. **13**, 2195 (1974).

73. DARDENNE, G., P.O. LARSEN, and E. WIECZORKOWSKA: Biosynthesis of p-Aminophenylalanine: Part of a General Scheme for the Biosynthesis of Chorismic Acid Derivatives. Biochim. Biophys. Acta **381**, 416 (1975).

74. DARDENNE, G., M. MARLIER, and A. WELTER: L-4-Hydroxy-3-methoxyphenylalanine à partir de *Cortinarius brunneus*. Phytochem. **16**, 1822 (1977).

75. DAVIS, J.S., ed.: Amino Acids and Peptides. London: Chapman and Hall. 1985.

76. DEFEUDIS, F.V.: Binding Studies with Muscimol: Relation to Synaptic γ-Aminobutyrate Receptors. Neuroscience **5**, 675 (1980).

77. DESPONTIN, J., M. MARLIER, and G. DARDENNE: L-cis-5-Hydroxypipecolic Acid from Seeds of *Gymnocladus dioicus*. Phytochem. **16**, 387 (1977).

78. DITTMER, K., H.L. GOERING, I. GOODMAN, and J.C. STANLEY: The Inhibition of Microbiological Growth by Allylglycine, Methallylglycine and Crotylglycine. J. Amer. Chem. Soc. **70**, 2499 (1948).

79. DONE, J., and L. FOWDEN: A New Amino-acid Amide in the Groundnut Plant (*Arachis hypogaea*): Evidence of the Occurrence of γ-Methyleneglutamine and γ-Methyleneglutamic Acid. Biochem. J. **51**, 451 (1952).

80. DÖPP, H., S. MAURER, A.N. SASAKI, and H. MUSSO: Fliegenpilzfarbstoffe VIII. Die Konstitution der Musca-aurine. Liebigs Ann. Chem. **1982**, 254.

81. DÖPP, H., and H. MUSSO: Die Konstitution des Muscaflavins aus *Amanita muscaria* und über Betalaminsäure [1]. Naturwiss. **60**, 477 (1973).

82. — —: Fliegenpilzfarbstoffe II. Isoliefung und Chromophore der Farbstoffe aus *Amanita muscaria*. Chem. Ber. **106**, 3473 (1973).

83. — —: Fine chromatographische Analysenmethode für Betalainfarbstoffe in Pilzen und höheren Pflanzen. Z. Naturforsch. **29C**, 640 (1974).

84. DOYLE, R., and B. LEVENBERG: Identification of Two New Unsaturated Amino Acids in the Mushroom, *Bankera fuligineoalba*. Biochemistry **7**, 2457 (1968).

85. — —: L-3-(3-Carboxyfuran-4-yl)alanine, a New Amino Acid from the Mushroom *Phyllotopsis nidulans*. Phytochem. **13**, 2813 (1974).

86. DRELL, W.: The Separation of Substituted *threo*- and *erythro*-Phenylserines by Paper Chromatography. The Configuration of Arterenol and Epinephrine. J. Amer. Chem. Soc. **77**, 5429 (1955).

87. DUNHILL, P.M., and L. FOWDEN: The Amino Acids of the Genus *Astragalus*. Phytochem. **6**, 1659 (1967).

88. DUNKELBLUM, E., H.E. MILLER, and A.S. DREIDING: On the Mechanism of Decarboxylation of Betanidine. A contribution to the interpretation of the biosynthesis of betalaines. Helv. Chim. Acta **55**, 642 (1972).

89. ELLINGTON, E.V., C.H. HASSALL, J.R. PLIMMER, and C.E. SEAFORTH: Amino-acids and Peptides. Part II. The Constitution of Hypoglycin A. J. Chem. Soc. (London) **1959**, 80.

90. ELLIS, B.E.: DOPA Ring-cleavage in the Biogenesis of Stizolobic Acid in *Mucuna deeringiana*. Phytochem. **15**, 489 (1976).

91. ENGVILD, K.C.: Chlorine-containing Natural Compounds in Higher Plants. Phytochem. **25**, 781 (1986).

92. ESAKI, N., H. TAKADA, M. MORIGUCHI, S.I. HATANAKA, H. TANAKA, and K. SODA: Mechanism-based Inactivation of Methionine γ-Lyase by L-2-Amino-4-chloro-4-pentenoate. Biochemistry **28**, 2111 (1989).

93. EUGSTER, C.H.: Wirkstoffe aus dem Fliegenpilz. Naturwiss. **55**, 305 (1968).

94. —: Chemie der Wirkstoffe aus dem Fliegenpilz, *Amanita muscaria*. Fortschr. Chem. organ. Naturstoffe **27**, 261 (1969).

95. EUGSTER, C.H., G.F.R. MÜLLER, and R. GOOD: Wirkstoffe aus *Amanita muscaria*: Ibotensäure und Muscazon. Tetrahedron Letters **1965**, 1813.

96. EUGSTER, C.H., and T. TAKEMOTO: Zur Nomenklatur der neuen Verbindungen aus *Amanita*-Arten. Helv. Chim. Acta **50**, 126 (1966).

97. EVANS, C.S., and E.A. BELL: "Uncommon" Amino Acids in 64 Species of Caesalpinieae. Phytochem. **17**, 1127 (1978).

98. EVANS, S.V., T.K.M. SHING, R.T. APLIN, L.E. FELLOWS, and G.W.J. FLEET: Sulphate Ester of *trans*-4-Hydroxypipecolic Acid in Seeds of *Peltophorum*. Phytochem. **24**, 2593 (1985).

99. FANG, S.-T. L.-C. LI, C.-I. NIU, and K.-F. TSÉNG: Chemical Studies on *Cucurbita moschata* I. The Isolation and Structural Studies of Cucurbitine, a New Amino Acid. Sci. Sin. (Peking) **10**, 845 (1961).

100. FAULSTICH, H., A. BUKU, H. BODENMÜLLER, and TH. WIELAND: Virotoxins: Actin-binding Cyclic Peptides of *Amanita virosa* Mushrooms. Biochemistry **19**, 334 (1980).

101. FAULSTICH, H., J. DÖLLING, K. MICHL, and TH. WIELAND: Synthese von α-Amino-γ-hydroxysäuren durch Photochlorierung. Liebigs Ann. Chem. **1973**, 560.

102. FISCHER, N., and A.S. DREIDING: Biosynthesis of Betalaines. On the Cleavage of the Aromatic Ring During the Enzymatic Transformation of DOPA into Betalamic Acid. Helv. Chim. Acta **55**, 649 (1972).

103. FLOSS, H.G., M.-D. TSAI, and R.W. WOODWARD: Stereochemistry of Biological Reactions at Proprochiral Centers. Top. Stereochem. **15**, 253 (1984).

104. FOWDEN, L.: The Enzymic Decarboxylation of γ-Methyleneglutamic Acid by Plant Extracts. J. Exp. Bot. **5**, 28 (1954).

105. —: Azetidine-2-carboxylic Acid: a New Cyclic Imino Acid Occurring in Plants. Biochem. J. **64**, 323 (1956).

106. —: The Acidic Amino Acids of Tulip: Isolation of γ-Ethylideneglutamic Acid. Biochem. J. **98**, 57 (1966).

107. —: Isolation of γ-Hydroxynorvaline from *Lathyrus odoratus* Seeds. Nature **209**, 807 (1966).

108. —: The Non-protein Amino Acids of Plants. In: Progress in Phytochemistry, Vol. 2 (L. REINHOLD and Y. LIWSCHITZ, eds.), pp. 203–266. London: Interscience. 1970.

109. —: Nonprotein Amino Acids. In: The Biochemistry of Plants, a Comprehensive Treatise, Vol. 7 (E.E. CONN, ed.), pp. 215–247. New York: Academic Press. 1981.

110. FOWDEN, L., and J. DONE: A Third Unsaturated Amino Acid in Groundnut: Evidence for the Occurrence of γ-Amino-α-methylenebutyric Acid. Biochem. J. **55**, 548 (1953).

111. FOWDEN, L., and P.J. LEA: Mechanism of Plant Avoidance of Autotoxicity by Secondary Metabolites, Especially by Nonprotein Amino Acids. In: Herbivores, Their Interaction with Secondary Plant Metabolites (G.A. ROSENTHAL and D.H. JANZEN, eds.), pp. 135–160. New York: Academic Press. 1979.

112. FOWDEN, L., P.J. LEA, and E.A. BELL: The Nonprotein Amino Acids in Plants. Adv. Enzymology **50**, 117 (1979).

113. FOWDEN, L., C.M. MACGIBBON, F.A. MELLON, and R.C. SHEPPARD: Newly Charac-terized Amino Acids from *Blighia unijugata*. Phytochem. **11**, 1105 (1972).

114. FOWDEN, L., and M. MAZELIS: Biosynthesis of 2-Amino-4-methylhex-4-enoic Acid in *Aesculus californica*: the Precursor Role of Isoleucine. Phytochem. **10**, 359 (1971).

115. FOWDEN, L., and H.M. PRATT: Cyclopropylamino Acids of the Genus *Acer*: Distribu-tion and Biosynthesis. Phytochem. **12**, 1677 (1973).

116. FOWDEN, L., H.M. PRATT, and A. SMITH: 4-Hydroxyisoleucine from Seed of *Trigonella foenum-graecum*. Phytochem. **12**, 1707 (1973).

117. FOWDEN, L., and A. SMITH: Newly Characterized Amino Acids from *Aesculus californica*. Phytochem. **7**, 809 (1968).

118. FOWDEN, L., A. SMITH, D.S. MILLINGTON, and R.C. SHEPPARD: Cyclopropane Amino Acids from *Aesculus* and *Blighia*. Phytochem. **8**, 437 (1969).

119. FOWDEN, L., and F.C. SWEWARD: Nitrogenous Compounds and Nitrogen Metabo-lism in the Liliaceae II. The Occurrence of Soluble Nitrogenous Compounds. Ann. Botany **21**, 53 (1957).

120. FOWDEN, L., and J.A. WEBB: The Incorporation of C^{14}-Labelled Substrates into Amino-acids of Groundnut Plants (*Arachis hypogaea*). Ann. Botany **22**, 73 (1958).

121. FRAHN, J.L., and R.J. ILLMAN: The Occurrence of D-Alanine and D-Alanyl-D-alanine in *Phalaris tuberosa*. Phytochem. **14**, 1464 (1975).

122. FRANCIS, M.M., and D.W.S. WESTLAKE: Biosynthesis of Chloramphenicol in *Streptomyces* Species 3022a: the Nature of the Arylamine Synthetase System. Canad. J. Microbiol. **25**, 1408 (1979).

123. FRIIS, P., P. HELBOE, and P.O. LARSEN: Synthesis and Resolution of Vinylglycine, a β,γ-Unsaturated α-Amino Acid. Acta Chem. Scand. **B28**, 317 (1974).

124. FRITZ, H., A.R. GAGNEUX, R. ZBINDEN, J.R. GEIGY, S.A. BASLE, and C.H. EUGSTER: The Structure of Muscazone. Tetrahedron Letters **1965**, 2075.

125. FUGMANN, B., and W. STEGLICH: Ungewöhnliche Inhaltsstoffe des Blätterpilzes *Lyophyllum connatum* (Agaricales). Angew. Chem. **96**, 71 (1984).

126. FUJII, S., and M. MARUOKA: 2-Amino-2-deoxy-D-erythrose in *Agaricus bisporus*. Biochim. Biophys. Acta **717**, 486 (1982).

127. FUKUDA, M., T. OGAWA, and K. SASAOKA: Optical Configuration of γ-Glutamylalanine in Pea Seedlings. Biochim. Biophys. Acta **304**, 363 (1973).

128. FURUYAMA, T., T. YAMASHITA, and S. SENOH: The Synthesis of L-Teanine. Bull. Chem. Soc. Japan **37**, 1078 (1964).

129. FUSHIYA, S., F. WATARI, T. TASHIRO, G. KUSANO, and S. NOZOE: A New Acidic Amino Acid from a Basidiomycetes, *Lactarius piperatus*. Chem. Pharm. Bull. Japan **36**, 1366 (1988).

130. GAGNEUX, A.R., F. HÄFLIGER, R. MEIER, J.R. GEIGY, S.A. BASLE, and C.H. EUGSTER: Synthesis of Ibotenic Acid. Tetrahedron Letters **1965**, 2081.

131. GALAT, A.: Ethyl Formylaminomalonate: an Intermediate in the Synthesis of Amino Acids. J. Amer. Chem. Soc. **69**, 965 (1947).

132. GALLINA, C., C. MARTA, C. COLOMBO, and A. ROMEO: Capreomycidine and 3-Guanidinoproline from Viromycidine: Synthesis of *cis*- and *trans*-3-Aminoprolines. Tetrahedron **27**, 4681 (1971).

133. GARTZ, J., and G. DREWITZ: Der erste Nachweis des Vorkommens von Psilocybin in Rißpilzen. Z. Mykol. **51**, 199 (1985).

134. GEBERT, U., TH. WIELAND, and H. BOEHRINGER: Über die Inhaltsstoffe des grünen Knollenblätterpilzes XXXIII. Die Konstitution von Amanin und Phallisin. Liebigs Ann. Chem. **705**, 227 (1967).

135. GEIPEL. H., J. GLOEDE, K.-P. HILGETAG, and H. GROSS: Über α-Aminosäuren I. Eine einfache Synthese für β-Hydroxy-α-aminosäuren. Chem. Ber. **98**, 1677 (1965).

136. GELLERT, E., B. HALPERN, and R. RUDZATS: Amino Acids and Steroids of a New Guinea *Boletus*. Phytochem. **12**, 689 (1073).

137. — — —: The Absolute Configuration of the New Amino Acid 2-Amino-4-methylhex-5-enoic Acid from a New Guinea *Boletus*. Phytochem. **17**, 802 (1978).

138. GEORGI, V., and TH. WIELAND: Über die Inhaltsstoffe des grünen Knollenblätterpilzes XXIX. Synthese von γ,δ-Dihydroxyisoleucin, der laktonisierenden Aminosäure von α- und β-Amanitin. Liebigs Ann. Chem. **700**, 149 (1966).

139. GERSHON, H., J.S. MEEK, and K. DITTMER: Propargylglycine: An Acetylenic Amino Acid Antagonist. J. Amer. Chem. Soc. **71**, 3573 (1949).

140. GIEREN, A., P. NARAYANAN, W. HOPPE, M. HASAN, K. MICHL, TH. WIELAND, H.O. SMITH, G. JUNG, and E. BREITMAIER: Über die Inhaltsstoffe des grünen Knollenblätterpilzes XLIV. Die Konfiguration der hydroxylierten Isoleucine der Amatoxine. Liebigs Ann. Chem. **1974**, 1561.

141. GIGLIOTTI, H.J., and B. LEVENBERG: Enzymatic Transfer of γ-Glutamyl Group Between Naturally Occurring Alline and Phenylhydrazine Derivatives in the Genus *Agaricus*. Biochim. Biophys. Acta **81**, 618 (1964).

142. GIGLIOTTI, H.J., and B. LEVENBERG: Studies on the γ-Glutamyltransferase of *Agaricus bisporus.* J. Biol. Chem. **239**, 2274 (1964).

143. GIRARDON, P., Y. SAUVAIRE, J.-C. BACCOU, and J.-M. BESSIERE: Identification de la 3-hydroxy-4,5-dimethyl-2(5*H*)-furanone dans l'arome des graines de fenugrec (*Trigonella foenum-graecum* L.). Lebensm.-Wiss. u. Technol. **19**, 44 (1986).

144. GMELIN, R., G. HASEMAIER, and G. STRAUSS: Über das Vorkommen von Djenkolsäure und einer C–S-Lyase in den Samen von *Albizzia lophantha* Benth. (Mimosaceae). Z. Naturforsch. **12b**, 687 (1957).

145. GMELIN, R., and P.O. LARSEN: L-γ-Ethylideneglutamic acid and L-γ-Methyleneglutamic Acid in Seeds of *Tetrapleura tetraptera* (Schum. et Thonn.) Taub. (Mimosaceae). Biochim. Biophys. Acta **136**, 572 (1967).

146. GMELIN, R., H.-H. LUXA, K. ROTH, and G. HÖFLE: Dipeptide Precursor of Garlic Odour in *Marasmius* Species. Phytochem. **15**, 1717 (1976).

147. GMELIN, R., M. N'GALAMULUME-TREVES, and G. HÖFLE: Epilentinsäure, ein neuer Aroma- und Geruchs-Precursor in *Tricholoma* Arten. Phytochem. **19**, 553 (1980).

148. GMELIN, R., R. SUSILO, and G.R. FENWICK: Cyclic Polysulphides from *Parkia speciosa.* Phytochem. **20**, 2521 (1981).

149. GODTFREDSEN, W.O., S. VANGEDAL, and D.W. THOMAS: Cycloheptamycin, a New Peptide Antibiotic, Structure Determination by Mass Spectrometry. Tetrahedron **26**, 4931 (1970).

150. GOOD, R., G.F.R. MÜLLER, and C.H. EUGSTER: Isolierung und Charakterisierung von Prämuscimol und Muscazon aus *Amanita muscaria* (L. ex Fr.) Hooker. Helv. Chim. Acta **48**, 927 (1965).

151. GORE, M.G., and P.M. JORDAN: Microbore Single-column Analysis of Pharmacologically Active Alkaloids from Fly Agaric Mushroom *Amanita muscaria.* J. Chromatogr. **243**, 323 (1982).

152. GÖTH, H., A.R. GAGNEUX, C.H. EUGSTER, and H. SCHMID: 2(3*H*)-Oxazolone durch Photolagerung von 3-Hydroxyisoxazolen. Synthese von Muscazon. Helv. Chim. Acta **50**, 137 (1967).

153. GRAY, D.O.: *trans*-4-Hydroxymethyl-D-proline from *Eriobotrya japonica.* Phytochem. **11**, 751 (1972).

154. GRAY, D.O., and L. FOWDEN: 4-Methyleneproline: a New Naturally Occurring Proline Derivative. Nature **193**, 1285 (1962).

155. GREENSTEIN, J.P., and M. WINITZ: Chemistry of the Amino Acids, Vol. 1, p. 86. New York: John Wiley & Sons. 1961.

156. HALL, R.H., and G.F. WRIGHT: Reaction of Acetyl Chloride with 1-Nitro-2-nitramino-2-propoxyimidazolidine. J. Amer. Chem. Soc. **73**, 2213 (1951).

157. HASHIMOTO, K., K. KONNO, H. SHIRAHAMA, and T. MATSUMOTO: Synthesis of Acromelic Acid B, a Toxic Principle of *Clitocybe acromelalga.* Chemistry Letters **1986**, 1399.

158. HASSALL, C.H., and D.I. JOHN: Amino-acids and Peptides. Part III. The Constitution of Hypoglycin B. J. Chem. Soc. (London) **1960**, 4112.

159. HASSALL, C.H., and K. REYLE: Hypoglycin A and B, Two Biologically Active Polypeptides from *Blighia sapida.* Biochem. J. **60**, 334 (1955).

160. HATANAKA, S.I.: A New Amino Acid Isolated from *Morchella esculenta* and Related Species. Phytochem. **8**, 1305 (1969).

161. —: Biochemical Studies on Nitrogen compounds of Fungi IV. L-Pipecolic Acid and *trans*-5-Hydroxy-L-pipecolic acid from *Stereum ostrea* (Blume et Nees) Fr. Sci. Pap. Coll. Gen. Educ. Univ. Tokyo **22**, 117 (1972).

162. —: Biochemical Studies on Nitrogen compounds of Fungi XIV. Formation of *cis*-3-

Amino-L-proline in Cultured Mycelium of *Morchella esculenta* Fr. Sci. Pap. Coll. Gen. Educ. Univ. Tokyo **26**, 33 (1976).

163. —: L-α-Amino-γ-nitraminobutyric Acid, a New Amino Acid, and Related Compounds Isolated from *Agaricus subrutilescens*. Trans. mycol. Soc. Japan **22**, 213 (1981).

164. HATANAKA, S.I., A. ATSUMI, K. FURUKAWA, and Y. ISHIDA: Chromatographic Separation of Synthetic Nopaline and Isonopaline and their Absolute Configuration. Phytochem. **21**, 225 (1982).

165. HATANAKA, S.I., H. IIZUMI, A. TSUJI, and R. GMELIN: L-2-Amino-4-methylpimelic acid: a New Amino Acid from *Lactarius* Species. Phytochem. **14**, 1559 (1975).

166. HATANAKA, S.I., and S. KANEKO: *cis*-5-Hydroxy-L-pipecolic acid from *Morus alba* and *Lathyrus japonicus*. Phytochem. **16**, 1041 (1977).

167. HATANAKA, S.I., S. KANEKO, Y. NIIMURA, F. KINOSHITA, and G. SOMA: L-2-Amino-4-chloro-4-pentenoic acid, a New Amino Acid from *Amanita pseudoporphyria*. Tetrahedron Letters **1974**, 3931.

168. HATANAKA, S.I., and H. KATAYAMA: L-γ-Propylideneglutamic Acid and Related Compounds from *Mycena pura*. Phytochem. **14**, 1434 (1975).

169. HATANAKA, S.I., and K. KAWAKAMI: Biochemical Studies on Nitrogen Compounds of Fungi XIX. Isolation and Identification of L-2-Amino-4,5-hexadienoic Acid from *Amanita neoovoidea* Hongo, Sci. Pap. Coll. Gen. Educ. Univ. Tokyo **30**, 147 (1980).

170. HATANAKA, S.I., K. KAWAKAMI, and Y. ISHIDA: Biochemical Studies on Nitrogen Compounds of Fungi XVIII. Isolation and Identification of γ-L-Glutamyl-β-alanine from *Phaeolepiota aurea* (Fr.) Maire. Sci. Pap. Coll. Gen. Educ. Univ. Tokyo **29**, 155 (1979).

171. HATANAKA, S.I., Y. MUROOKA, K. SAITO, and Y. TAKEUCHI: *E*-2-(*S*)-Amino-3-methyl-3-pentenoic acid from *Coniogramme intermedia*. Phytochem. **21**, 453 (1982).

172. HATANAKA, S.I., and Y. NIIMURA: L-3-(3-Carboxy-4-furyl)alanine from *Tricholomopsis rutilans*. Phytochem. **14**, 1436 (1975).

173. HATANAKA, S.I., Y. NIIMURA, and K. TAKISHIMA: Non-protein Amino Acids of Unsaturated Norleucine-type in *Amanita pseudoporphyria*. Trans. mycol. Soc. Japan **26**, 61 (1985).

174. HATANAKA, S.I., Y. NIIMURA, and K. TANIGUCHI: L-2-Aminohex-4-ynoic acid: a New Amino Acid from *Tricholomopsis rutilans*. Phytochem. **11**, 3327 (1972).

175. — — —: Biochemische Studien über Stickstoffverbindungen in Pilzen V. Eine weitere neue Aminosäure vom Acetylen-Typ aus *Tricholomopsis rutilans*. Z. Naturforsch. **28C**, 475 (1973).

176. HATANAKA, S.I., and K. TAKISHIMA: α-Methylene-γ-aminobutyric Acid from *Mycena pura*. Phytochem. **16**, 1820 (1977).

177. — —: Biochemical Studies on Nitrogen Compounds of Fungi XX. L-*threo*- and L-*erythro*-γ-Methylglutamic acid from *Mycena pura* (Fr.) Kummer. Sci. Pap. Coll. Gen. Educ. Univ. Tokyo **31**, 33 (1981).

178. HATANAKA, S.I., and H. TERAKAWA: Biochemical Studies on Nitrogen Compounds of Fungi I. Distribution of Some Non-protein Amino Acids 1. Bot Mag. Tokyo **81**, 259 (1968).

179. HATANAKA, S.I., H. AKATSUKA, J. FURUKAWA, E. NAGASAWA, and S. OKUDA: unpublished.

180. HATANAKA, S.I., T. AOKI, J. FURUKAWA, and S. OKUDA: unpublished.

181. HATANAKA, S.I., Y. NIIMURA, K. TAKISHIMA, and J. SUGIYAMA: unpublished.

182. HATFIELD, G.M., and J.P. SCHAUMBERG: Isolation and Structural Studies of Coprine, the Disulfiram-like Constituent of *Coprinus atramentarius*. Lloydia **38**, 489 (1975).

183. HATFIELD, G.M., L.J. VALDES, and A.H. SMITH: The Occurrence of Psilocybin in *Gymnopilus* Species. Lloydia **41**, 140 (1978).

184. HATTORI, S., and A. KOMAMINE: Stizolobic Acid: a New Amino Acid in *Stizolobium hassjoo*. Nature **183**, 1116 (1959).

185. HEGARTY, M.P.: The Isolation and Identification of 5-Hydroxypiperidine-2-carboxylic Acid from *Leucaena glauca* Benth. Austral. J. Chem. **10**, 484 (1957).

186. HERRMANN, M.: Der Rettichhelmling—*Mycena pura* (Pers. ex Fr.) Kumm.—ist giftig! Mykol. Mitt. bl. **17**, 17 (1973).

187. HILBIG, S., T. ANDREAS, W. STEGLICH, and T. ANKE: Zur Chemie und antibiologischen Aktivität des Carbolegerlings (*Agaricus xanthoderma*). Angew. Chem. **97**, 1063 (1985).

188. HIGTON, A.A., and A.D. ROBERTS: Dictionary of Antibiotics and Related Substances (B.W. BYCROFT, ed.) London: Chapman and Hall. 1988.

189. HIRAYAMA, F., K. KONNO, H. SHIRAHAMA, and T. MATSUMOTO: 4-Aminopyridine-2,3-dicarboxylic Acid from *Clitocybe acromelalga*. Phytochem. **28**, 1133 (1989).

190. HÖFLE, G., R. GMELIN, H.-H. LUXA, M. N'GALAMULUME-TREVES, and S.I. HATANAKA: Struktur der Lentinsäure: 2-(γ-Glutamylamino)-4,6,8,10,10-pentaoxo-4,6,8,10-tetrathiaundecansäure. Tetrahedron Letters **1976**, 3129.

191. HOFMANN, A., R. HEIM, A. BRACK, H. KOBEL, A. FREY, H. OTT, T. PETRZILK, and F. TROXLER: Psilocybin und Psilocin, zwei psychotrope Wirkstoffe aus mexikanischen Rauschpilzen (*Psilocybe*). Helv. Chim. Acta **42**, 1557 (1959).

192. HOMA, A.D., and E.E. DEKKER: Decarboxylation of γ-Hydroxyglutamate Decarboxylase of *Escherichia coli* (ATCC 11246). Biochemistry **6**, 2626 (1967).

193. HULME, A.C., and W. ARTHINGTON: Methylproline in Young Apple Fruits. Nature **173**, 588 (1954).

194. HUNT, S.: The Non-protein Amino Acids. In: Chemistry and Biochemistry of the Amino Acids (G.C. BARRETT, ed.). pp. 55–138. London: Chapman and Hall. 1985.

195. IBRAHIM, S.A., P.L. LEA, and L. FOWDEN: Purification and Properties of 4-Methyleneglutaminase from the Leaves of Peanut (*Arachis hypogaea*). Phytochem. **23**, 1545 (1984).

196. ICHIHARA, A., H. HASEGAWA, H. SATO, M. KOYAMA, and S. SAKAMURA: The Structure of a New Amino Acid from *Fagopyrum esculentum* Moench. Tetrahedron Letters **1973**, 37.

197. IKEDA, M., Y. NAGANUMA, K. OHTA, T. SASSA, and Y. MIURA: Isolation and Identification of a Plant-growth Inhibitor, Azetidine-2-carboxylic Acid, from *Clavaria miyabeana* S. Ito and its Occurrence in the Family Clavariaceae. Nippon Nogeikagaku Kaishi **51**, 519 (1977).

198. IKUTANI, Y., T. OKUDA, and S. AKABORI: β-Hydroxyleucine I. Synthesis by Means of Copper Complex and Separation of the Diastereomeric Racemates. Bull. Chem. Soc. Japan **33**, 582 (1960).

199. IMAZEKI, R., and T. HONGO: Colored Illustrations of Fungi of Japan, Vol. II, p. 103. Osaka: Hoikusha. 1957.

200. IMPELLIZZERI, G., S. MANGIAFICO, G. ORIENTE, M. PIATTELLI, S. SCIUTO, E. FATTORUSSO, S. MAGNO, C. SANTACROCE, and D. SICA: Amino Acids and Low-molecular-weight Carbohydrate of Some Marine Red Algae. Phytochem. **14**, 1549 (1975).

201. IMPELLIZZERI, G., and M. PIATTELLI: Biosynthesis of Indicaxanthin in *Opuntia ficus-indica* fruits. Phytochem. **11**, 2499 (1972).

202. IMPELLIZZERI, G., M. PIATTELLI, S. SCIUTO, and E. FATTORUSSO: Pyrrolidine-2,4-dicarboxylic Acid, a New Naturally Occurring Imino Acid. Phytochem. **16**, 1601 (1977).

203. INATOMI, H., F. INUKAI, and T. MURAKAMI: Isolierung und Identifizierung von
     erythro-3-Hydroxy-L-asparaginsäure von den unreifen Samen von *Astragalus sinicus*
     L. Chem. Pharm. Bull. Japan **19**, 216 (1971).
204. IRREVERRE, F., K. MORITA, S. ISHII, and B. WITKOP: Occurrence of *cis*- and *trans*-3-
     Hydroxy-L-proline in Acid Hydrolyzate of Telomycin. Biochem. Biophys. Res.
     Commun. **9**, 69 (1962).
205. IRREVERRE, F., K. MORITA, A.V. ROBERTSON, and B. WITKOP: Isolation, Configura-
     tion, and Synthesis of Natural *cis*- and *trans*-3-Hydroxyprolines. J. Amer. Chem. Soc.
     85, 2824 (1963).
206. ISHIDA, M., and H. SHINOZAKI: Acromelic Acid is a Much More Potent Excitant than
     Kainic Acid or Domoic Acid in the Isolated Rat Spinal Cord. Brain Res. **474**, 386
     (1988).
207. ITO, Y., Y. OHASHI, S. KAWABE, H. ABE, and T. OKUDA: β-Hydroxy-L-valine, a
     Constitutional Amino Acid of Antibiotics YA-56 X and Y. J. Antibiotics **25**, 360 (1972).
208. IWAMI, K., K. YASUMOTO, and H. MITSUDA: Enzymatic Cleavage of Cysteine Sulfoxide
     in *Lentinus edodes*. Agric. Biol. Chem. **39**, 1947 (1975).
209. IWAMI, K., K. YASUMOTO, K. NAKAMURA, and H. MITSUDA: Properties of γ-
     Glutamyltransferase from *Lentinus edodes*. Agric. Biol. Chem. **39**, 1933 (1975).
210. —————: Reactivity of *Lentinus* γ-Glutamyltransferase with Lentinic Acid as the
     Principal Endogenous Substrate. Agric. Biol. Chem. **39**, 1941 (1975).
211. IWASAKI, H., T. KAMIYA, O. OKA, and J. UEYANAGI: Synthesis of Tricholomic Acid, a
     Flycidal Amino Acid I. Chem. Pharm. Bull. (Japan) **13**, 753 (1965).
212. —————: Synthesis of Tricholomic Acid II. Chem. Pharm. Bull. (Japan) **14**, 1307
     (1966).
213. JADOT, J., J. CASIMIR, and F. ALDERWEIRELDT: Charactérisation de la L-*threo*-β-
     hydroxyleucine à partir de *Deutzia gracilis*. Biochim. Biophys. Acta **78**, 500 (1963).
214. JADOT, J., J. CASIMIR, and A. LOFFET: Séparation et charactérisation de l'acid 2(*S*),
     4(*R*)-γ-hydroxy-γ-methylglutamique à partir de *Ledenbergia rosea-aenea* et de l'acid
     2(*S*), 4(*S*)-γ-hydroxy-γ-methylglutamique à partir de *Pandanus veitchii*. Biochim.
     Biophys. Acta **136**, 79 (1967).
215. JADOT, J., J. CASIMIR, and G. MAGHUIN: Identification de la L(+)cystationine dans
     Boletus erythropus. Bull. soc. roy. sci. Liège **40**, 355 (1971).
216. JADOT, J., J. CASIMIR, and M. RENARD: Séparation et charactérisation du L(+)-γ-(*p*-
     hydroxy)anilide de l'acide glutamique à partir du *Agaricus hortensis*. Biochim.
     Biophys. Acta **43**, 322 (1960).
217. JADOT, J., J. CASIMIR, and R. WARIN: Establissement de la formule de structure d' un
     nouvel acide amine soufre de *Xerocomus subtomentosus*. Bull. soc. chim. Belges **78**, 299
     (1969).
218. JIRACEK, V., J. SÜSS, and J. KOCOUREK: Sugar and Free and Bound Amino Acids in
     Polygala vulgaris. Planta Med. **10**, 298 (1962).
219. JOHNSTON, M., D. JANKOOWSKI, P. MARCOTTE, H. TANAKA, N. ESAKI, K. SODA, and
     C. WALSH: Suicide Inactivation of Bacterial Cystathionine γ-Synthase and Methionine
     γ-Lyase During Processing of L-Propargylglycine. Biochemistry **18**, 4690 (1979).
220. JOHNSTON, M., R. RAINES, M. CHANG, N. ESAKI, K. SODA, and C. WALSH: Mechanistic
     Studies on Reactions of Bacterial Methionine γ-Lyase with Olefinic Amino Acids.
     Biochemistry **20**, 4325 (1981).
221. JOUNATHAN, E.S., and E. FRIEDEN: Studies on Amylase Synthesis by Pigeon Pancrease
     Slices. J. Biol. Chem. **220**, 801 (1956).
222. KALYANKAR, G.D., and F.F. SNELL: Differentiation of α-Amino Acids and Amines by
     Non-enzymatic Transamination on Paper Chromatogram. Nature **180**, 1069 (1959).

223. KAMEOKA, H., and Y. DEMIZU: 3,5-Diethyl-1,2,4-trithiolane from *Allium cepa*. Phytochem. **18**, 1397 (1979).

224. KARLE, I.L., J.W. DALY, and B. WITKOP: 2,3-*cis*-3,4-*trans*-3,4-Dihydroxy-L-proline: Mass Spectrometry and X-Ray Analysis. Science **164**, 1401 (1969).

225. KASAI, T., Y. KISHI, M. SANO, and S. SAKAMURA: Isolation and Identification of 3-(2-Furoyl)alanine and L-Pipecolic Acid from Green Gram Seeds. Agric. Biol. Chem. **37**, 2923 (1973).

226. KASAI, T., and P.O. LARSEN: Chemistry and Biochemistry of γ-Glutamyl Derivatives from Plants Including Mushrooms (Basidiomycetes). Fortschr. Chem. organ. Naturstoffe **39**, 173 (1980).

227. KASAI, T., P.O. LARSEN, and H. SØRENSEN: Free Amino Acids and γ-Glutamyl Peptides in Fagaceae. Phytochem. **17**, 1911 (1978).

228. KATO, K.: Synthesis of Cleonine, Amino(1-hydroxycyclopropyl)acetic Acid, a Novel Amino Acid Contained in Cleomycin. Tetrahedron Letters **1980**, 4925.

229. KAZT, E., K.T. MASON, and A.B. MAUGER: The Presence of α-Amino-β, γ-dihydroxybutyric Acid in Hydrolysates of Actinomycin Z_1. J. Antibiotics **27**, 952 (1974).

230. KIMLER, L., R.A. LARSON, L. MESSENGER, J.B. MOORE, and T.J. MABRY: Betalamic Acid, a New Naturally Occuring Pigment. Chem. Commun. **1971**, 1329.

231. KIMOTO, M., T. OGAWA, and S. SASAOKA: Effect of 1-Aminoproline on Methionine Metabolism in Rats. Arch. Biochem. Biophys. **206**, 336 (1981).

232. KING, F.F., T.J. KING, and A.J. WARWICK: The Chemistry of Extractives from Hardwoods. Pt. III. Baikiain, an Amino Acid Present in *Baikiaea plurijuga*. J. Chem. Soc. (London) **1950**, 3590.

233. KING, G.S., E.S. WAIGHT, P.G. MANTLE, and C.A. SZCZYRBAK: The Structure of Clavicepitic Acid, an Azepinoindole Derivative from *Claviceps fusiformis*. J. Chem. Soc. (London) Perkin Trans. I. **1977**, 2099.

234. KLOSTERMAN, H.J., G.L. LAMOUREUX, and J.L. PARSONS: Isolation, Characterization, and Synthesis of Linatine. A Vitamin B_6 Antagonist from Flaxseed (*Linum usitatissimum*). Biochemistry **6**, 170 (1967).

235. KÖGL, F., and H. ERXLEBEN: Untersuchungen über Pilzfarbstoffe VIII. Über den roten Farbstoff des Fliegenpilzes. Liebigs Ann. Chem. **479**, 11 (1930).

236. KOIKE, Y., K. WADA, G. KUSANO, S. NOZOE, and K. YOKOYAMA: Isolation of Psilocybin from *Psilocybe argentipes* and its Determination in Specimens of Some Mushrooms. J. Nat. Prod. **44**, 362 (1981).

237. KOLLONITSCH, J., A. ROSEGAY, and G. DONDOURAS: Reactions in Strong Acids II. New Concept in Amino Acid Chemistry: C-derivatization of amino acids. J. Amer. Chem. Soc. **86**, 1857 (1964).

238. KOMAMINE, A.: Metabolism of Aromatic Amino Acids in Plants I. On 3,4-Dihydroxyphenylalanine in *Stizolobium hassjoo*. Bot. Mag. Tokyo **75**, 228 (1962).

239. KONNO, K., K. HASHIMOTO, Y. OHFUNE, H. SHIRAHAMA, and T. MATSUMOTO: Synthesis of Acromelic Acid A, a Toxic Principle of *Clitocybe acromelalga*. Tetrahedron Letters **27**, 607 (1986).

240. — — — — —: Acromelic Acids A and B. Potent Neuroexitatory Amino Acids Isolated from *Clitocybe acromelalga*. J. Amer. Chem. Soc. **110**, 4807 (1988).

241. KONNO, K., K. HAYANO, H. SHIRAHAMA, H. SAITO, and T. MATSUMOTO: Structure and Synthesis of Clitidine, a New Toxic Pyridine Nucleotide from *Clitocybe acromelalga*. Tetrahedron Letters **1977**, 481.

242. — — — — — : Clitidine, a New Toxic Pyridine Nucleotide from *Clitocybe acromelalga*. Tetrahedron **38**, 3281 (1982).

243. KONNO, K., H. SHIRAHAMA, and T. MATSUMOTO: Isolation and Structure of Clithion-

eine, a New Amino Acid Betaine from *Clitocybe acromelalga*. Tetrahedron Letters **22**, 1617 (1981).

244. — — — : Isolation and Structure of Acromelic acid A and B. New Kainoids of *Clitocybe acromelalga*. Tetrahedron Letters **24**, 939 (1983).

245. — — —: Clitioneine, an Amino Acid Betaine from *Clitocybe acromelalga*. Phytochem. **23**, 1003 (1984).

246. KRISTENSEN, I., and P.O. LARSEN; Azetidine-2-carboxylic acid Derivatives from Seeds of *Fagus silvatica* L. and a Revised Structure for Nicotianamine. Phytochem. **13**, 2791 (1974).

247. KRISTENSEN, I., P.O. LARSEN, and C.E. OLESEN: Two Diastereomers of 5-Hydroxy-6-methylpipecolic Acid from Seeds of *Fagus silvatica* L. Tetrahedron **32**, 2799 (1976).

248. KUSANO, G., H. OGAWA, A. TAKAHASHI, S. NOZOE, and K. YOKOYAMA: A New Amino Acid, (2S,3R)-(−)-3-Hydroxybaikiain from *Russula subnigricans* Hongo. Chem. Pharm. Bull. (Japan) **35**, 3482 (1987).

249. LAMBEIN, F., Y.-H. KUO, and R. VAN PARIJS: Isoxazolin-5-ones: Chemistry and Biology of a New Class of Plant Products. Heterocycles **4**, 567 (1976).

250. LAMBEIN, F., N. SCHAMP, L. VANDENDRIESSSCHE, and R. VAN PARIJS: A new UV-Sensitive Heterocyclic Amino Acid from Pea Seedlings: 2-Alanyl-3-isoxazolin-5-one. Biochem. Biophys. Res. Commun. **37**, 375 (1969).

251. LAMBEIN, F., and R. VAN PARIJS: New Isoxazolinone Amino Acids from *Lathyrus odoratus* Seedlings. Biochem. Biophys. Res. Commun. **61**, 155 (1974).

252. LAMOTTE, J., B. OLEKSYN, L. DUPONT, O. DIDEBERG, H. CAMPSTEYN, and M. VERMEIRE: The Crystal and Molecular Structure of 3-[(5S)-5-Carboxy-2-oxotetrahydrofur-5-yl]-(2S)-alanine (Lycoperdic acid). Acta Crystallogr. **B34**, 3635 (1978).

253. LANCINI, G.C., A. DIENA, and E. LAZZARI: The Synthesis of Alanosine [L-2-Amino-3-(N-nitrosohydroxyamino)propionic acid]. Tetrahedron Letters **1966**, 1769.

254. LARSEN, P.O.: Amino Acids and γ-Glutamyl Derivatives in Seeds of *Lunaria annua* L. Part III. Acta Chem. Scand. **21**, 1592 (1967).

255. — : Free Amino Acids in Cruciferae and Resedaceae. Danish Atomic Energy Commission Research Establishment Risö, Report No. 189 (1969).

256. LARSEN, P.O., and A. KJAER: Paper-chromatographic Differentiation Between α-Monoamino Acids and Other Ninhydrin-positive Substances. Biochim. Biophys. Acta **38**, 149 (1960).

257. LEETE, E.: The Biosynthesis of Azetidine-2-carboxylic Acid. J. Amer. Chem. Soc. **86**, 3162 (1964).

258. LEETE, E., G.E. DAVIS, C.R. HUTCHINSON, K.W. WOO, and M.R. CHEDEKEL: Biosynthesis of Azetidine-2-carboxylic Acid in *Convallaria majalis*. Phytochem. **13**, 427 (1974).

259. LEETE, E., L.L. LOUTERS, and H.S.P. RAO: Biosynthesis of Azetidine-2-carboxylic Acid in *Convallaria majalis*: Studies With N-15 Labelled Precursors. Phytochem. **25**, 2753 (1986).

260. LEUNG, A.Y., and A.G. PAUL: Baeocystin, a Mono-methyl Analog of Psilocybin from *Psilocybe baeocystis* Saprophyte Culture. J. Pharm. Sci. **56**, 146 (1967).

261. — —: Baeocystin and Norbaeocystin: New Analogs of Psilocybin from *Psilocybe baeocystis*. J. Pharm. Sci. **57**, 1667 (1968).

262. LEVENBERG, B.: Structure and Enzymatic Cleavage of Agaritine, a Phenylhydrazide of L-Glutamic Acid Isolated from Agaricaceae. J. Amer. Chem. Soc. **83**, 503 (1961).

263. — : Isolation and Structure of Agaritine, a γ-Glutamyl-substituted Arylhydrazine Derivative from Agaricaceae. J. Biol. Chem. **239**, 2267 (1964).

264. LEVENBERG, B.: Isolation and Characterization of β-Methylene-L-(+)-norvaline from *Lactarius helvus*. J. Biol. Chem. **243**, 6009 (1968).

265. LEWIS, D.E., and F.W. DUNN: β-3-Furylalanine a New Analog of Phenylalanine. Arch. Biochem. Biophys. **107**, 363 (1964).

266. LICHTENSTEIN, N.: Preparation of γ-Alkylamides of Glutamic acid. J. Amer,. Chem. Soc. **64**, 1021 (1942).

267. LINCOFF, G., and D.H. MITCHEL: Toxic and Hallucinogenic Mushroom Poisoning—a Handbook for Physicians and Mushroom Hunters. New York: Van Nostrand Rheinhold. 1977.

268. LINDBERG, P., R. BERGMAN, and B. WICKBERG: Isolation and Structure of Coprine, a Novel Physiologically Active Cyclopropanone Derivative from *Coprinus atramentarius* and its Synthesis *via* 1-Aminocyclopropanol. Chem. Commun. **1975**, 946.

269. ———: Isolation and Structure of Coprine, the *in vivo* Aldehyde Dehydrogenase Inhibitor in *Coprinus atramentarius*; Synthesis of Coprine and Related Cyclopropanone Derivatives. J. Chem. Soc. (London) Perkin I, **1977**, 684.

270. LINDLAR, H.: Ein neuer Katalysator für selektive Hydrierung. Helv. Chim. Acta **35**, 446 (1952).

271. LINKO, P., and A.I. VIRTANEN: On the Biosynthesis of Some Recently Discovered Derivatives of Glutamic Acid in Plants. Acta Chem. Scand. **12**, 68 (1958).

272. LUND, U.: Estimation of Muscimol and Ibotenic Acid in *Amanita muscaria* Using High-performance Liquid Chromatography. Arch. Pharm. Chem., Sci. Ed. **7**, 115 (1979).

273. MABRY, T.J.: The Betacyanins and Betaxanthins. In: Comparative Phytochemistry (T. Swain, ed.), pp. 231–244. London: Academic Press. 1966.

274. —: Betacyanins. In: Encyclopedia of Plant Physiology, Vol. **8**, Secondary Plant Products (E.A. BELL and B.V. CHARLWOOD, eds.), pp. 513–533. Berlin Heidelberg New York: Springer. 1980.

275. MABRY, T.J., A. TAYLOR, and B.L. TURNER: The Betacyanins and their Distribution. Phytochem. **2**, 61 (1963).

276. MARCUS, A., and L.M. SHANNON: γ-Methyl-γ-hydroxy-α-ketoglutaric Aldolase. J. Biol. Chem. **237**, 3348 (1962).

277. MARLIER, M., G. DARDENNE, and J. CASIMIR: 2S-Carboxy-4R,5S-dihydroxypiperidine et 2S-carboxy-4S,5S-dihydroxypiperidine à partir de *Derris elliptica*. Phytochem. **15**, 183 (1976).

278. ———: 2S,4R-Carboxy-2-acetylamino-4-piperidine dans les feuilles de *Calliandra haematocephala*. Phytochem. **18**, 479 (1979).

279. MARTIN, D.G., C.G. CHIDESTER, S.A. MIZSAK, D.J. DUCHAMP, L. BACZYNSKYJ, W.C. KRUEGER, R.J. WNUK, and P. A. MEULMAN: The Isolation, Structure, and Absolute Configuration of U-43, 795, A New Antitumor Agent. J. Antibiotics **28**, 91 (1975).

280. MARTIN, D.G., D.J. DUCHAMP, and C.G. CHIDESTER: The Isolation, Structure, and Absolute Configuration of U-42, 126 a Novel Antitumor Antibiotic. Tetrahedron Letters **1973**, 2549.

281. MATSUMOYO, N.: Isolation and Identification of S-2-Aminoethyl-L-cysteine from *Rozites caperata* and 2-Amino-3-butenoic Acid from *Rhodophyllus crassipes* and their Antibacterial Activity. J. Med. Soc. Toho Japan **31**, 249 (1984).

282. MATSUMOTO, T., W. TRUEB, R. GWINNER, and C.H. EUGSTER: Isolierung von (−)-R-4-Hydroxy-pyrrolidon-(2) und einigen weiteren Verbindungen aus *Amanita muscaria*. Helv. Chim. Acta **52**, 716 (1969).

283. MATZINGER, P., PH. CATALFOMO, and C.H. EUGSTER: Isolierung von (2S, 4S)-(+)-γ-Hydroxynorvalin aus *Boletus satanas* Lenz. Helv. Chim. Acta **55**, 1478 (1972).

284. MEEK, J.S., and J.W. ROWE: The Synthesis of β-Cyclopropyl-α-aminopropionic Acid. J. Amer. Chem. Soc. 77, 6675 (1955).

285. MEISTER, A., and S.S. TATE: Glutathione and Related γ-Glutamyl Compounds: Biosynthesis and Utilization. Ann. Rev. Biochem. 45, 559 (1976).

286. MICHAEL, E., B. HENNING, and H. KREISEL: Handbuch für Pilzfreunde, Vol. I. Jena: Gustav Fisher. 1978.

287. MILLER, S.L.: A Production of Amino Acids Under Possible Primitive Earth Conditions. Science 117, 528 (1953).

288. MILLER, H.E., H. RÖSLER, A. WOHLPART, H. WYLER, M.E. WILCOX, H. FROHOFER, T.J. MABRY, and A.S. DREIDING: Biogenese der Betalaine. Biotransformation von DOPA und Tyrosin in den Betalaminsäureteil des Betanins. Helv. Chim. Acta 51, 1470 (1968).

289. MILLINGTON, D.S., and L. FOWDEN: Addendum. Spectroscopic Characterization of Two New Amino Acids from Aesculus californica. Phytochem. 7, 1027 (1968).

290. MINALE, L., M. PIATTELLI, and R.A. NICOLAUS: Pigments of Centrospermae IV. On the Biogenesis of Indicaxanthin and Betanin in Opuntia ficus-indica Mill. Phytochem. 4, 593 (1965).

291. MINATO, S.: Isolation of Anthglutin, an Inhibitor of γ-Glutamyl Transpeptidase from Penicillium oxalicum. Arch Biochem. Biophys. 192, 235 (1979).

292. MIX, H.: Synthese aliphatischer 2-Amino-3-hydroxysäuren und Diastereomerentrennung mit Hilfe von Kupferchelaten. Z. physiol. Chem. 327, 41 (1961).

293. MORIGUCHI, M., Y. HARA, and S.I. HATANAKA: Antibacterial Activity of L-2-Amino-4-chloro-4-pentenoic Acid Isolated from Amanita pseudoporphyria Hongo. J. Antibiotics 40, 904 (1987).

294. MORIGUCHI, M., S. HOSHINO, and S.I. HATANAKA: Dehalogenation and Deamination of L-2-Amino-4-chloro-4-pentenoic Acid by Proteus mirabilis. Agric. Biol. Chem. 51, 3295 (1987).

295. MORIGUCHI, M., K. KIMURA, S.I. HATANAKA: Isolation and Identification of a New γ-L-Glutamylpeptide, γ-L-Glutamyl-cis-3-amino-L-proline from the Cultured Mycelia of Morchella esculenta. Trans. mycol. Soc. Japan 24, 191 (1983).

296. MORIGUCHI, M., S. SADA, and S.I. HATANAKA: Isolation of cis-3-Amino-L-proline from Cultured Mycelia of Morchella esculenta Fr. Appl. Env. Microbiol. 38, 1018 (1979).

297. MORIGUCHI, M., M. YAMADA, S. SUENAGA, H. TANAKA, A. WAKASUGI, and S.I. HATANAKA: Partial Purification and Properties of γ-Glutamyltranspeptidase from Mycelia of Morchella esculenta. Arch. Microbiol. 144, 15 (1986).

298. MORITA, K., and K. KOBAYASHI: Isolation and Synthesis of Lenthionine, an Odorous Substance of Shiitake, an Edible Mushroom. Tetrahedron Letters 1966, 573.

299. ——: Isolation, Structure, and Synthesis of Lenthionine and its Analogs. Chem. Pharm. Bull. Japan 15, 988 (1967).

300. MORRIS, C.J., J.F. THOMPSON, S. ASEN, and F. IRREVERRE: The Isolation of γ-Glutamyl-β-alanine from Iris Bulbs. J. Biol. Chem. 237, 2180 (1962).

301. MORRIS, L.J.: Mechanisms and Stereochemistry in Fatty Acid Metabolism — The Fifth Colworth Medical Lecture. Biochem. J. 118, 681 (1970).

302. MOSER, M.: Die Röhrlinge und Blätterpilze (Polyporales, Boletales, Agaricales, Russulales). In: Kleine Kryptogamenflora IIb/2, Basidiomyceten, 2. Teil, p 156. Stuttgart: Gustav Fisher. 1983.

303. MÜLLER, G.F.R., and C.H. EUGSTER: Muscimol, ein pharmakodynamisch wirksamer Stoff aus Amanita muscaria. Helv. Chim. Acta 48, 910 (1965).

304. MURAKAMI, N., J. FURUKAWA, S. OKUDA, and S.I. HATANAKA: Stereochemistry of 2-Aminopimelic Acid and Related Amino Acids in Three Species of Asplenium. Phytochem. 24, 2291 (1985).

305. MURAKOSHI, I., F. IKEGAMI, F. KATO, and J. HAGINIWA: Enzymic Alanylation of an Isoxazolinone Glucoside by Legume Seedling Extracts. Phytochem. **14**, 1515 (1975).

306. MURAKOSHI, I., Y.-Y. MEI, M. GOO, and J. HAGINIWA: Isolation of (−)-*trans*-4-Hydroxypipecolic Acid from the Young Leaves of *Acacia mollissima.* Yakugaku Zasshi **89**, 1723 (1969).

307. MURTHY, Y.K.S., J.E. THIEMANN, C. CORONELLI, and P. SENSI: Alanosine, a New Antiviral and Antitumour Agent Isolated from a *Streptomyces.* Nature **211**, 1198 (1966).

308. MUSSO, H.: The Pigments of Fly Agaric, *Amanita muscaria.* Tetrahedron **35**, 2843 (1979).

309. — : Über die Farbstoffe des Fliegenpilzes. Aufgaben und Ziele der Naturstoffchemie heute. Naturwiss. **69**, 326 (1982).

310. MUTO, T., and R. SUGAWARA: 1,3-Diolein, a House Fly Attractant in the Mushroom, *Amanita muscaria* (L.) Fr. In: Control of Insect Behavior by Natural Products (D.L. WOOD, R.M. SILVERSTEIN, and M. NAKAJIMA, eds.), pp. 189–208. New York: Academic Press. 1970.

311. NAKAJIMA, T., and B. VOLCANI: 3,4-Dihydroxyproline: a New Amino Acid in Diatom Cell Walls. Science **164**, 1400 (1969).

312. NAKANO, K., Y. INAMASU, S. HAGIHARA, and F. OBO: Isolation and Properties of L-Amino Acid Oxidase in Habu snake (*Trimeresurus fravoviridis*) Venom. Acta Medica Univ. Kagoshima **14**, 229 (1972).

313. NARAYANAN, S., M.R.S. IYENGAR, P.L. GANJU, and S. RENGAJU. γ-Chloronorvaline, a Leucine Analog from *Streptomyces.* J. Antibiotics **33**, 1249 (1980).

314. NIIMURA, Y., and S.I. HATANAKA: L-*threo-* and L-*erythro*-2-Amino-3-hydroxyhex-4-ynoic acid: New Amino Acids from *Tricholomopsis rutilans.* Phytochem. **13**, 175 (1974).

315. — — : Biochemical Studies on Nitrogen Compounds of Fungi XIII. Metabolism of L-norleucine in *Tricholomopsis rutilans* (Fr.) Sing. Sci. Pap. Coll. Gen. Educ. Univ. Tokyo **26**, 27 (1976).

316. — — : Two γ-Glutamylpeptides of Acetylenic Amino Acids in *Tricholomopsis rutilans.* Phytochem. **16**, 1435 (1977).

317. NIINURA, Y., F. KINOSHITA, and S.I. HATANAKA: Non-protein Amino Acids in Five Species in the Genus *Tricholomopsis.* Trans. mycol. Soc. Japan **15**, 218 (1974).

318. NISHIKAWA, M., K. KAMIYA, S. KOBAYASHI, K. MORITA, and Y. TOMIIE: The X-Ray Analysis of Lenthionine, an Odorous Substance of Shiitake, an Edible Mushroom. Chem. Pharm. Bull. Japan **15**, 756 (1967).

319. NOMA, M., M. NOGUCHI, and E. TAMAKI: A New Amino Acid, Nicotianamine, from Tobacco Leaves. Tetrahedron Letters **1971**, 2017.

320. — — — : Isolation and Characterization of D-Alanyl-D-alanine from Tobacco Leaves. Agric. Biol. Chem. **37**, 2439 (1973).

321. NORRIS, R.D., and L. FOWDEN: Substrate Discrimination by Prolyl-tRNA Synthetase from Various Higher Plants. Phytochem. **11**, 2921 (1972).

322. NULU, J.R., and E.A. BELL: Configuration of L-γ-Ethylideneglutamic Acid from *Guilandina crista.* Phytochem. **11**, 2573 (1972).

323. OGAWA, T., M. FUKUDA, and K. SASAOKA: Occurrence of *N*-Malonyl-D-alanine in Pea Seedlings. Biochim. Biophys. Acta **297**, 60 (1973).

324. OGAWA, T., Y. OKA, and K. SASAOKA: D$_s$-*erythro*-2-Amino-3,4-dihydroxybutanoic Acid, a Constituent in the Edible Mushroom, *Lyophyllum ulmarium.* Phytochem. **23**, 684 (1984).

325. — — — : D$_s$-*erythro*-2-Amino-4-ethoxy-3-hydroxybutanoic Acid from the Fruiting Bodies of the Edible Mushroom, *Lyophyllum ulmarium.* Phytochem. **24**, 1837 (1985).

326. — — —: Amino Acid Profiles of Common Cultivated Mushrooms Including the Identification of N-(N-γ-L-Glutamyl-3-sulfo-L-alanyl)glycine in *Flammulina velutipes*. J. Food Sci. **52**, 135 (1987).

327. OHASHI, Y., H. ABE, and Y. ITO: β-Hydroxy-L-valine and 4-Amino-3,6-dihydroxy-2-methylhexanoic Acid, Constitutional Amino Acids of the Antibiotic YA-56. Agric. Biol. Chem. **37**, 2283 (1973).

328. OHENOLA, E., J. JOKIRANTA, T. MÄKINEN, A. KAIKKONEN, and M.M. AIRAKSINEN: The Occurrence of Psilocybin and Psilocin in Finnish Fungi. J. Nat. Prod. **50**, 741 (1987).

329. OHTA, T., S. NAKAJIMA, S.I. HATANAKA, M. YAMAMOTO, Y. SHIMMEN, S. NISHIMURA, Z. YAMAIZUMI, and S. NOZOE: A Chlorohydrin Amino Acid from *Amanita abrupta*. Phytochem. **26**, 565 (1987).

330. OHTA, T., S. NAKAJIMA, Z. SATO, T. AOKI, S.I. HATANAKA, and S. NOZOE: Cyclopropylalanine, an Antifungal Amino Acid of the Mushroom *Amanita virgineoides* Bas. Chemistry Letters **1986**, 511.

331. OKA, Y., T. OGAWA, and K. SASAOKA: Occurrence of L-Saccharopine and γ-Glutamylglycine in the Mushroom, *Agaricus bisporus*. Agric. Biol. Chem. **43**, 1995 (1979).

332. — — —: Occurrence of N-(γ-L-Glutamyl)ethanolamine in the Mushroom, *Agaricus bisporus*. Agric. Biol. Chem. **44**, 1959 (1980).

333. — — —: First Evidence for the Occurrence of N^δ-Acetyl-L-ornithine and Quantitation of the Free Amino Acids in the Cultured Mushroom, *Pleurotus ostreatus*. J. Nutr. Sci. Vitaminol. **30**, 27 (1984).

334. OKA, Y., H. TSUJI, T. OGAWA, and K. SASAOKA: Quantitative Determination of the Free Amino Acids and their Derivatives in the Common Edible Mushroom, *Agaricus bisporus*. J. Nutr. Sci. Vitaminol. **27**, 253 (1981).

335. ONDA, M., H. FUKUSHIMA, and M. AKAGAWA: A Flycidal Constituent of *Amanita pantherina* (DC.) Fr. Chem. Pharm. Bull. Japan **12**, 751 (1964).

336. OZAWA, M., H. INATOMI, Y. SUYAMA, and F. INUKAI: Studies on the Free Amino Acids in Plants. Part VI. Isolation of γ-Methyl- and γ-Methyleneglutamic Acids from Edible Lily Bulbs. Meiji Daigaku Nogakubu Kenkyu Hokoku **22**, 7 (1967).

337. PANT, R., and H.M. FALES: Occurrence of a New Amino Acid in *Crotalaria* Seeds. Phytochem. **13**, 1626 (1974).

338. PARKER, E.D., C.G. SKINNER, and W. SHIVE: Biological Specificities of 4,5-Dehydro analogues of Isoleucine and Alloisoleucine. J. Biol. Chem. **236**, 3267 (1961).

339. PEARSON, A.A.: The genus *Lactarius*. The Naturalist, July-September, 81 (1950).

340. PETERSON, P.J., and L. FOWDEN: Purification, Properties and Comparative Specificities of the Enzymes Prolyl-transfer Ribonucleic Acid Synthetase from *Phaseolus aureus* and *Polygonatum multiflorum*. Biochem. J. **97**, 112 (1965).

341. — —: The Biosynthesis of γ-Substituted Glutamic Acids in *Gleditsia triacanthos*. Phytochem. **11**, 663 (1972).

342. PFLEIDERER, G., W. GRUBER, and TH. WIELAND: Eine enzymatische Bestimmung der L-Asparaginsäure. Biochem. Z. **326**, 446 (1954/55).

343. PIATTELLI, M.: The Betalains: Structure, Biosynthesis, and Chemical Taxonomy. In: The Biochemistry of Plants, a Comprehensive Treatise, Vol. 7 (E.E. CONN, ed.), pp. 557–575. New York: Academic Press. 1981.

344. PIATTELLI, M., and L. MINALE: Pigments of Centrospermae I. Betacyanins from *Phyllocactus hybridus* Hort. and *Opuntia ficus-indica* Mill. Phytochem. **3**, 307 (1964).

345. — — : Structura dell'amarantina e dell'isoamarantina II. Ann. Chim. **56**, 1060 (1966).

346. PIATTELLI, M., L. MINALE, and R.A. NICOLAUS: Pigments of Centrospermae V. Betaxanthins from *Mirabilis jalapa* L. Phytochem. **4**, 817 (1965).

347. PIATTELLI, M., L. MINALE, and G. PROTA: Isolation, Structure and Absolute Configuration of Indicaxanthin. Tetrahedron **20**, 2325 (1964).

348. ———: Structure della betanica. Ann. Chim. **54**, 955 (1964).

349. ———: Pigments of Centrospermae III. Betaxanthins from *Beta vulgaris* L. Phytochem. **4**, 121 (1965).

350. PILBEAM, D.J., and E.A, BELL: A Reappraisal of the Free Amino Acids in Seeds of *Crotalaria juncea* (Leguminosae). Phytochem. **18**, 320 (1979).

351. POLLARD, J.K., E. SONDHEIMER, and F.C. STEWARD: New Hydroxyamino Acids in Plants and their Identification. Nature **182**, 1356 (1958).

352. POTGIETER, H.C., N.M.J. VERMEULEN, D.J.J. POTGIETER, and H.F. STRAUSS: A Toxic Amino Acid, 2(*S*)3(*R*)-2-Amino-3-hydroxypent-4-ynoic Acid from the Fungus *Sclerotium rolfsii*. Phytochem. **16**, 1757 (1977).

353. POWELL, G.K., and E.E. DEKKER: Purification and Properties of 4-Methylene-L-glutamine Aminohydrolase from Peanut Leaves. J. Biol. Chem. **258**, 8677 (1983).

354. PRZYBYLSKA, J., and F.M. STRONG: Identification of γ-Methylglutamic Acid in *Lathyrus maritimus*. Phytochem. **7**, 471 (1968).

355. REHR, S.S., E.A. BELL, D.E. JANZEN, and P.P. FEENY: Insecticidal Amino Acids in Legume Seeds. Biochem. System. **1**, 63 (1973).

356. REHR, S.S., D.H. JANZEN, and P.P. FEENY: L-DOPA in Legume Seeds: a Chemical Barrier to Insect Attack. Science **181**, 81 (1973).

357. REPKE, D.B., D.T. LESLIE, and N.G. KISH: GLC-Mass Spectral Analysis of Fungal Metabolites. J. Pharm. Sci. **67**, 485 (1978).

358. RHUGENDA-BANGA, N., A. WELTER, J. JADOT, and J. CASIMIR: Un nouvel acide amine isolé de *Lycoperdon perlatum*. Phytochem. **18**, 482 (1979).

359. RING, D., Y. WOLMAN, N. FRIEDMANN, and S.L. MILLER: Prebiotic Synthesis of Hydrophobic and Protein Amino Acid. Proc. Nat. Acad. Sci. (USA) **69**, 765 (1972).

360. ROBBERS, J.E., and H.G. FLOSS: Clavicepitic Acid, a New 4-Substituted Indole Amino Acid Obtained from Submerged Cultures of Ergot Fungus. Tetrahedron Letters **1969**, 1857.

361. ROBBERS, J.E., V.E. TYLER, and G.M. OLA'H: Additional Evidence Supporting the Occurrence of Psilocybin in *Panaeolus foenisecii*. Lloydia **32**, 399 (1969).

362. ROMEO, J.T., L.A. SWAIN, and A.B. BLEECKER: *cis*-4-Hydroxypipecolic Acid and 2,4-*cis*-4,5-*trans*-4,5-Dihydroxypipecolic Acid from *Calliandra*. Phytochem. **22**, 1615 (1983).

363. ROSENTHAL, G.A.: Plant Nonprotein Amino and Imino Acids, Biological, Biochemical, and Toxicological Properties. New York: Academic Press. 1982.

364. ROSENTHAL, G.A., and E.A. BELL: Naturally Occurring, Toxic Nonprotein Amino Acids. In: Herbivores (G.A. ROSENTHAL and E.A. BELL, eds.), pp. 353–385. New York: Academic Press. 1977.

365. ROSSETTI, V., and M. SURIA: γ-Methyleneglutamic Acid from *Lilium candidum* Bulbs. Phytochem. **11**, 859 (1972).

366. RUDZATS, R., E. GELLERT, and B. HALPERN: Constituents of a New Guinea *Boletus*, Isolation and Identification of a New Unsaturated α-Amino Acid. Biochem. Biophys. Res. Commun. **47**, 290 (1972).

367. RULE, C.J., B.A. WURZBURG, and B. GANEM: 3*R*,4*R*-Dihydroxy-L-proline: a Potent and Specific β-D-Glucuronidase Inhibitor. Tetrahedron Letters **1985**, 5379.

368. SAITO, K., and A. KOMAMINE: Biosynthesis of Stizolobinic acid and Stizolobic Acid in Higher Plants—An Enzyme System(s) Catalyzing the Conversion of Dihydroxyphenylalanine into Stizolobinic Acid and Stizolobic Acid from Etiolated Seedlings of *Stizolobium hassjoo*. Eur. J. Biochem. **68**, 237 (1976).

369. ——: Biosynthesis of Stizolobinic Acid and Stizolobic Acid in Higher Plants—Stizolobinic Acid Synthase and Stizolobic Acid Synthase, New Enzymes Which Catalyze the Reaction Sequences Leading to the Formation of Stizolobinic Acid and Stizolobic Acid from 3,4-Dihydroxyphenylalanine in *Stizolobium hassjoo*. Eur. J. Biochem. **82**, 385 (1978).

370. SAITO, K., A. KOMAMINE, and S.I. HATANAKA: Biosynthesis of Stizolobic Acid and Stizolobinic Acid in *Amanita pantherina*. Z. Naturforsch. **33C**, 793 (1978).

371. SAITO, K., A. KOMAMINE, and S. SENOH: Biosynthesis of Stizolobinic Acid and Stizolobic Acid in the Etiolated Seedlings of *Stizolobium hassjoo*. Z. Naturforsch. **30C**, 659 (1975).

372. ———: Further Studies on the Biosynthesis of Stizolobinic Acid and Stizolobic Acid in the Etiolated Seedlings of *Stizolobium hassjoo*. Z. Naturforsch. **31C**, 15 (1976).

373. SAKATO, Y., T. HASHIZUME, and Y. KISHIMOTO: Studies on the Chemical Constituents of Tea. Part III. On a New Amide Theanine. Nippon Nogeikagaku Kaishi **23**, 262 (1950).

374. ———: Studies on the Chemical Constituents of Tea. Part V. Synthesis of Teanine. Nippon Nogeikagaku Kaishi **23**, 269 (1950).

375. SASAOKA, K., and M. KITO: Synthesis of Theanine by Tea Seedling Homogenate. Agric. Biol. Chem. **28**, 313 (1964).

376. SASAOKA, K., M. KITO, and Y. ONISHI: Synthesis of Theanine by Pea Seed Acetone Powder Extract. Agric. Biol. Chem. **28**, 318 (1964).

377. ———: Synthesis of Theanine by Pigeon Liver Powder Extract. Agric. Biol. Chem. **28**, 325 (1964).

378. ———: Some Properties of the Theanine Synthesizing Enzyme in Tea Seedlings. Agric. Biol. Chem. **29**, 984 (1965).

379. SASAOKA, K., T. OGAWA, K. MORITOKI, and M. KIMOTO: Antivitamin B-6 Effect of 1-Aminoproline on Rats. Biochim. Biophys. Acta **428**, 396 (1976).

380. SASAOKA, K., T. OGAWA, H. TSUJI, and N. BANDO: Incorporation of Radioactive Shikimic Acid into *N*-(γ-L-Glutamyl)-4-hydroxyaniline in *Agaricus bisporus*. Biochim. Biophys. Acta **630**, 137 (1980).

381. SAUPE, S.G.: Occurrence of Psilocybin/Psilocin in *Pluteus salicinus* (Pluteaceae). Mycologia **73**, 781 (1981).

382. SCANNELL, J.P., D.L. PRUESS, T.C. DEMNY, F. WEISS, T. WILLIAMS, and A. STEMPEL: Antimetabolites Produced by Microogranisms II. L-2-Amino-4-pentynoic acid. J. Antibiotics **24**, 239 (1971).

383. SCHLEIFER, K.H., R. PLAPP, and O. KANDLER: Identification of *threo*-3-Hydroxyglutamic Acid in the Cell Wall of *Microbacterium lacticum*. Biochem. Biophys. Res. Commun. **28**, 566 (1967).

384. SCHWIMMER, S., and A. KJAER: Purification and Specificity of the C–S-Lyase of *Albizzia lophanta*. Biochim. Biophys. Acta **42**, 316 (1960).

385. SENOH, S., S. IMAMOTO, Y. MAENO, K. YAMASHITA, M. MATSUI, T. TOKUYAMA, T. SAKAN, A. KOMAMINE, and S. HATTORI: α-Pyrone-6-carboxylic Acid Derivatives II. Synthesis of DL-Stizolobinic Acid, DL-Stizolobic Acid and DL-β-(6-Carboxy-α'-pyron-5-yl)alanine. Tetrahedron Letters **1964**, 3431.

386. ————————: Two Novel Amino Acids in *Stizolobium hassjoo*. Tetrahedron Letters **1964**, 3437.

387. SHEEHAN, J.C., K. MAEDA, A.K. SEN, and J.A. STOCK: The Isolation, Characterization and Synthesis of *erythro*-β-Hydroxy-L-leucine, a New Amino Acid from the Antibiotic Telomycin. J. Amer. Chem. Soc. **84**, 1303 (1962).

388. SHEWRY, P.R., and L. FOWDEN: 4,5-Dihydroxypipecolic Acids in the Seeds of *Julbernardia, Isoberlinia* and *Brachystegia*. Phytochem. **15**, 1981 (1976).

389. SHINOZAKI, H., M. ISHIDA, and T. OKAMOTO: Acromelic Acid, a Novel Excitatory Amino Acid from a Poisonous Mushroom: Effects on the Crayfish Neuromuscular Junction. Brain Res. **399**, 395 (1986).

390. SHRIFT, A.: Aspects of Selenium Metabolism in Higher Plants. Ann. Rev. Plant Physiol. **20**, 475 (1969).

391. SINGER, R.: The Agaricales in Modern Taxonomy, 4th ed. Koenigstein: Koeltz Scientific Books. 1986.

392. p. 451 of the reference 391.

393. SKINNER, C.G., J. EDELSON, and W. SHIVE: A Conformation of Methionine Essential for Its Biological Utilization. J. Amer. Chem. Soc. **83**, 2281 (1961).

394. STADELMANN, R.J., E. MÜLLER, and C.H. EUGSTER: Über die Verbreitung der stereomeren Muscarine innerhalb der Ordnung der Agaricales. Helv. Chim. Acta **59**, 2432 (1976).

395. STEGLICH, W., and F. ESSER: L-3,4-Dihydroxyphenylalanin aus *Strobilomyces floccopus.* Phytochem. **12**, 1817 (1973).

396. STEGLICH, W., and R. PREUSS: L-3,4-Dihydroxyphenylalanine from Carpophores of *Hygrocybe conica* and *H. Ovina.* Phytochem. **14**, 1119 (1975).

397. STIJVE, T.: High-performance Thin-layer Chromatographic Determination of the Toxic Principles of Some Poisonous Mushrooms. Mitt. Geb. Lebensmittelunters. Hyg. **72**, 44 (1981).

398. STRACK, D., D. SCHMIDT, H. REZNIK, W. BOLAND, L. GROTJAHN, and V. WRAY: Humilixanthin, a New Betaxanthin from *Rivina humilis.* Phytochem. **26**, 2285 (1987).

399. STÜSSI, H., and D.M. RAST: The Biosynthesis and Possible Function of γ-Glutaminyl-4-hydroxybenzene in *Agaricus bisporus.* Phytochem. **20**, 2347 (1981).

400. SU, E.F.W., and B. LEVENBERG: Utilization of the Intact Carbon Chain of Methionine for Biosynthesis of Azetidine-2-carboxylic Acid in *Convallaria majalis.* Acta Chem. Scand. **21**, 493 (1967).

401. SULSER, H., M. HABEGGER, and W. BÜCHI: Synthese und Geschmacksprüfungen von 3,4-disubstituierten 2-Hydroxy-2-buten-1,4-oliden. Z. Lebensm.-Unters. Forsch. **148**, 215 (1972).

402. SUNG, M.-L., and L. FOWDEN: *trans*-3-Hydroxy-L-proline: a Constituent of *Delonix regia.* Phytochem. **7**, 2061 (1968).

403. — —: Acetylenic Amino Acids from *Euphoria longan.* Phytochem. **8**, 1277 (1969).

404. — —: Imino Acid Biosynthesis in *Delonix regia.* Phytochem. **10**, 1523 (1971).

405. SUNG, M.-L., L. FOWDEN, D.S. MILLINGTON, and R.C. SHEPPARD: Acetylenic Amino Acids from *Euphoria longan.* Phytochem. **8**, 1227 (1969).

406. SWAIN, L.A., and J.T. ROMEO: Metabolism of Pipecolic Acid Derivatives in *Calliandra* and *Zapoteca.* Phytochem. **27**, 397 (1988).

407. SZENT-GYORGYI, A., R.H. CHUNG, M.J. BOYAJIAN, M. TISHLER, B.H. ARISON, E.F. SCHOENEWALDT, and J.J. WITTICK: Agaridoxin, a Mushroom Metabolite. Isolation, Structure, and Synthesis. J. Org. Chem. **41**, 1603 (1976).

408. TAKAHASHI, K., M. TADENUMA, and S. SATA: 3-Hydroxy-4,5-dimethyl-2(5*H*)-furanone, a Burnt Flavoring Compound from Aged Sake. Agric. Biol. Chem. **40**, 325. (1976).

409. TAKANO, S., Y. IWABUCHI, and K. OGASAWARA: A Concise Enantioselective Synthesis of Acromelic Acid A. J. Amer. Chem. Soc. **109**, 5523 (1987).

410. TAKEMOTO, T., and T. NAKAJIMA: Studies on the Constituents of Indigenous Fungi I. Isolation of the Flycidal Constituent from *Tricholoma muscarium.* Yakugaku Zasshi **84**, 1183 (1964).

411. — —: Structure of Tricholomic acid. Yakugaku Zasshi **84**, 1230 (1964).

412. TAKEMOTO, T., T. NAKAJIMA, and R. SAKUMA: Isolation of a Flycidal Constituent "Ibotenic Acid" from *Amanita muscaria* and *A. pantherina.* Yakugaku Zasshi **84**, 1233 (1964).

413. TAKEMOTO, T., T. NAKAJIMA, and T. YOKOBE: Structure of Ibotenic Acid. Yakugaku Zasshi **84**, 1232 (1964).

414. TAKEMOTO, T., T. YOKOBE, and T. NAKAJIMA: Studies on the Constituents of Indigenous Fungi II. Isolation of the Flycidal Constituent from *Amanita strobiliformis.* Yakugaku Zasshi **84**, 1186 (1964).

415. TAKEO, T.: L-Alanine as a Precursor of Ethylamine in *Camellia sinensis.* Phytochem. **13**, 1401 (1974).

416. —: L-Alanine Decarboxylase in *Camellia sinensis.* Phytochem. **17**, 313 (1978).

417. TAKITA, T., and H. NAGANAWA: L-2-Amino-4-hexenoic Acid in Ilamycins. J. Antibiotics **A16**, 246 (1963).

418. TAKITA, T., K. OHI, Y. OKAMI, K. MAEDA, and H. UMEZAWA: New Antibiotics, Ilamycins. J. Antibiotics **A15**, 46 (1962).

419. TEBBETT, I. R., and B. CADDY: Mushroom Toxins of the Genus *Cortinarius.* Experientia **40**, 441 (1984).

420. THEOBALD, W., O. BÜCH, H.A. KUNZ, P. KRUPP, E.G. STENGER, and H. HEIMANN: Pharmakologische und experimentalpsychologische Untersuchungen mit 2 Inhaltsstoffen des Fliegenpilzes (*Amanita muscaria*). Arzneim. Forsch. **18**, 311 (1968).

421. THOMPSON, G.A., A.H. DATKO, and H. MUDD: Methionine Synthesis in *Lemna.* Plant Physiol. **70**, 1347 (1982).

422. THOMPSON, J.F., C.J. MORRIS, and G.E. HUNT: The Identification of L-α-Amino-δ-hydroxyvaleric Acid and L-Homoserine in Jack Bean Seeds (*Canavalia ensiformis*). J. Biol. Chem. **239**, 1122 (1964).

423. TOKIMOTO, Y., A. KOBAYASHI, T. YAMANISHI, and S. MURAKI: Studies on the "Sugary Flavor" of Raw Cane Sugar III. Key Compound of the Sugary Flavor. Proc. Japan Acad. **B56**, 457 (1980).

424. TOTTMAR, O., and P. LINDBERG: Effects on Rat Liver Acetaldehyde Dehydrogenase *in vitro* and *in vivo* by Coprine, the Disulfiram-like Constituent of *Coprinus atramentarius.* Acta Pharmacol. Toxicol. **40**, 476 (1977).

425. TOUZÉ-SOULET, J.M., and CH. MONTANT: Étude de quelques formes combinées nouvelles de l'acide glutamique chez *Boletus edulis* Fr. ex Bull. Bull. soc. chim. biol. (Paris) **44**, 451 (1962).

426. TSCHIERSCH, B.: Über γ-Methyleneglutamin und γ-Methyleneglutaminsäure in Keimlingen von *Amorpha fruticosa* L. Phytochem. **1**, 103 (1962).

427. TSUJI, H., N. BANDO, T. OGAWA, and K. SASAOKA: Identification of Two Metabolites of Radioactive Shikimic Acid in *Agaricus bisporus.* Agric. Biol. Chem. **45**, 541 (1981).

428. — — — —: Studies on the Biosynthesis of *N*-(γ-L-Glutamyl)-4-hydroxyaniline in *Agaricus bisporus*: Identification of the Position in Shikimic Acid at Which the Amination Occurs. Biochim. Biophys. Acta **677**, 326 (1981).

429. TSUJI, H., T. OGAWA, N. BANDO, and K. SASAOKA: Purification and Properties of 4-Aminobenzoate Hydroxylase, a New Monooxydase from *Agaricus bisporus.* J. Biol. Chem. **261**, 13203 (1986).

430. — — — —: Stereospecificity of NAD(P)H Oxidation Catalyzed by 4-Aminobenzoate Hydroxylase from *Agaricus bisporus.* Biochim. Biophys. Acta **991**, 380 (1989).

431. TURNER, W.B.: Fungal Metabolites. London: Academic Press. 1971.

432. TURNER, W.B., and D.C. ALDRIDGE: Fungal Metabolites II. London: Academic Press. 1983.

433. TYLER, V.E. JR., and M.H. MALONE: An Investigation of the Culture, Constituents, and Physiological Activity of *Panaeolus campanulatus.* J. Amer. Pharm. Assoc., Sci. Ed. **49**, 23 (1960).

434. UMEZAWA, H., Y. MURAOKA, A. FUJII, H. NAGANAWA, and T. TAKITA: Chemistry of Bleomycin XXVII. Cleomycin, a New Family of Bleomycin-phleomycin Group. J. Antibiotics **33**, 1079 (1980).

435. VANDERHAEGHE, H., and G. PARMENTIER: The Structure of Factor S of Staphylomycin. J. Amer. Chem. Soc. **82**, 4414 (1960).

436. VAN HARREVELD, A.: Effects of L-Proline and Some of its Analogs on Retinal Spreading Depression. J. Neurobiol. **10**, 355 (1979).

437. VERVIER, R., and J. CASIMIR: Isolement et caractérisation d'un nouvel acide aminé, la β-méthylène-L-(+)-norleucine, dans les carpophores d'*Amanita vaginata* var. *fulva.* Phytochem. **9**, 2059 (1970).

438. VIRTANEN, A.I., and A.-M. BERG: New Aminodicarboxylic Acids and Corresponding α-Keto Acids in *Phyllitis scolopendrium.* Acta Chem. Scand. **9**, 553 (1955).

439. VIRTANEN, A.I., and P.K. HIETALA: γ-Hydroxyglutamic Acid in Green Plants. Acta Chem. Scand. **9**, 175 (1955).

440. VON ARDENNE, R., H. DÖPP, H. MUSSO, and W. STEGLICH: Über das Vorkommen von Muscaflavin bei Hygrocyben (Agaricales) und seine Dihydroazepin-Struktur. Z. Naturforsch. **29C**, 637 (1974).

441. WARIN, R., J. JADOT, and J. CASIMIR: Séparation et caractérisation de la S-(amino-2'-éthyl)cystéine et de la S-(acétamido-2'-ethyl)cystéine de *Rozites caperata.* Bull. soc. roy. sci. Liège **38**, 280 (1969).

442. WATSON, R., and L. FOWDEN: Amino Acids of *Caesalpinia tinctoria* and Some Allied Species. Phytochem. **12**, 617 (1973).

443. WEAVER, R.F., W.L. BYRNE, and F.S. VOGEL: Formation of the Dormant Spore in the Common Meadow Mushroom: Appearance of Respiratory Inhibitor(s). Fed. Proc. (Amer. Soc. Exp. Biol.) **27**, 248 (1968).

444. WEAVER, R.F., K.V. RAJAGOPALAN, P. HANDLER, and W. L. BYRNE: γ-L-Glutamyl-3,4-benzoquinone, Structural Studies and Enzymatic Synthesis. J. Biol. Chem. **246**, 2015 (1971).

445. WEAVER, R.F., K.V. RAJAGOPALAN, P. HANDLER, P.W. JEFFS, W.L. BYRNE, and D. ROSENTHAL: Isolation of γ-L-Glutaminyl-4-hydroxybenzene and γ-L-Glutaminyl-3,4-benzoquinone: a Natural Sulfhydryl Reagent, from Sporulating Gill Tissues of the Mushroom *Agaricus bisporus.* Proc. Nat. Acad. Sci. (USA) **67**, 1050 (1970).

446. WEAVER, R.F., K.V. RAJAGOPALAN, P. HANDLER, D. ROSENTHAL, and P.W. JEFFS: Isolation from the Mushroom *Agaricus bisporus* and Chemical Synthesis of γ-L-Glutaminyl-4-hydroxybenzene. J. Biol. Chem. **246**, 2010 (1971).

447. WEEKS, R.A., R. SINGER, and W.L. HEARM: A New Psilocybian Species of *Copelandia.* Lloydia **42**, 469 (1979).

448. WELTER, A., J. JADOT, G. DARDENNE, M. MARLIER, and J. CASIMIR: L-γ-Glutamyl-2-amino-3-hexanone dans *Russula ochroleuca.* Phytochem. **15**, 1984 (1974).

449. WELTER, A., M. MARLIER, and G. DARDENNE: Nouveau acides amines libres de *Afzelia bella*: *trans*-Hydroxy-4-L-proline et *trans*-carboxy-4-L-proline. Phytochem. **17**, 131 (1978).

450. WESTLEY, J.W., D.L. PRUESS, L.A. VOLPE, T.C. DEMNY, and A. STEMPEL: Antimetabolites Produced by Microorganisms IV. L-*threo*-α-Amino-β,γ-dihydroxybutyric acid. J. Antibiotics **24**, 330 (1971).

451. WEYGAND, F., and F. MEYER: Synthese von γ,δ,δ'-Trihydroxy-L-leucin. Chem. Ber. **101**, 2065 (1968).

452. WICKSON, M., and G.H.N. TOWERS: $C^{14}O_2$ Assimilation in *Lilium regale* With Reference to γ-Methyleneglutamic Acid. Canad. J. Biochem. Physiol. **34**, 502 (1956).

453. WIELAND, TH.: Struktur und Wirkung der Amatoxine. Naturwiss. **59**, 225 (1972).

454. —: Peptides of Poisonous Amanita Mushrooms. New York: Springer. 1986.

455. WIELAND, TH., and A. BUKU,: Über die Inhaltsstoffe des grünen Knollenblätterpilzes XXXVIII. Die Konstitution von ε-Amanitin und Amanullin. Liebigs Ann. Chem. **717**, 215 (1968).

456. WIELAND, TH., and J. DÖLLING: Die Entdeckung von γ-Hydroxyleucin im Hydrolysat von Gelatine. Naturwiss. **53**, 526. (1966).

457. WIELAND, TH., and A. FAHMEIER: Über die Inhaltsstoffe des grünen Knollenblätterpilzes XL. Oxydation und Reduktion an der γ,δ-Dihydroxy-isoleucin-Seitenkette des O-Methyl-α-amanitins. Methyl-aldoamanitin, ein ungiftiges Abbauprodukt. Liebigs Ann. Chem. **736**, 95 (1970).

458. WIELAND, TH., M. HASAN, and P. PFAENDER: Die absoluten Konfigurationen der in den Phytotoxinen enthaltenen γ-Hydroxyaminosäuren und der γ-Hydroxynorvaline. Liebigs Ann. Chem. **717**, 205 (1968).

459. WIELAND, TH., and H. KRANZ: Die isomeren γ-Lactone des γ,δ-Dihydroxyleucins. Chem. Ber. **91**, 2619 (1958).

460. WIELAND, TH., G. LÜBEN, H. OTTENHEYM, J. FAESEL, J.X. DE VRIES, A. PROX, and J. SCHMID: The Discovery, Isolation, Elucidation of Structure, and Synthesis of Antamanide. Angew. Chem. internat. Edit. **7**, 204 (1968).

461. WIELAND, TH., K. MANNES, and A. SCHÖPF: Über die Giftstoffe des grünen Knollenblätterpilzes XV. Die Konstitution des Phalloins. Liebigs Ann. Chem. **617**, 152 (1958).

462. WIELAND, TH., and H. WEHRT: Über die Inhaltsstoffe des grünen Knollenblätterpilzes XXVI. Die Bausteine des γ-Amanitins. Liebigs Ann. Chem. **700**, 120 (1966).

463. WILDING, M.D., and M.A. STAHMANN: Hydroxyaspartic Acid from Alfalfa and Clover Roots. Phytochem. **1**, 241 (1962).

464. WILSON, M.F., and E.A. BELL: Amino Acids and β-Aminopropionitrile as Inhibitors of Seed Germination and Growth. Phytochem. **17**, 403 (1978).

465. WISEMAN, J.S., and R.H. ABELES: Mechanism of Inhibition of Aldehyde Dehydrogenase by Cyclopropanone Hydrate and the Mushroom Toxin Coprine: Biochemistry **18**, 427 (1979).

466. WITKOP, B., and C.M. FOLTZ: The Configuration of 5-Hydroxypipecolic Acid from Dates. J. Amer. Chem. Soc. **79**, 192 (1957).

467. WOLMAN, Y., W.J. HAVERLAND, and S.L. MILLER: Nonprotein Amino Acids from Spark Discharges and Their Comparison With the Murchison Meteolite Amino Acids. Proc. Nat. Acad. Sci. (USA) **69**, 809 (1972).

468. WRATTEN, S.J., and D.J. FAULKNER: Cyclic Polysulfides from the Red Alga *Chondria californica*. J. Org. Chem. **41**, 2465 (1976).

469. WYLER, H., T.J. MABRY, and A.S. DREIDING: Über die Konstitution des Randenfarbstoffs Betanin: Zur Struktur des Betanidins. Helv. Chim. Acta **46**, 1745 (1963).

470. YAMAURA, Y., M. FUKUHARA, E. TAKABATAKE, N. ITO, and T. HASHIMOTO: Hepatotoxic Action of a Poisonous Mushroom, *Amanita abrupta* in Mice and its Toxic Component. Toxicology **38**, 161 (1986).

471. YASUMOTO, K., K. IWAMI, and H. MITSUDA: A New Sulfur-containing Peptide from *Lentinus edodes* Acting as a Precursor for Lenthionine. Agric. Biol. Chem. **35**, 2059 (1971).

472. — — —: Enzyme-catalyzed Evolution of Lenthionine from Lentinic acid. Agric. Biol. Chem. **35**, 2070 (1971).

473. YASUMOTO, K., K. IWAMI, T. YONEZAWA, and H. MITSUDA: Anion Activation of γ-Glutamyltransferase from Fruiting Bodies of *Lentinus edodes*. Phytochem. **16**, 1351 (1977).
474. ZACHARIUS, R.M., J.K. POLLARD, and F.C. STEWARD: γ-Methyleneglutamine and γ-Methyleneglutamic acid in the Tulip (*Tulipa gesneriana*). J. Amer. Chem. Soc. **76**, 1961 (1954).

(*Received February 18, 1991*)

Cembranoids, Pseudopteranoids, and Cubitanoids of Natural Occurrence

I. Wahlberg and A.-M. Eklund, Reserca AB, Stockholm, Sweden

Contents

I. Introduction

The structures of (+)-cembrene (**1**), a macrocyclic diterpene hydro-carbon isolated from pine oleoresins (*1, 2*), and of the two epimeric cembratrienediols **35** and **36**, which were obtained from tobacco (*3*), were reported in 1962. Since these pioneering studies, the structures of some 300 naturally occurring cembranoids have been resolved. These include both hydrocarbons and oxygen-containing compounds, the latter possessing *e.g.* epoxy, oxo, hydroxy and lactone functions. Many are polyoxygenated.

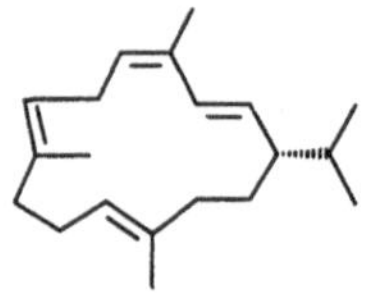

| 1 | 35 | 36 |

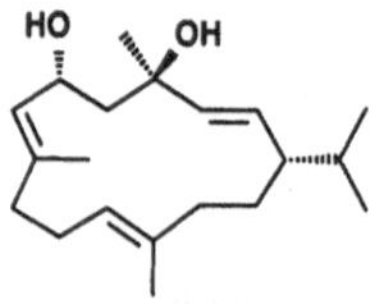

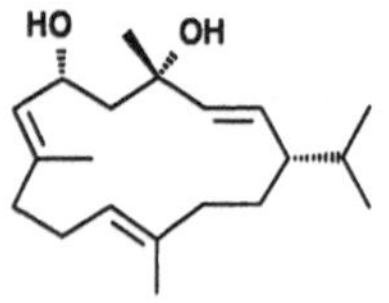

Cembranic diterpenoids are present both in the plant and the animal kingdom. The majority of the cembranoids of plant origin derives from tobacco, the remainder having been isolated from several different plant species. Only two cembrenes have hitherto been encountered in the insect phylum, while the coelenterata phylum, comprising mainly marine in-vertebrates, is the by far predominant source of cembranoids in the animal kingdom.

A review on cembranoids of natural origin was published by WEIN-HEIMER, CHANG, and MATSON in 1979 (*4*). Our intention in the present article is to give a comprehensive compilation of the cembranic diterpen-oids appearing in the literature through December 1990, and to discuss their biogenetic relationships and chemistry with special emphasis on results obtained after 1979. Also included are carbomonocyclic pseudo-pteranes, which are viewed as rearranged cembranoids, while com-pounds such as the briaranes and trinervitanes, which are viewed as being formed from cembrane precursors by secondary carbon-carbon bond closures, will be dealt with in a forthcoming article.

It is often difficult to predict how the introduction of a substituent will influence the conformation of the fourteen-membered ring of a certain cembranoid in solution. As a consequence, the use of NMR methods for stereochemical assignments has, at least in the past, been severely hampered, and the stereostructures of quite a number of cem-branoids have been resolved by X-ray diffraction methods. We have therefore included references to X-ray, spectral and conformational studies in Tables 1–7.

The total syntheses of several cembranoids have been accomplished during the past decade. References to the pertinent articles are given in the Tables 1–7. Since the literature has recently been reviewed (5), the methodology used is not discussed here.

Besides the challenge to construct the macrocyclic cembrane skeleton with stereochemical control, the synthetic efforts in this field are explained by the fact that many cembranoids show important biological and pharmacological effects. Some of these effects are listed in Tables 2–6.

A. Structural Representation and Nomenclature

In the present article we are using the structural representation and nomenclature suggested by WEINHEIMER *et al.* (4). This means that the cembrane skeleton, which is composed of a fourteen-membered carbocyclic ring substituted by an isopropyl group at C-1 and by three methyl groups at C-4, C-8, and C-12, should be drawn as shown in **A**.

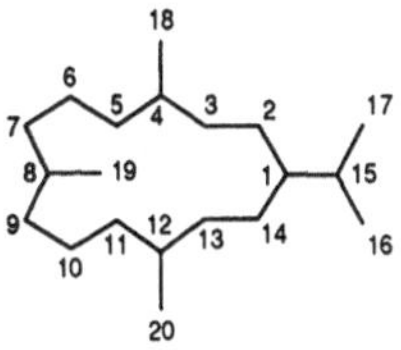

A 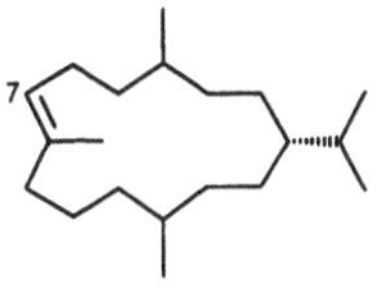   B 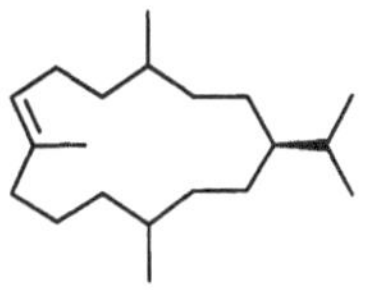   C

The numbering is anti-clockwise, the 1,8-axis of skeletal symmetry is in the horizontal and the 7,8 double bond, or a vestigial indication of its original location, is an internal point of reference with C-7 situated above the 1,8-axis. Compounds having the isopropyl substituent at C-1, or its equivalent, oriented downward belong to the α-series (**B**), while those having this substituent oriented upward belong to the β-series (**C**). In drawing formulae a dotted line is used to represent a substituent situated below the general plane of the macrocyclic ring and a thickened line for a substituent situated above the plane of the ring. When applying these rules here, it has been necessary to redraw many formulae or to change stereochemical notations in certain formulae as compared with their presentations in the original articles. It should be added, however, that even if these rules are followed, graphical representation of cembranoid structures is not always straightforward. In the interest of clarity, we have therefore used a semi-systematic cembrane nomenclature and the *R*- and *S*-system for configurational assignment. We urge authors publishing the structures of new cembranoids to use this mode of representation in order to avoid unnecessary confusion.

II. Cembranoids from Plants

A. Plants Other than Tobacco

It has long been known that various conifer oleoresins are sources of cembrane hydrocarbons and alcohols such as cembrene (**1**) (*1, 2, 6–13*), cembrene-A (**2**) (*14, 15*), isocembrene (**3**) (*16*), thunbergol (**5**) (*8, 11, 12, 16–18*), 4-epiisocembrol (**6**) (*17, 19*) and cembrol (**7**) (*20*) (Table 1). Thunbergol (**5**) and 4-epiisocembrol (**6**) are also constituents of *Nicotiana sylvestris* (*17*) (and of *N. tabacum*, Table 2). WEINHEIMER and coworkers suggested that these two alcohols have (4R)- and (4S)-configurations, respectively (*4*), assignments which have now been reinforced by chemical and ^{13}C NMR studies (*17, 21*).

An alcohol (**8**), enantiomeric (or isomeric) to cembrol (**7**) and giving rise to (−)-(1R)-cembrene (**109**) on dehydration, has been isolated from *Helipterum venustum* (*22*). Poilaneic acid (**13**), another compound of the β-series, is the first cembranoid obtained from a *Croton* species (*C. poilanei*) (*23*).

Mukulol (**4**), having a (2R)-hydroxy substituent, and cembrene-A (**2**) are well-known constituents of the gum resin of *Commiphora mukol*, which is a crude ayurvedic drug (*14, 24–26*). The chemically more

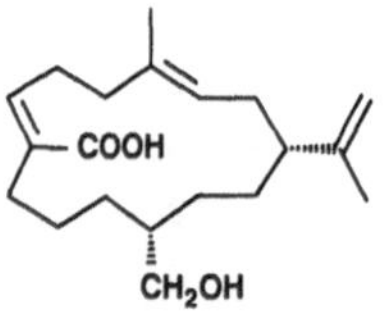

14

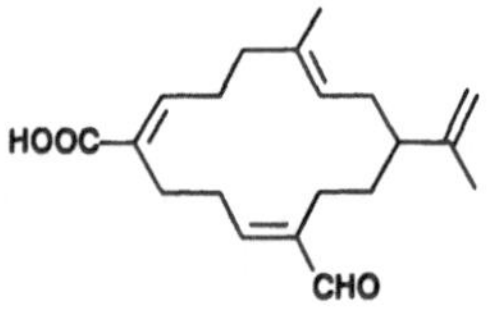

15

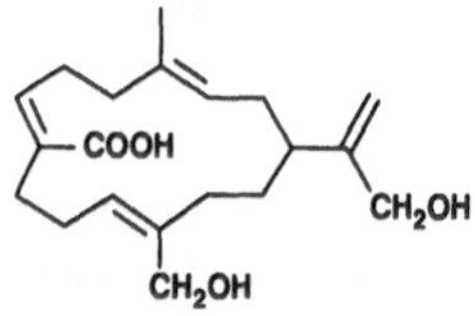

16

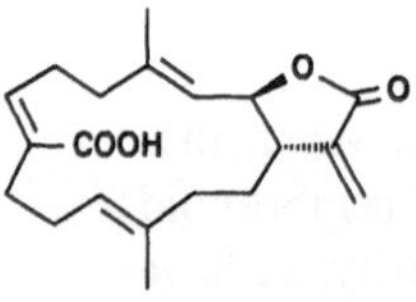

27

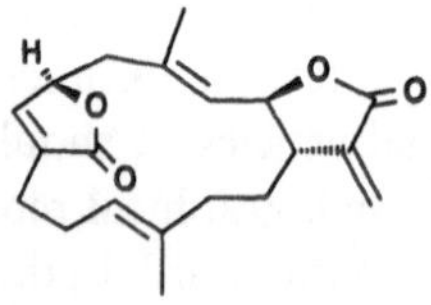

28

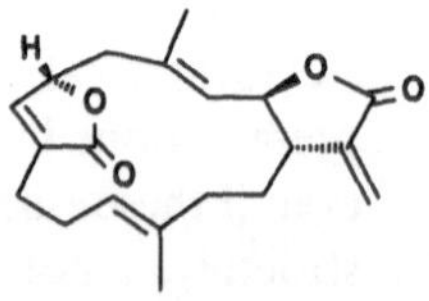

29

elaborate cembrane lactones anisomelic acid (**27**), ovatodiolide (**28**) and isoovatodiolide (**29**) are present in three *Anisomeles* species (*27–30*).

Three acids, **14**, **15**, and **16**, have been found in *Cleome viscosa* (*31, 32*), a plant used in herbal folk medicine. Of these, cleomaldeic acid (**15**) contains two double bonds which have Z-geometries with respect to the carbon ring.

The cembranoids isolated from the resin coatings of the leaves of various *Eremophila* species form an interesting group (*33–36*). The 11,12 double bond has a Z-geometry in all compounds and, when present, the 7,8 double bond has a Z-geometry with respect to the carbon ring. All compounds have an oxygen-containing substituent at C-15: a hydroxy group in **17**, **18**, and **24** and a 3,15 ether bridge in **19–23** and **25–26**.

It is interesting to note that while **19** has a (3R)-configuration, all other 3,15 ether bridged compounds have (3S)-configurations. As suggested by JEFFERIES and coworkers (*35*), this difference may be accounted for by differences in the stereochemistry of the biosynthetic steps leading to the formation of **19** on the one hand and of the (3S)-compounds on the other. A probable mode of biogenesis is shown in Scheme 1.

Epoxidation of the 3,4 double bond in a precursor such as **316** gives the (3S,4R)- and (3R,4S)-epoxides **317** and **318**. An intramolecular substitution by the hydroxy group at C-15 on C-3 of epoxide **317** affords **19**, while an analogous process occurring with the (3R,4S)-epoxide **318** followed by functionalization at C-18, a pinacolic rearrangement and oxidation lead to the generation of **20**, **21**, and **22**. Subsequent esterification converts **22** into **23** and **21** into **26**.

The validity of this pathway is supported by the fact that acyclic diterpenoids having Z-oriented double bonds (with respect to the carbon chain) are present in various *Eremophila* species (*37*). Moreover, compounds **17** and **18**, do have a 3,4 double bond of Z-geometry.

JEFFERIES et al. (*35*) also point out that an alternative mode of formation of the 3,15 ether bridged compounds would involve a β-attack of the hydroxy group at C-15 on an α,β-unsaturated carbonyl function. The (3S)-compounds, *e.g.* **20** and **21** would then originate from an intermediate having a (3E) double bond (with respect to the carbon ring) and the (3R)-compound **19** from an intermediate having a (3Z) double bond. This alternative is in harmony with the discovery of compounds having either a Z or an E terminal double bond among the acyclic diterpenoids in *Eremophila* species (*37*).

Frankincense producing *Boswellia carteri* is noted for the presence of a group of cembranoids possessing a 1,12 epoxide group: incensole (**10**), incensole oxide (**11**) and isoincensole oxide (**12**) (*38–42*). With the isolation of cembrenol (**9**) (*39*), a (1S)-hydroxy compound from this

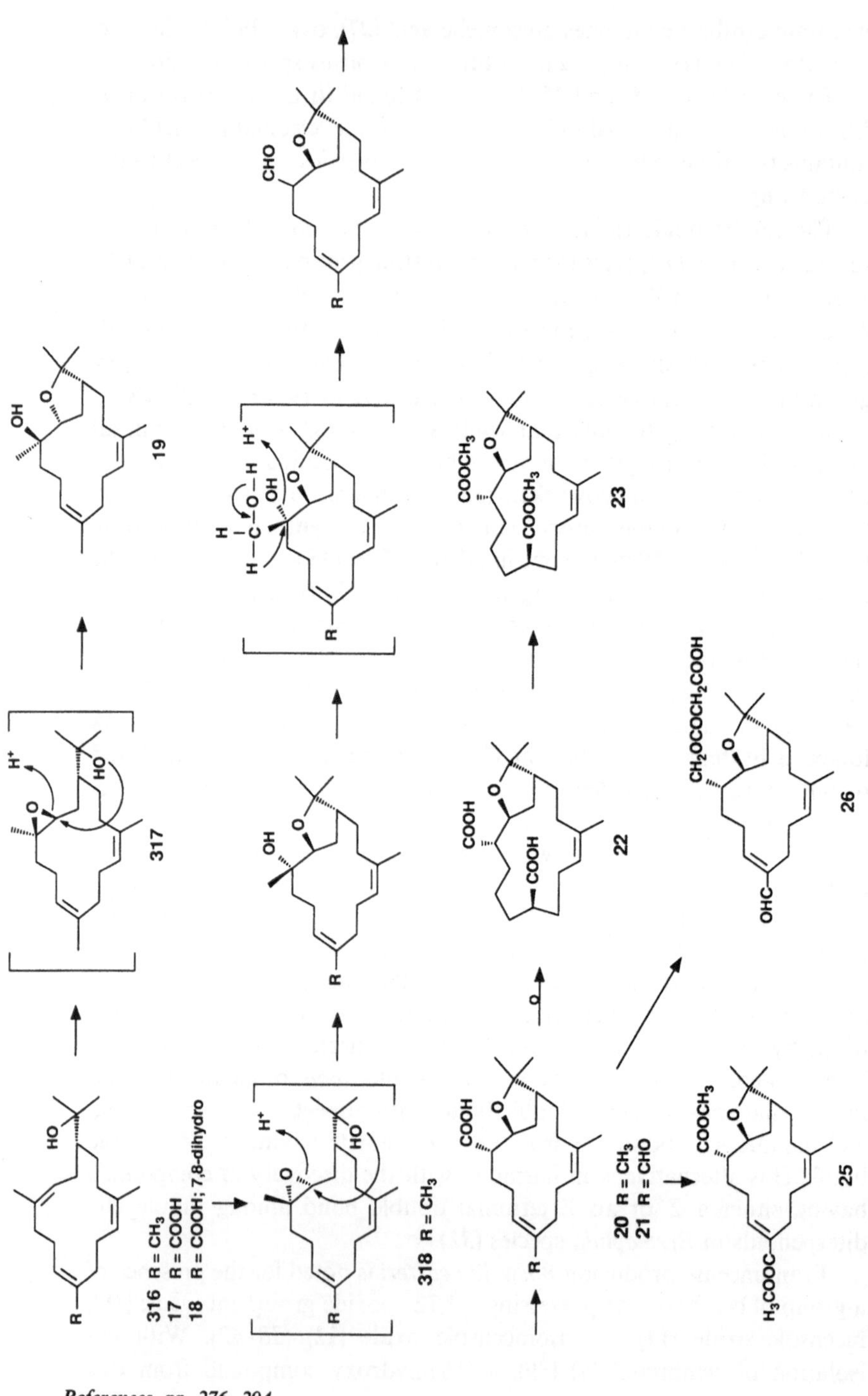

Scheme 1. Proposed biogenesis of compounds 19–23, 25 and 26

material* and the determination of the relative configuration of **11** by X-ray analysis (*43*), it is reasonable to assume that the biogenesis of the 1,12 ether bridged compounds takes place as outlined in Scheme 2.

The existence of this pathway is borne out by results from chemical studies. Thus, cembrenol (**9**) has been shown to give rise to incensole (**10**) and incensole oxide (**11**) by treatment with *m*-chloroperoxybenzoic acid (*44*). Incensole (**10**), in turn, reacts with *p*-nitroperoxybenzoic acid to form incensole oxide (**11**) and isoincensole oxide (**12**) (*42*).

B. Tobacco

The group of tobacco cembranoids has grown extensively during the past decade and now includes more than seventy compounds. These have a close biogenetic relationship as demonstrated by the fact that all compounds for which absolute configurations have been determined have (1*S*)-chirality, *i.e.* they belong to the α-series. All but one (**31**) have a 2,3 double bond of *E*-geometry and most possess a hydroxy substituent at C-4. They are therefore often divided into two groups: one that comprises compounds having a (4*R*)-configuration and the other compounds having a (4*S*)-configuration. Additional oxygenation has hitherto been encountered at C-6, C-7, C-8, C-10, C-11, C-12, C-13, and C-20.

The two 4,6-diols **35** and **36** are the major tobacco cembranoids. Although their gross structures were reported in the early 1960's (*3*), the assignment of a (1*S*,4*R*,6*R*)-configuration to diol **35** had to await results from X-ray diffraction and chemical studies completed in 1975 (*72, 73*). Diol **36** was identified as the (4*S*)-epimer of **35** by chemical correlation with the (4*S*,6*R*,11*S*)-triol **69**, whose stereostructure was unequivocally established by X-ray analysis in 1982 (*74*).

The biosynthesis of these two diols has been shown to take place in the glandular heads of the trichomes present on the surface of the tobacco leaf and in the tobacco flower (*75*). They are then secreted into the cuticular wax, where they co-occur with compounds such as aliphatic hydrocarbons, fatty alcohols, wax esters and, for some tobacco varieties, also with diterpenoids of the labdane class and with sucrose esters (*76*).

Labelling studies have shown that [1-^{14}C]-acetate, [2-^{14}C]-mevalonic acid, *all*-(*E*)-[2-^{14}C]-geranylgeraniol and (R/S)-[3-^{14}C]-cembrene are all incorporated, although at different levels, into the two diols (**35, 36**). On the basis of these results CROMBIE *et al.* (*77*) have suggested that the diols **35** and **36** are generated as illustrated in Scheme 3. Geranylger-

---

* The formula given for (1*S*)-cembrenol (**9**) in the original article (*39*) represents the (1*R*)-compound.

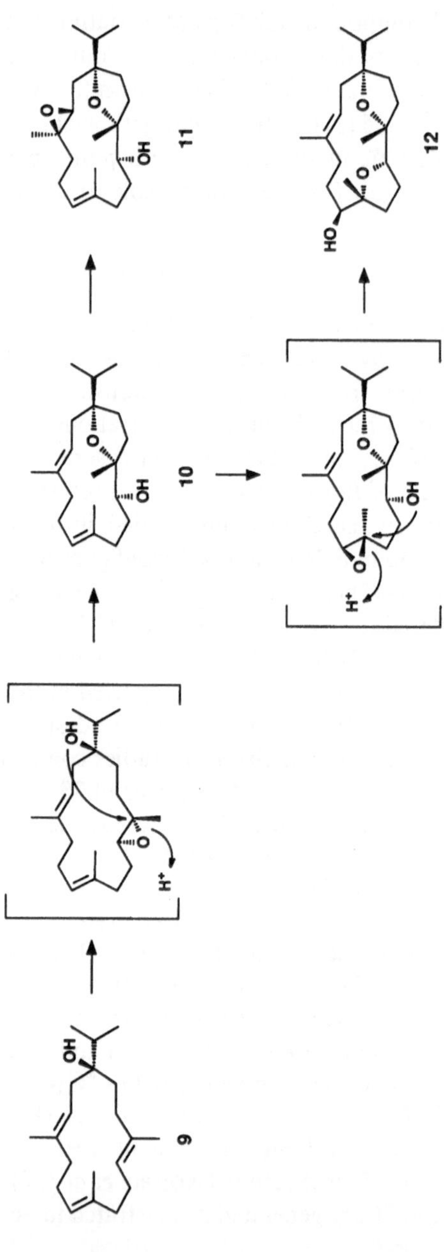

Scheme 2. Proposed biogenesis of incensole (10), incensole oxide (11) and isoincensole oxide (12) from cembrenol (9)

Table 1. *Cembranoids from Plants Other than Tobacco*

Structure no.	Compound Name	Source	X-ray and spectral studies	Synthesis
1	(+)-Cembrene (1S,2E,4Z,7E,11E)-Cembra-2,4,7,11-tetraene	*Picea obovata* (8) *Pinus albicaulis* (1) *Pinus armandi* (6, 7) *Pinus sibirica* (13) *Pinus thunbergii* (2, 45) *Larix sibirica* (9, 46) *Larix sukaczawii* (9) *Larix czekanovskii* (9) *Chamaecyparis obtusa* (10) *Pseudotsuga menziesii* (11, 12)	X-ray analysis of **1** (relative configuration) (7)	(±)-**1** (47, 48, 49) (+)-**1** (50)
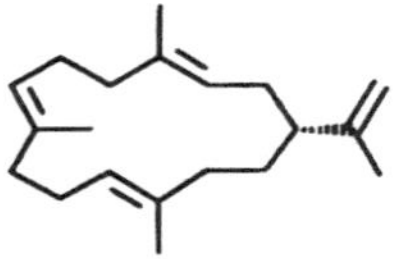 **2**	(−)-Cembrene-A (Neocembrene) (Neocembrene-A) (1R,3E,7E,11E)-Cembra-3,7,11,15-tetraene	*Commiphora mukol* (14) *Picea obovata* (15) *Pinus koraiensis* (15)		(±)-**2** (51–54) (−)-**2** (50, 55, 56) (+)-**2** (55)
3	Isocembrene (1S,2E,7E,11E)-Cembra-2,4(18),7,11-tetraene	*Pinus sibirica* (16)		

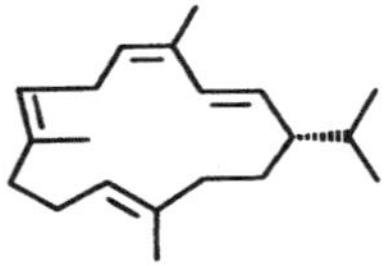

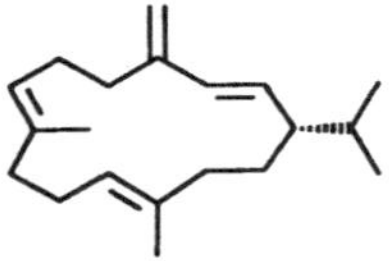

Table 1 (continued)

Structure no.	Compound Name	Source	X-ray and spectral studies	Synthesis
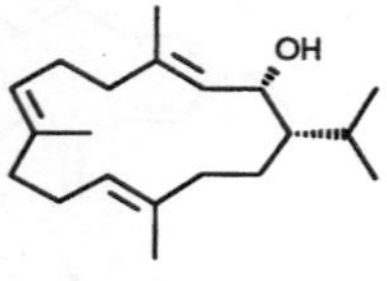 4	Mukulol (1S,2R,3E,7E,11E)-Cembra-3,7,11-trien-2-ol	Commiphora mukol (14, 24–26)	X-ray analysis of (±)-15,16-dehydro-epimukulol acetate (relative configuration) (57)	(±)-4 (49, 58, 59)
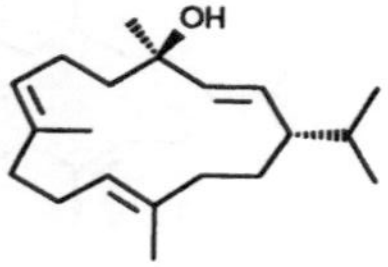 5	Thunbergol (Isocembrol) (1S,2E,4R,7E,11E)-Cembra-2,7,11-trien-4-ol	Pseudotsuga menziesii (11, 12, 17) Pinus sibirica (16, 21) Pinus koraiensis (18) Picea ajanensis (60) Picea obovata (8) Nicotiana sylvestris (17)		(±)-5 (61, 62)
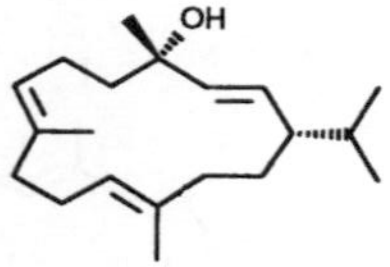 6	4-Epiisocembrol (Epithunbergol) (1S,2E,4S,7E,11E)-Cembra-2,7,11-trien-4-ol	Pinus koraiensis (19) Pinus sibirica (19) Pseudotsuga menziesii (17) Nicotiana sylvestris (17)		(±)-6 (61, 62)
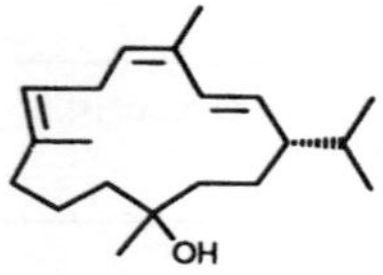 7	Cembrol (1S,2E,4Z,7E)-Cembra-2,4,7-trien-12-ol	Pinus sibirica (20)		

Structure	Name	Source	Ref.	Notes	Synthesis
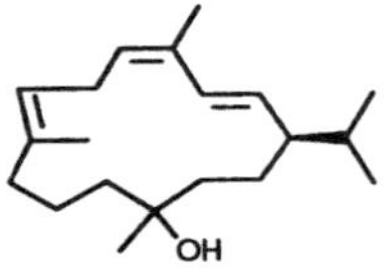 8	(1R,2E,4Z,7E)-Cembra-2,4,7-trien-12-ol	*Helipterum venustum*	(22)		
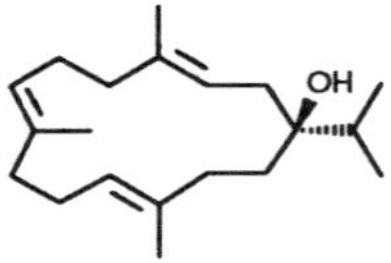 9	Cembrenol (Serratol) (1S,3E,7E,11E)-Cembra-3,7,11-trien-1-ol	*Boswellia carteri* *Boswellia serrata*	(39) (63)		(±)-9 (64)
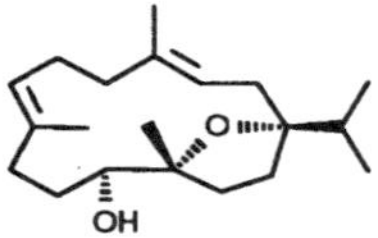 10	Incensole (1S,3E,7E,11S,12R)-1,12-Epoxycembra-3,7-dien-11-ol	*Boswellia carteri*	(39, 41)	^{13}C NMR study (65)	(±)-10 (66, 67)
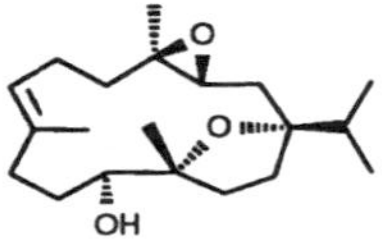 11	Incensole oxide (1S,3R,4R,7E,11S,12R)-1,12:3,4-Diepoxycembr-7-en-11-ol	*Boswellia carteri*	(40)	X-ray analysis of 11 (relative configuration) (43) ^{13}C NMR study (65)	
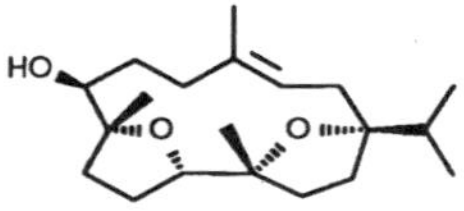 12	Isoincensole oxide (1S,3E,7S,8R,11S,12R)-1,12:8,11-Diepoxycembr-3-en-7-ol	*Boswellia carteri*	(38, 42)	^{13}C NMR study (65)	(±)-12 (67)

Table 1 (*continued*)

Structure no.	Compound Name	Source	X-ray and spectral studies	Synthesis
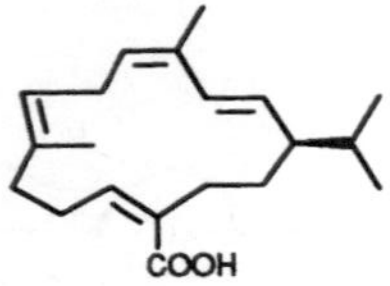 13	Poilaneic acid (1R,2E,4Z,7E,11Z)-Cembra-2,4,7,11-tetraen-20-oic acid	*Croton poilanei*	(23)	
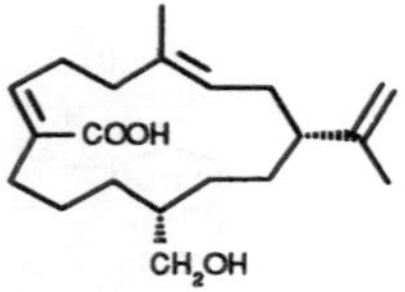 14	(1R*,3E,7Z,12R*)-20-Hydroxycembra-3,7,15-trien-19-oic acid	*Cleome viscosa*	(31) X-ray analysis of 14 (relative configuration) (31)	
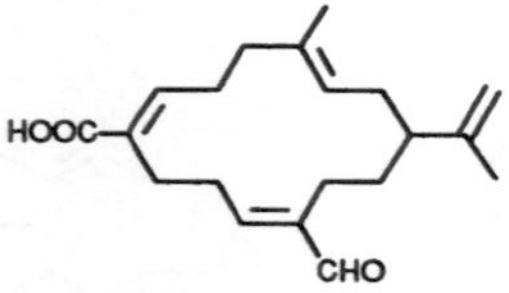 15	Cleomaldeic acid (3E,7E,11E)-20-Oxocembra-3,7,11,15-tetraen-19-oic acid	*Cleome viscosa*	(32)	
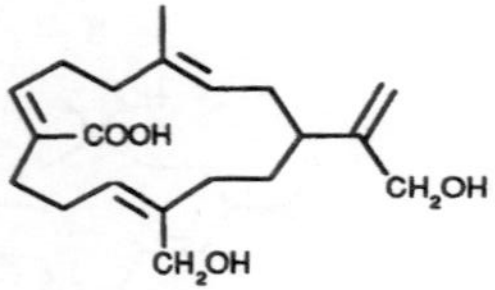 16	(3E,7Z,11Z)-17,20-Dihydroxycembra-3,7,11,15-tetraen-19-oic acid	*Cleome viscosa*	(31)	

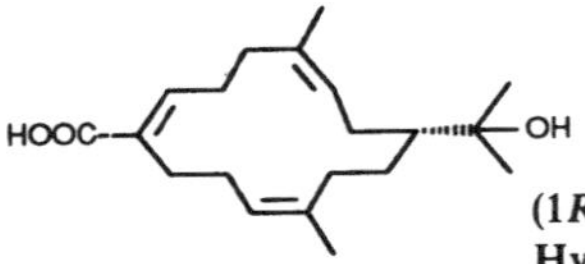

17 — (1R,3Z,7E,11Z)-15-Hydroxycembra-3,7,11-trien-19-oic acid — *Eremophila dempsteri* — (33)

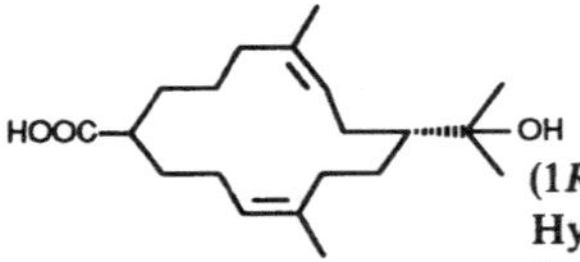

18 — (1R,3Z,11Z)-15-Hydroxycembra-3,11-dien-19-oic acid — *Eremophila dempsteri* — (33)

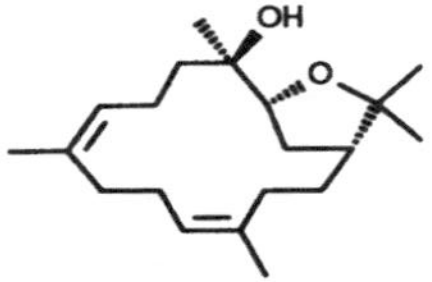

19 — (1R,3R,4R,7Z,11Z)-3,15-Epoxycembra-7,11-dien-4-ol — *Eremophila georgei* — (36) X-ray analysis of **19** (relative configuration) (68)

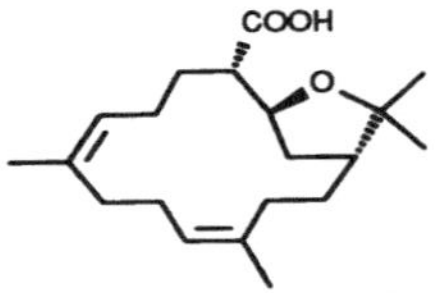

20 — (1R,3S,4S,7Z,11Z)-3,15-Epoxycembra-7,11-dien-18-oic acid — *Eremophila granitica* — (35)

Table 1 (continued)

Structure no.	Compound Name	Source	X-ray and spectral studies	Synthesis
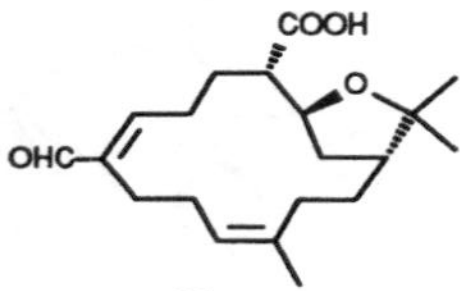 **21**	(1R,3S,4S,7E,11Z)-3,15-Epoxy-19-oxocembra-7,11-dien-18-oic acid	*Eremophila granitica* *Eremophila abietina*	(35) (35)	
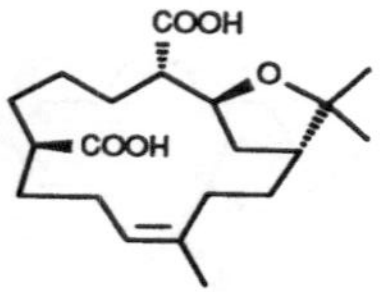 **22**	(1R*,3S*,4S*,8R*,11Z)-3,15-Epoxycembr-11-ene-18,19-dioic acid	*Eremophila abietina*	(35) X-ray analysis of the corresponding diol (relative configuration) (35)	
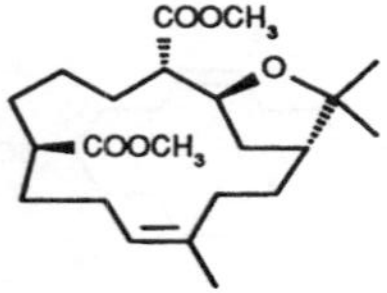 **23**	Dimethyl (1R*,3S*,4S*,8R*,11Z)-3,15-Epoxycembr-11-ene-18,19-dioate	*Eremophila abietina*	(35) X-ray analysis of the corresponding diol (relative configuration) (35)	
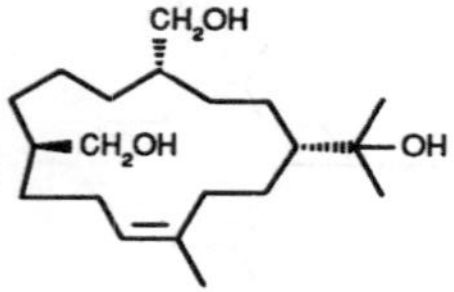 **24**	(1S*,4S*,8R*,11Z)-Cembr-11-ene-15,18,19-triol	*Eremophila clarkei*	(34) X-ray analysis of **24** (relative configuration) (69)	

References, pp. 276–294

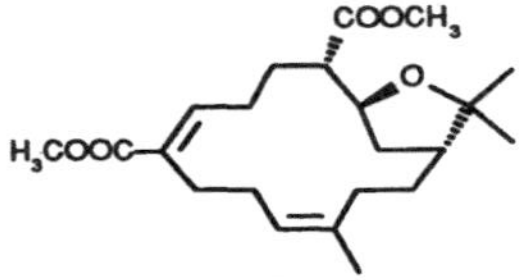

	Name	Species	Ref.	Remarks
25	Dimethyl (1R,3S,4S,7E,11Z)-3,15-Epoxycembra-7,11-diene-18,19-dioate	Eremophila abietina	(35)	
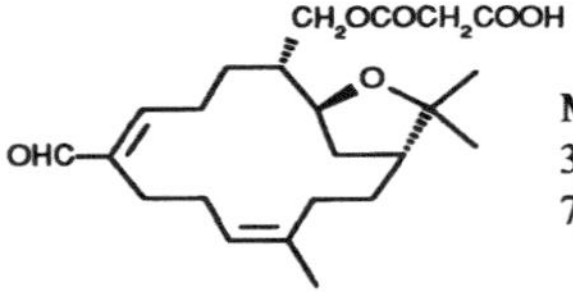 26	Mono (1R,3S,4R,7E,11Z)-3,15-Epoxy-19-oxocembra-7,11-dien-18-yl malonate	Eremophila platycalyx Eremophila fraseri	(33) (33)	
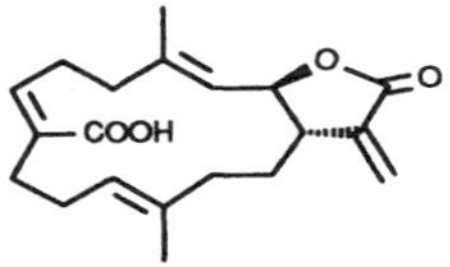 27	Anisomelic acid (1S*,2R*,3E,7Z,11E)-16,2-Olidylcembra-3,7,11,15(17)-tetraen-19-oic acid	Anisomeles malabarica	(29)	(±)-27 (70, 71)
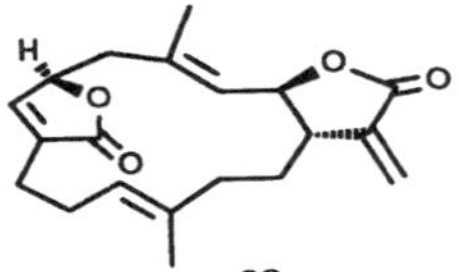 28	Ovatodiolide (1S*,2R*,3E,6R*,7Z,11E)-Cembra-3,7,11,15(17)-tetraene-16,2:19,6-diolide	Anisomeles ovata Anisomeles malabarica Anisomeles indica	(27, 28, 30) (29) (28)	X-ray analysis of 28 (relative configuration) (27, 28)
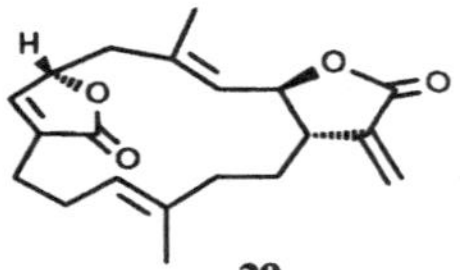 29	Isoovatodiolide (1S*,2R*,3E,6S*,7Z,11E)-Cembra-3,7,11,15(17)-tetraene-16,2:19,6-diolide	Anisomeles ovata Anisomeles indica	(28) (28)	X-ray analysis of 29 (relative configuration) (27, 28)

aniol (**319**), formed *via* the mevalonate pathway, is cyclized to (1*S*)-cembrene (**1**). This process is probably best accounted for as occurring through (1*R*,3*S*)-casbene (**320**) *via* hydroxylation at C-5, solvolysis and reduction of the carbonium ion formed. (1*S*)-Cembrene (**1**), which is concluded to be an experimentally satisfactory precursor for the diols **35** and **36**, is proposed to undergo stereospecific enzymic hydroxylation at the allylic C-6 position prior to the hydroxylation of the 4,5 double bond. The latter reaction is either non-stereospecific or takes place *via* assistance of two different enzymes. As a consequence, compound **30**, which possesses a 4,5 double bond and a hydroxy substituent of *S*-configuration at C-6, may be an intermediate. This is then not true for thunbergol (**5**) and its (4*S*)-epimer (**6**), and the mode of formation of **31**, another tobacco constituent, remains unclear.

An insight into the subsequent metabolism of the two 4,6-diols **35** and **36** has been provided by biomimetic experiments. The results confirm that the two diols are the principal precursors of most of the other tobacco cembranoids and show that reactions such as oxidations, reductions, acid- and base-induced rearrangements, and eliminations are involved in these bioconversions. Cleavages of carbon-carbon bonds occur and *seco*-cembranoids have been encountered.

The biodegradation is quite extensive during the curing (drying) and ageing of the tobacco leaf. Thus with the use of $[U-^{14}C]$-**35** and -**36** in a model curing experiment BEVAN *et al.* (*78*) found that only about 10% of the original (4*S*,6*R*)-diol **36** was retained during curing, while the chemically more stable (4*R*,6*R*)-diol **35** was retained to some 40%. Simultaneously, a wide array of carboacyclic products having less than twenty carbon atoms is generated. Since many of these have desirable aroma properties, a high capability of synthesizing cembranoids is regarded as a quality criterion for a tobacco cultivar.

The presence of cembranoids in tobacco is of importance not solely from a flavour point of view. It has been found that the two 4,6-diols (**35, 36**) exert an inhibitory effect on the germination of the spores of *Peronospora tabacina* (*79*) and that they have insect repelling properties (*80, 81*). They also act as plant growth regulators (*82*), while the 11-hydroperoxide **77** inhibits the action of indole-3-acetic acid (*83*). Of considerable interest is the recent discovery that the 4,6-diols (**35, 36**) show anti-tumour promoter activity (*84, 85*). This is also true for some of the other tobacco cembranoids (*86*) (see Table 2).

Reviews dealing with the tobacco cembranoids and their biogenesis, as formulated with the aid of results from biomimetic experiments, have been published in 1984 (*87*) and 1987 (*88*). In this article we will follow an essentially similar outline, but have added recent results.

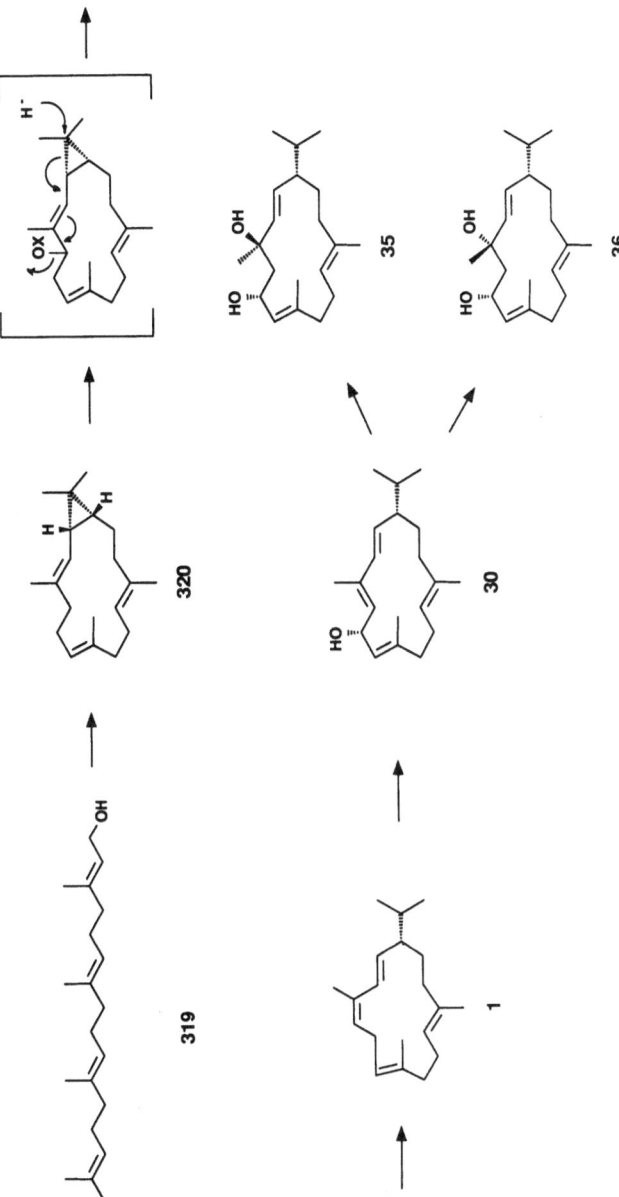

Scheme 3. Proposed biosynthesis of cembrene (1) and the 4,6-diols **35** and **36** from geranylgeraniol (**319**)

Table 2. *Cembranoids from Tobacco (Nicotiana tabacum)*

Structure no.	Compound Name	References	X-ray studies	Synthesis	Biological activity
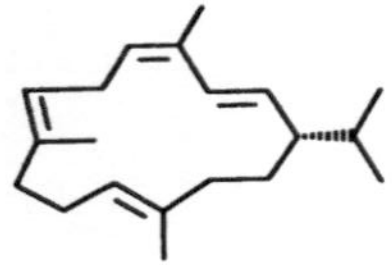	(+)-Cembrene (1S,2E,4Z,7E,11E)-Cembra-2,4,7,11-tetraene	(113)	X-ray analysis of **1** (relative configuration) (7)	(±)-**1** (47–49) (+)-**1** (50)	
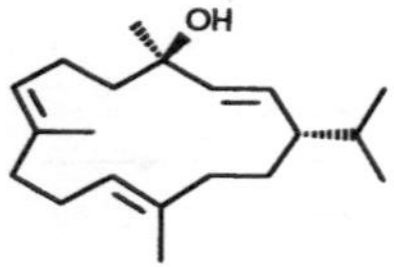	Thunbergol (Isocembrol) (1S,2E,4R,7E,11E)-Cembra-2,7,11-trien-4-ol	(114)		(±)-**5** (61, 62)	
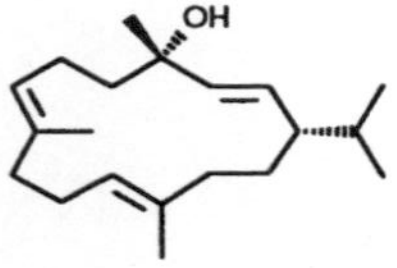	4-Epiisocembrol (Epithunbergol) (1S,2E,4S,7E,11E)-Cembra-2,7,11-trien-4-ol	(114)		(±)-**6** (61, 62)	
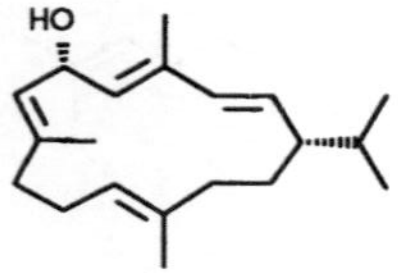	(1S,2E,4E,6S,7E,11E)-Cembra-2,4,7,11-tetraen-6-ol	(115)			

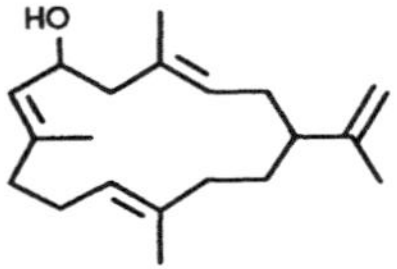

31 — (3E,7E,11E)-Cembra-3,7,11,15-tetraen-6-ol (116)

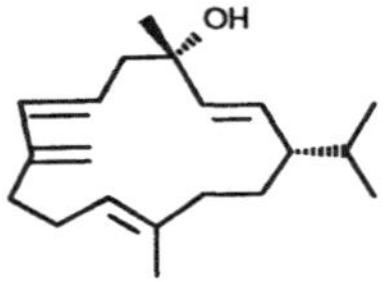

32 — (1S,2E,4S,6E,11E)-Cembra-2,6,8(19),11-tetraen-4-ol (111) — Tumor promoter inhibitory effect (86)

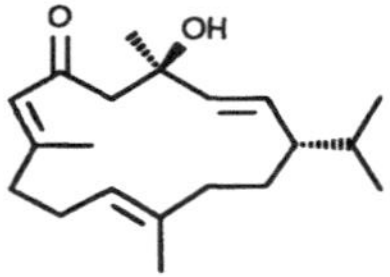

33 — (1S,2E,4R,7E,11E)-4-Hydroxycembra-2,7,11-trien-6-one (107) — Tumor promoter inhibitory effect (86)

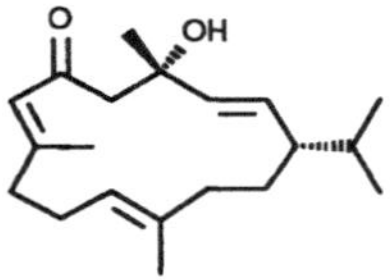

34 — (1S,2E,4S,7E,11E)-4-Hydroxycembra-2,7,11-trien-6-one (107) — X-ray analysis of **34** (relative configuration) (78) — Tumor promoter inhibitory effect (86)

Table 2 (continued)

Structure no.	Compound Name	References	X-ray studies	Synthesis	Biological activity
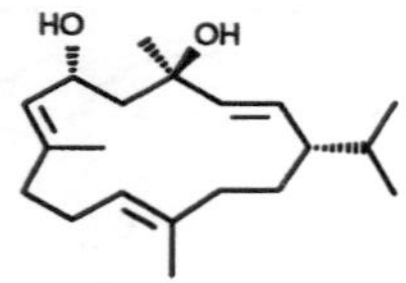 35	(1S,2E,4R,6R,7E,11E)-Cembra-2,7,11-triene-4,6-diol	(3, 72, 73)	X-ray analysis of 35 (relative configuration) (72, 78)	(±)-35 (62, 117, 118)	Tumor promoter inhibitory effect (84, 86) Insect repelling properties (80, 81) Plant growth inhibitory properties (82, 119, 120) Inhibitory effect on aldose reductase (121)
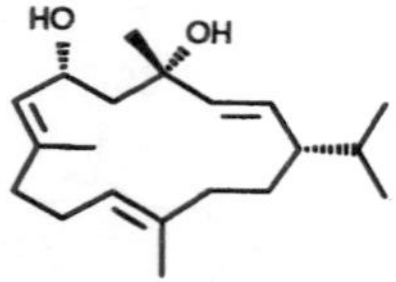 36	(1S,2E,4S,6R,7E,11E)-Cembra-2,7,11-triene-4,6-diol	(3, 74)		(±)-36 (62, 117) (+)-36 (122)	Tumor promoter inhibitory effect (84, 86) Insect repelling properties (80, 81) Plant growth inhibiting properties (82, 119, 120) Inhibitory effect on aldose reductase (121)
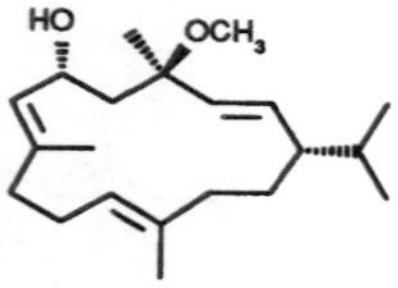 37	(1S,2E,4R,6R,7E,11E)-4-Methoxycembra-2,7,11-trien-6-ol	(112)			

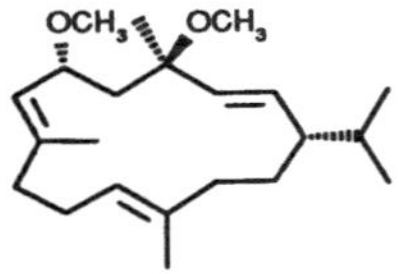

(1S,2E,4R,6R,7E,11E)-Cembra-2,7,11-triene-4,6-dimethoxide   (112)

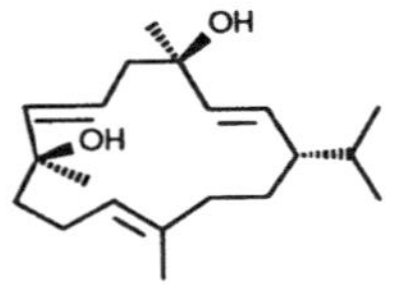

38

(1S,2E,4R,6E,8S,11E)-Cembra-2,6,11-triene-4,8-diol   (109, 110) X-ray analysis of a 11,12-epoxide of **39** (relative configuration) (110)

39

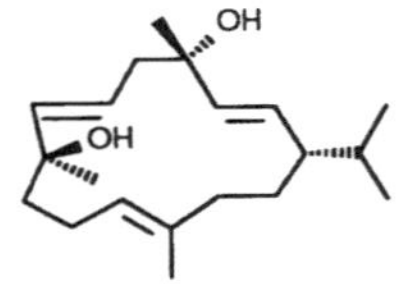

(1S,2E,4S,6E,8S,11E)-Cembra-2,6,11-triene-4,8-diol   (109, 110) X-ray analysis of **40** (relative configuration) (110)

40

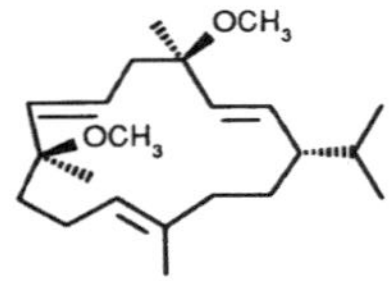

(1S,2E,4R,6E,8S,11E)-Cembra-2,6,11-triene-4,8-dimethoxide   (112)

41

Table 2 (continued)

Compound Structure no.	Name	References	X-ray studies	Synthesis	Biological activity
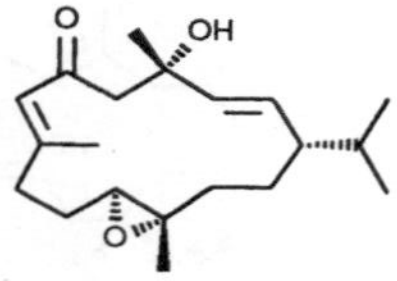 42	(1S,2E,4S,7E,11S,12S)-11,12-Epoxy-4-hydroxy-cembra-2,7-dien-6-one	(95)			
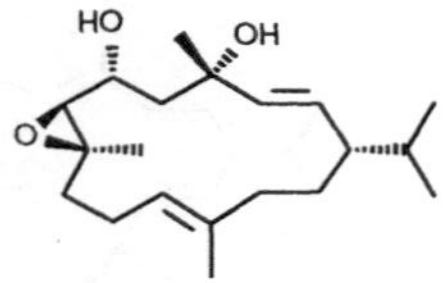 43	(1S,2E,4S,6R,7S,8S,11E)-7,8-Epoxycembra-2,11-diene-4,6-diol	(101)	X-ray analysis of **43** (relative configuration) (101)		
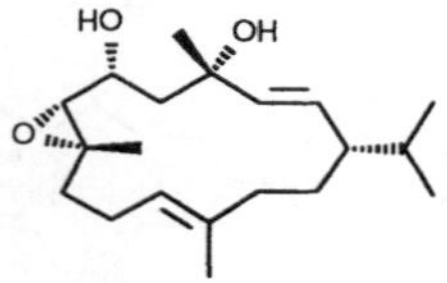 44	(1S,2E,4S,6R,7R,8R,11E)-7,8-Epoxycembra-2,11-diene-4,6-diol	(101)	X-ray analysis of **44** (relative configuration) (101)		
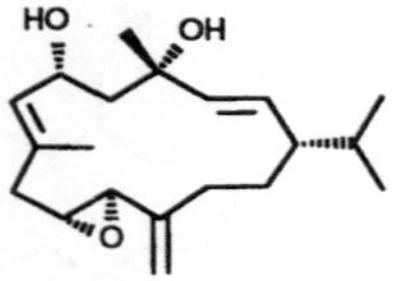 45	(1S,2E,4S,6R,7E,10R,11R)-10,11-Epoxycembra-2,7,12(20)-triene-4,6-diol	(99)	X-ray analysis of **45** (relative configuration) (99)		

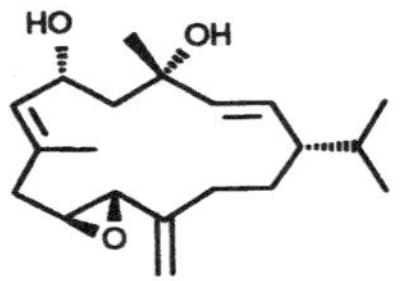 **46**	(1*S*,2*E*,4*S*,6*R*,7*E*,10*S*,11*S*)-10,11-Epoxycembra-2,7,12(20)-triene-4,6-diol	(*99*)	
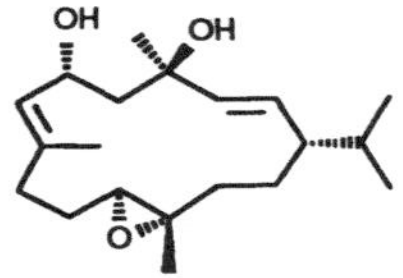 **47**	(1*S*,2*E*,4*R*,6*R*,7*E*,11*S*,12*S*)-11,12-Epoxycembra-2,7-diene-4,6-diol	(*98*)	X-ray analysis of the monoacetate of **47** (relative configuration) (*98*)
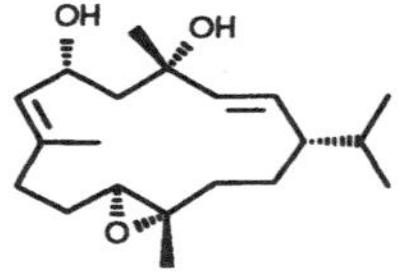 **48**	(1*S*,2*E*,4*S*,6*R*,7*E*,11*S*,12*S*)-11,12-Epoxycembra-2,7-diene-4,6-diol	(*74*)	
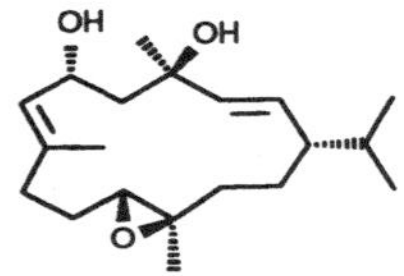 **49**	(1*S*,2*E*,4*R*,6*R*,7*E*,11*R*,12*R*)-11,12-Epoxycembra-2,7-diene-4,6-diol	(*99*)	

Table 2 (*continued*)

Structure no.	Compound Name	References	X-ray studies	Synthesis	Biological activity
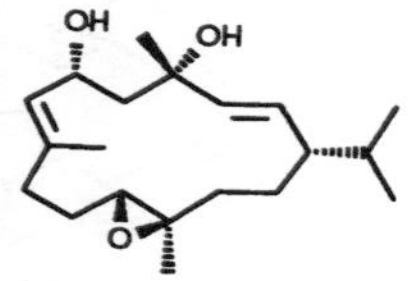 **50**	(1S,2E,4S,6R,7E,11R, 12R)-11,12-Epoxycembra-2,7-diene-4,6-diol	(99)			
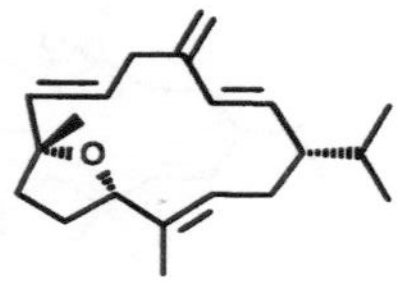 **51**	(1S,2E,6E,8R,11S,12E)-Cembra-2,4(18),6,12-tetraen-8,11-epoxide	(111)			
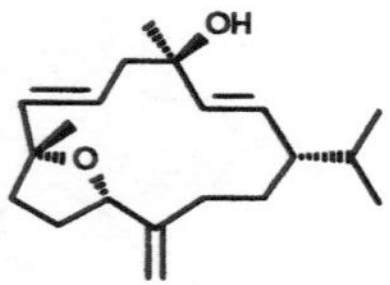 **52**	(1S,2E,4R,6E,8R,11S)-8,11-Epoxycembra-2,6,12(20)-trien-4-ol	(98, 123)			
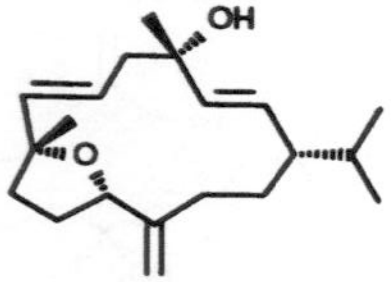 **53**	(1S,2E,4S,6E,8R,11S)-8,11-Epoxycembra-2,6,12(20)-trien-4-ol	(123)	X-ray analysis of **53** (relative configuration) (73)		

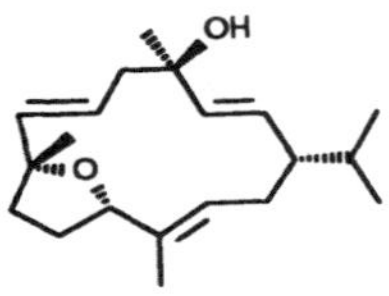

(1S,2E,4R,6E,8R,11S,12E)-8,11-Epoxycembra-2,6,12-trien-4-ol   (98, 111)

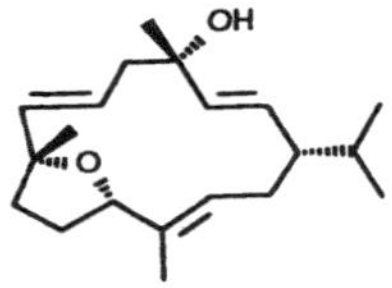

(1S,2E,4S,6E,8R,11S,12E)-8,11-Epoxycembra-2,6,12-trien-4-ol   (98, 123)

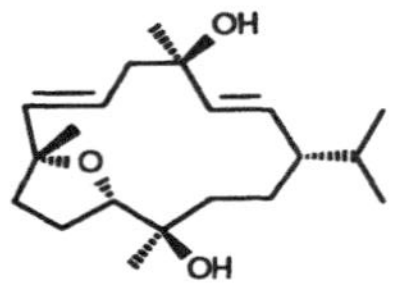

(1S,2E,4R,6E,8R,11S,12R)-8,11-Epoxycembra-2,6-diene-4,12-diol   (124)

(1S,2E,4S,6E,8R,11S,12R)-8,11-Epoxycembra-2,6-diene-4,12-diol   (124)

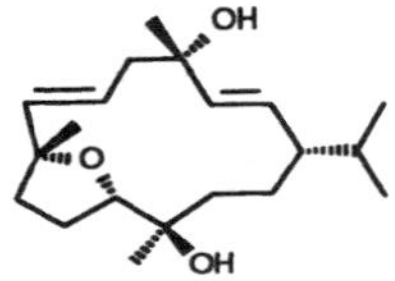

X-ray analysis of **57** (relative configuration) (124)

54

55

56

57

Table 2 (continued)

Structure no.	Compound Name	References	X-ray studies	Synthesis	Biological activity
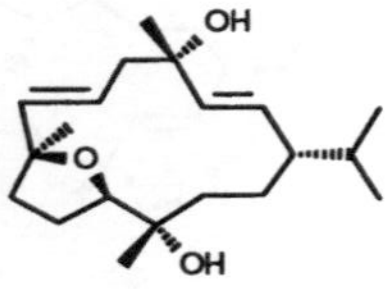 58	(1S,2E,4S,6E,8S,11R,12S)-8,11-Epoxycembra-2,6-diene-4,12-diol	(100)	X-ray analysis of the (4R)-epimer of **58** (relative configuration) (100)		
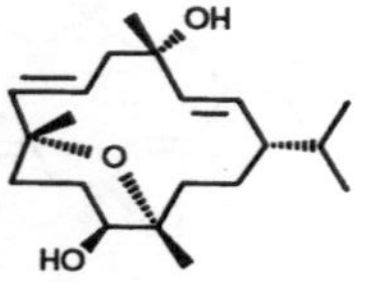 59	(1S,2E,4S,6E,8R,11S,12R)-8,12-Epoxycembra-2,6-diene-4,11-diol	(125)	X-ray analysis of the monoacetate of **59** (relative configuration) (125)		
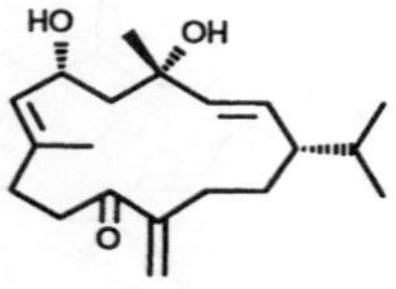 60	(1S,2E,4S,6R,7E)-4,6-Dihydroxycembra-2,7,12(20)-trien-11-one	(96)			
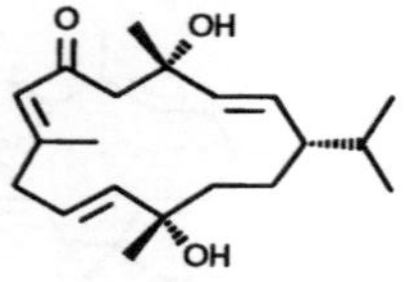 61	(1S,2E,4S,7E,10E,12S)-4,12-Dihydroxycembra-2,7,10-trien-6-one	(95)			

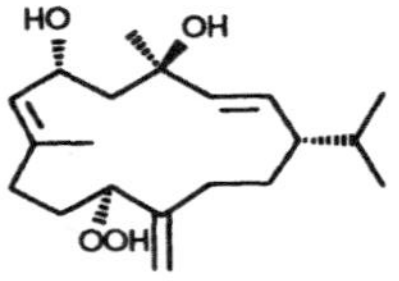

Tumor promoter
inhibitory effect (86)

Tumor promoter
inhibitory effect (86)

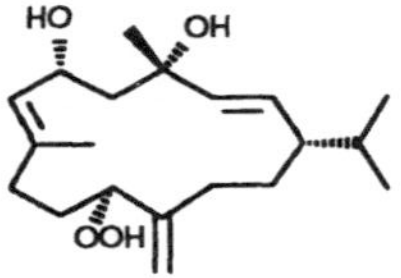

(89)

(1S,2E,4R,6R,7E,11S)-11-
Hydroperoxycembra-
2,7,12(20)-triene-
4,6-diol

(89)

(1S,2E,4S,6R,7E,11S)-11-
Hydroperoxycembra-
2,7,12(20)-triene-
4,6-diol

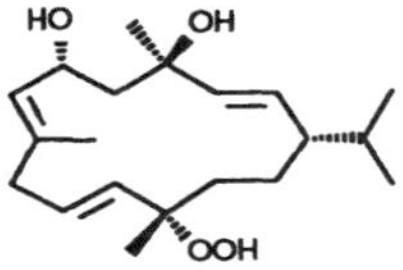

(89)

(1S,2E,4R,6R,7E,10E,
12S)-12-Hydroperoxy-
cembra-2,7,10-triene-
4,6-diol

(89)

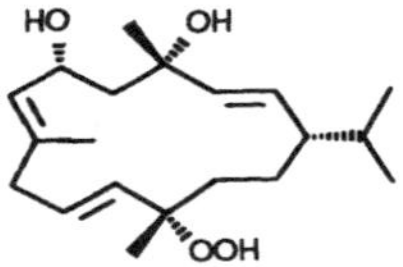

(1S,2E,4S,6R,7E,10E,12S)-
12-Hydroperoxycembra-
2,7,10-triene-4,6-diol

62

63

64

65

Table 2 (*continued*)

Structure no.	Compound Name	References	X-ray studies	Synthesis	Biological activity
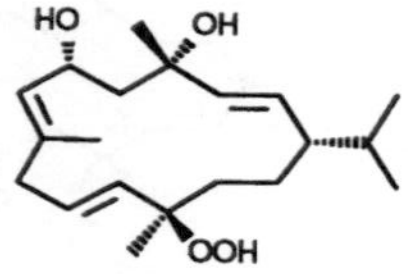 **66**	(1S,2E,4S,6R,7E,10E, 12R)-12-Hydroperoxy-cembra-2,7,10-triene-4,6-diol	(89)			
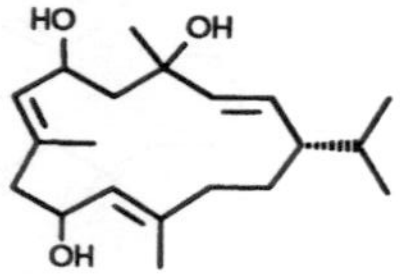 **67**	(1S,2E,7E,11E)-Cembra-2,7,11-triene-4,6,10-triol	(103)			
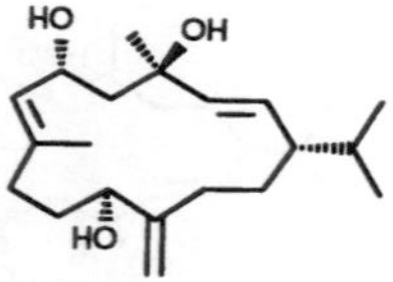 **68**	(1S,2E,4R,6R,7E,11S)-Cembra-2,7,12(20)-triene-4,6,11-triol	(90)			
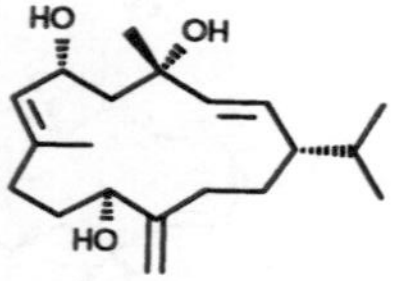 **69**	(1S,2E,4S,6R,7E,11S)-Cembra-2,7,12(20)-triene-4,6,11-triol	(74, 90)	X-ray analysis of **69** (relative config-uration) (74)		

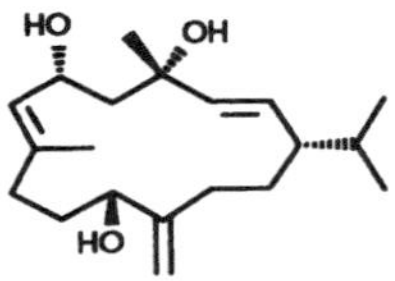

(90)

(1S,2E,4S,6R,7E,11R)-
Cembra-2,7,12(20)-
triene-4,6,11-triol

70

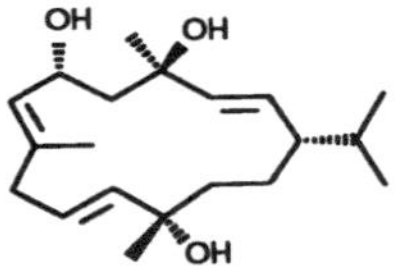

(90)

(1S,2E,4R,6R,7E,10E,
12S)-Cembra-2,7,10-
triene-4,6,12-triol

71

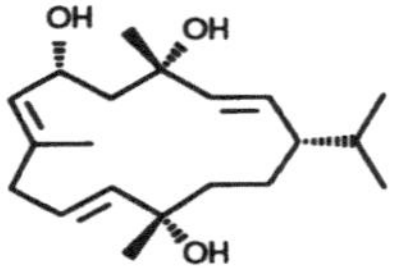

(74, 90)

(1S,2E,4S,6R,7E,10E,12S)-
Cembra-2,7,10-triene-
4,6,12-triol

72

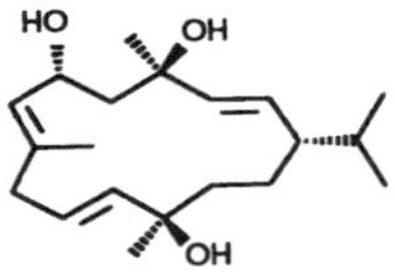

(90)

(1S,2E,4R,6R,7E,10E,
12R)-Cembra-2,7,10-
triene-4,6,12-triol

73

Table 2 (continued)

Structure no.	Compound Name	References	X-ray studies	Synthesis	Biological activity
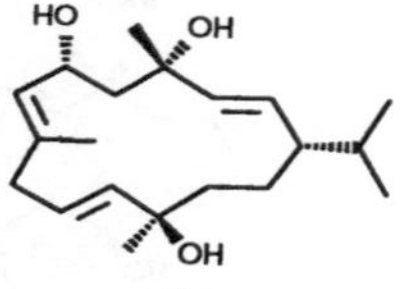 74	(1S,2E,4S,6R,7E,10E, 12R)-Cembra-2,7,10-triene-4,6,12-triol	(90)			
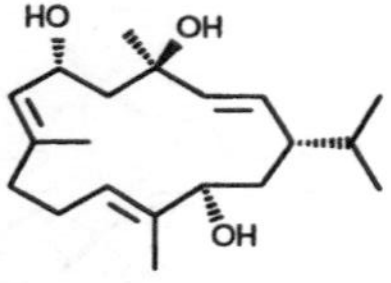 75	(1S,2E,4R,6R,7E,11E, 13R)-Cembra-2,7,11-triene-4,6,13-triol	(104)	X-ray analysis of the 6-O-acetate of 75 (relative configuration) (104)		
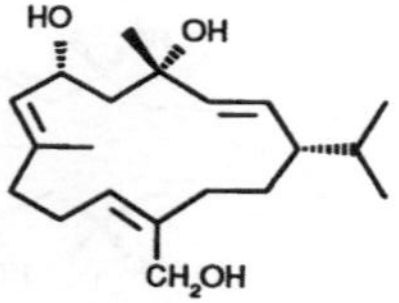 76	(1S,2E,4S,6R,7E,11Z)-Cembra-2,7,11-triene-4,6,20-triol	(104)			
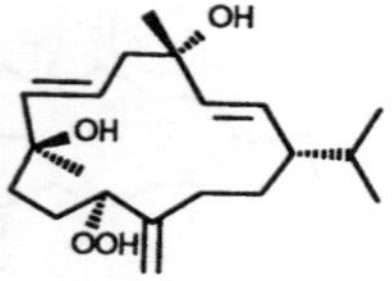 77	(1S*,2E,4S*,6E,8S*,11S*)-11-Hydroperoxycembra-2,6,12(20)-triene-4,8-diol	(83)	X-ray analysis of 77 (relative configuration) (83)		IAA inhibitory activity (83)

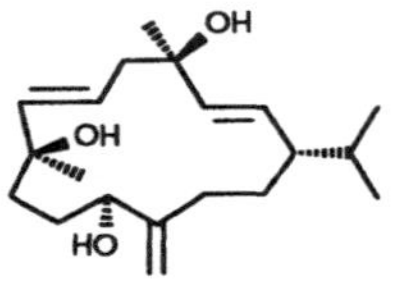

(90)

(1S,2E,4R,6E,8S,11S)-
Cembra-2,6,12(20)-
triene-4,8,11-triol

78

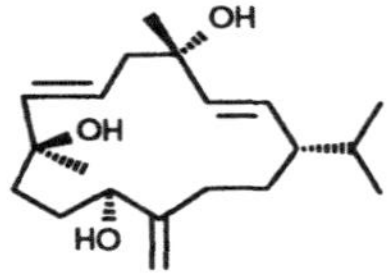

(94, 126)

(1S,2E,4S,6E,8S,11S)-
Cembra-2,6,12(20)-
triene-4,8,11-triol

79

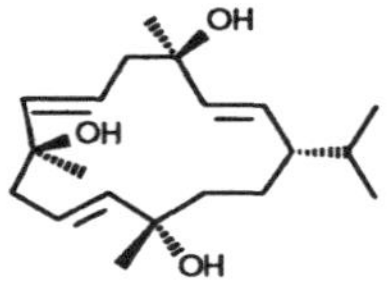

(94)

(1S,2E,4R,6E,8S,10E,12S)-
Cembra-2,6,10-triene-
4,8,12-triol

80

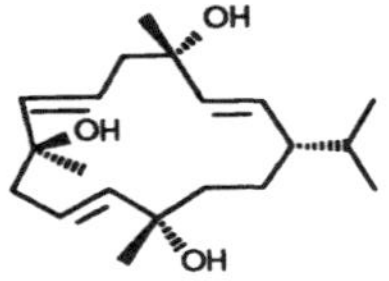

(94)

(1S,2E,4S,6E,8S,10E,12S)-
Cembra-2,6,10-triene-
4,8,12-triol

81

Table 2 (continued)

Compound Structure no.	Name	References	X-ray studies	Synthesis	Biological activity
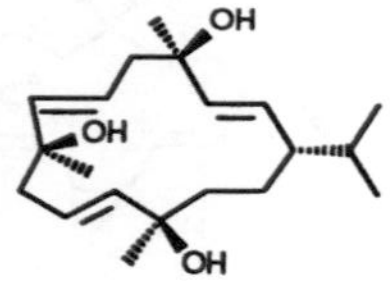 82	(1S,2E,4R,6E,8S,10E,12R)-Cembra-2,6,10-triene-4,8,12-triol	(94)			
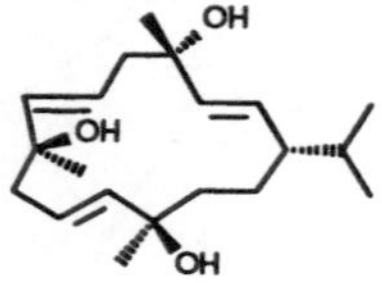 83	(1S,2E,4S,6E,8S,10E,12R)-Cembra-2,6,10-triene-4,8,12-triol	(94)			
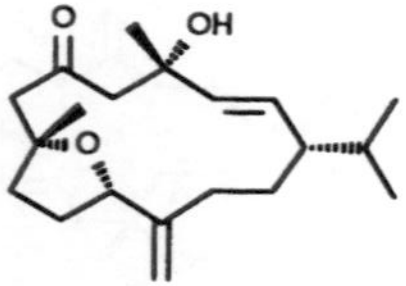 84	(1S,2E,4S,8R,11S)-8,11-Epoxy-4-hydroxycembra-2,12(20)-dien-6-one	(95)			
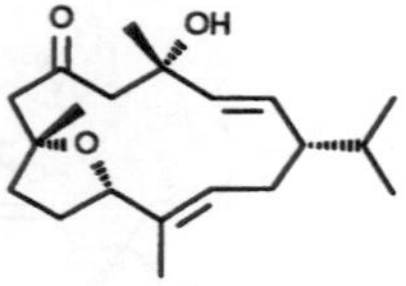 85	(1S,2E,4S,8R,11S,12E)-8,11-Epoxy-4-hydroxy-cembra-2,12-dien-6-one	(95)	X-ray analysis of 85 (relative configuration) (95)		

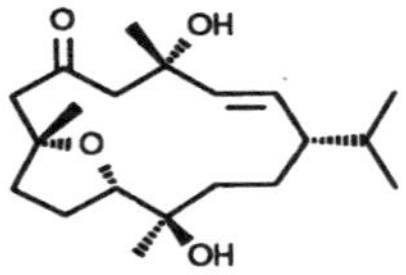

(1S,2E,4S,8R,11S,12R)-
4,12-Dihydroxy-8,11-
epoxycembr-2-en-6-one

(95)

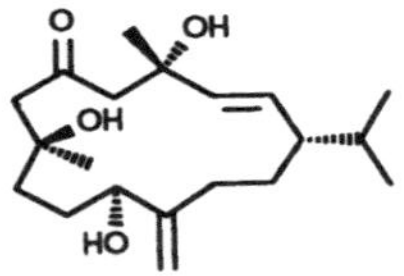

86

(1S,2E,4S,8S,11S)-4,8,11-
Trihydroxycembra-
2,12(20)-dien-6-one

(97)

87

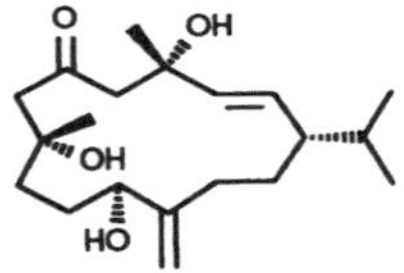

(1S,2E,4S,8R,11S)-4,8,11-
Trihydroxycembra-
2,12(20)-dien-6-one

(97)

88

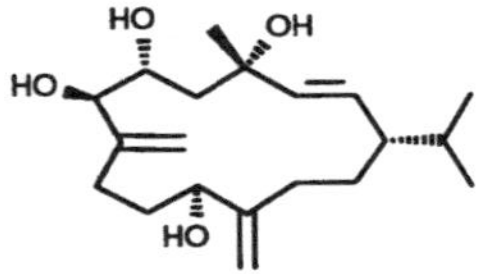

(1S,2E,4S,6R,7R,11S)-
Cembra-2,8(19),12(20)-
triene-4,6,7,11-tetraol

(97)

X-ray analysis of 89
(relative configuration)
(97)

89

Table 2 (*continued*)

Structure no.	Compound Name	References	X-ray studies	Synthesis	Biological activity
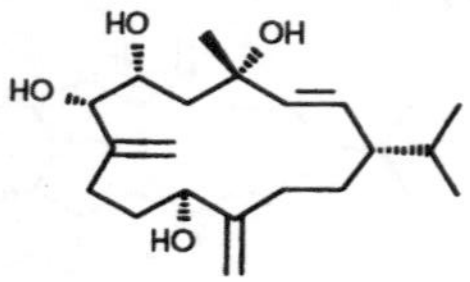 **90**	(1S,2E,4S,6R,7S,11S)-Cembra-2,8(19),12(20)-triene-4,6,7,11-tetraol	(97)			
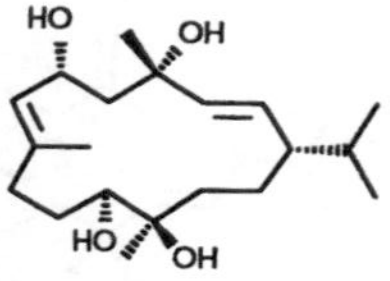 **91**	(1S,2E,4S,6R,7E,11S,12R)-Cembra-2,7-diene-4,6,11,12-tetraol	(97)			
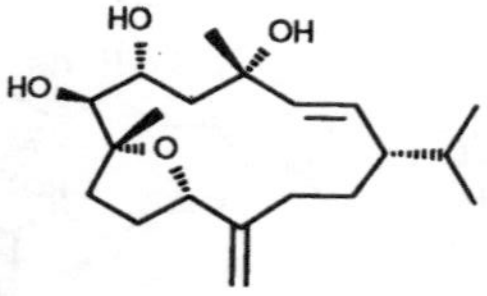 **92**	(1S,2E,4S,6R,7S,8R,11S)-8,11-Epoxycembra-2,12(20)-diene-4,6,7-triol	(102)			
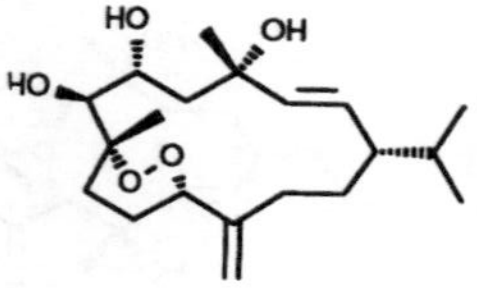 **93**	(1S,2E,4S,6R,7S,8R,11S)-8,11-Epidioxycembra-2,12(20)-diene-4,6,7-triol	(102)	X-ray analysis of the 6-benzoate of **93** (relative configuration) (102)		

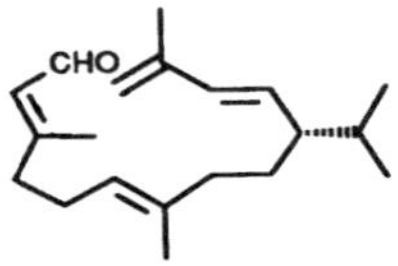

(108)

(2E,6E,10S,11E)-10-
Isopropyl-3,7,13-
trimethyltetradeca-
2,6,11,13-tetraenal

(107)

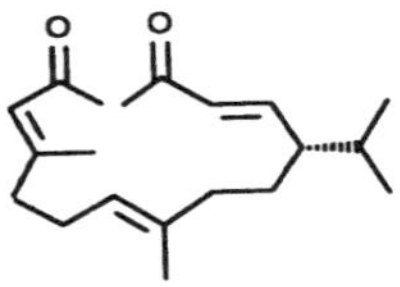

(3E,7E,11S,12E)-4,8-
Dimethyl-11-isopropyl-
pentadeca-3,7,12-triene-
2,14-dione

(91)

(4E,6R,8S,9E,11S)-4,8-
Dimethyl-11-isopropyl-
6,8-dihydroxy-14-oxo-
pentadeca-4,9-dienal

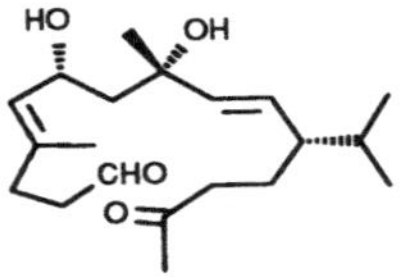

(92, 93)

(4E,6R,8R,9E,11S)-6,8-
Dihydroxy-4,8-dimethyl-
11-isopropyl-14-oxo-
pentadeca-4,9-dienoic acid

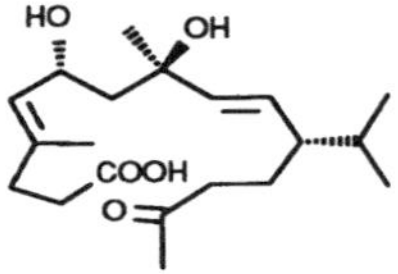

94

95

96

97

Table 2 (*continued*)

| Compound | | | | | |
Structure no.	Name	References	X-ray studies	Synthesis	Biological activity
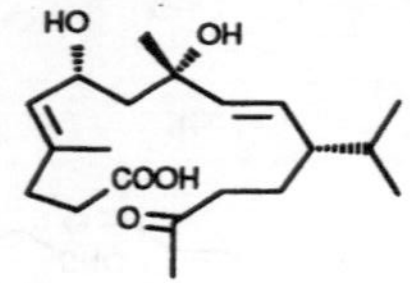 **98**	(4E,6R,8S,9E,11S)-6,8-Dihydroxy-4,8-dimethyl-11-isopropyl-14-oxo-pentadeca-4,9-dienoic acid	(92, 93)			
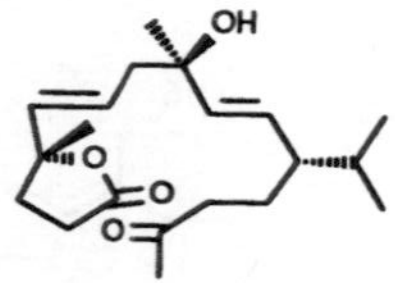 **99**	(4R,5E,8R,9E,11S)-4,8-Dimethyl-8-hydroxy-11-isopropyl-14-oxo-pentadeca-5,9-dien-4-olide	(93, 114)			
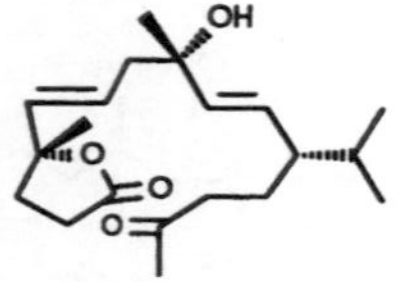 **100**	(4R,5E,8S,9E,11S)-4,8-Dimethyl-8-hydroxy-11-isopropyl-14-oxo-pentadeca-5,9-dien-4-olide	(93, 114)			

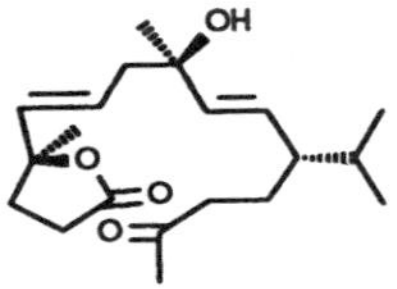

(4S,5E,8R,9E,11S)-4,8-
Dimethyl-8-hydroxy-11-
isopropyl-14-oxo-
pentadeca-5,9-dien-4-olide

(93, 114)

101

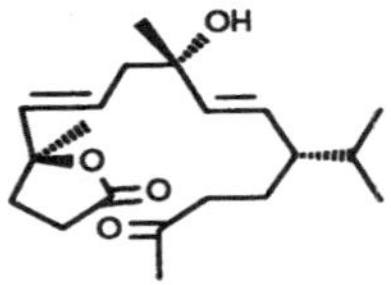

(4S,5E,8S,9E,11S)-4,8-
Dimethyl-8-hydroxy-11-
isopropyl-14-oxo-
pentadeca-5,9-dien-4-olide

(93, 114)

102

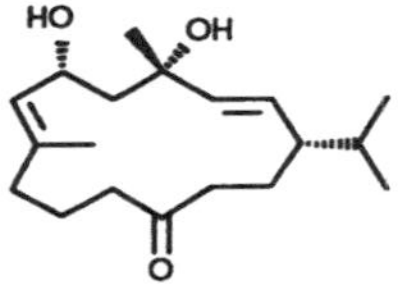

(1S,2E,4S,6R,7E)-4,6-
Dihydroxy-20-nor-
cembra-2,7-dien-12-one

(127)

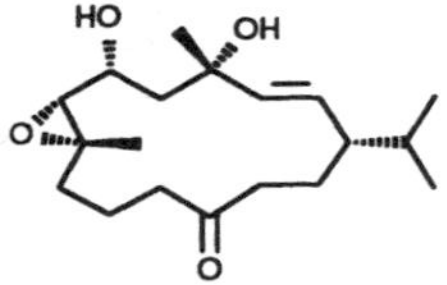

103

(1S,2E,4S,6R,7R,8R)-4,6-
Dihydroxy-7,8-epoxy-
20-norcembr-2-en-12-one

(127)

104

1. Oxidation of the 11,12 Double Bond

A large number of the tobacco cembranoids is evidently formed from the 4,6-diols **35** and **36** by reactions initiated by oxidation of the 11,12 double bond. One route is shown in Scheme 4. It involves the conversion of **35** and **36** into 11- and 12-hydroperoxides (**62–66**) either by sensitized photo-oxygenation or, more probably, by an oxygenase-catalyzed reaction. Subsequent reduction accounts for the generation of the 4,6,11- and 4,6,12-triols (**68–74**). These hydroperoxides and triols have all (6R)-chirality but differ with respect to the configuration of C-4, C-11, or C-12 (*74, 89, 90*).

Experimental results reinforce the validity of these biogenetic pathways. Thus, model experiments involving treatment of the 4,6-diols (**35, 36**) with singlet oxygen and subsequent reduction of the hydroperoxides formed show that the attack occurs preferentially at the trisubstituted 11,12 double bond. Only *syn*-ene products are formed, and a stereoselectivity favouring the generation of the (11S)- and (12S)-triols (**68, 69, 71, 72**) over that of their (11R)- and (12R)-counterparts (*e.g.* **70, 73, 74**) is observed. In keeping with the *cis*-cyclic mechanism of the ene reaction this stereoselectivity has been accounted for by the fact that one of the possible conformers about the 10,11 bond is more populated or reacts faster than the other. It is noteworthy that the (11S)- and (12S)-triols are more abundant in tobacco than the corresponding (11R)- and (12R)-triols (*90*).

The hydroperoxides **62–66** are evidently prone to undergo scission of the 11,12 bond. This view is supported by the fact that the *seco*-aldehyde **96** and the *seco*-acids **97** and **98** isolated from tobacco all have the required (4E,6R,9E,11S)-stereochemistry (*91–93*).

The four *seco*-lactones **99–102** obtained from tobacco are diastereoisomers with respect to the chiralities of C-4 and C-8 (*93*). Their biogenesis may then be explained by a non-stereospecific allylic rearrangement involving the 4,5-en-6-ol systems in the *seco*-acids **97** and **98** followed by lactonization. The allylic rearrangement involved in the conversion of the 4,6,11- and 4,6,12-triols **68, 69** and **71–74** into the newly discovered 4,8,11- and 4,8,12-triols **78–83**, on the other hand, shows a high stereospecificity, as implied by the fact that all six triols of natural occurrence (**78–83**) have (8S)-configurations (*90, 94*). It is worthy of mention that the 11-hydroperoxide **77**, which may derive from **63** and may be a precursor of **79** also has (8S)-chirality (*83*).

Oxidation of the (4S,6R,12S)-triol **72** is one of the two most plausible routes to the 6-oxo compound **61** (*95*). There are also two routes to the 11-oxo compound **60** (*96*), one involving dehydration of the 11-hydro-

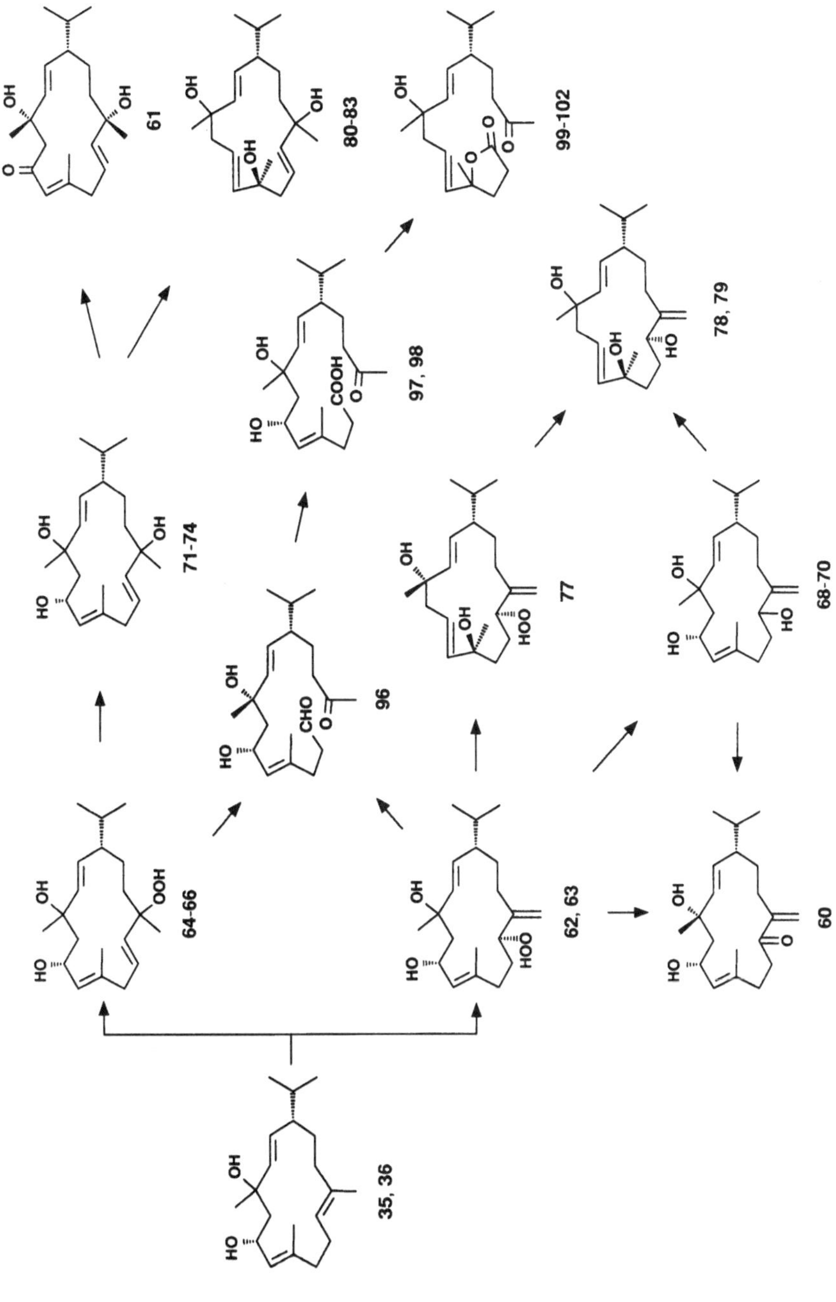

Scheme 4. Proposed biogenesis of compounds **60–66**, **68–74**, **77–83** and **96–102** from the 4,6-diols **35** and **36**

peroxide **63** and the other regiospecific oxidation of a (4S,6R,11)-triol (**69, 70**).

Besides being converted to (8R,11S)-epoxides (**52, 53**) in an acid-catalyzed reaction (see Scheme 6), the (4,6R,11S)-triols **68** and **69** are susceptible to oxidation of the 7,8 double bond. Thus, as illustrated in Scheme 5, the (4S,6R,11S)-triol **69** gives rise to four ene products on sensitized photo-oxygenation: the (4S,6R,7R,11S)-tetraol and its (7S)-epimer (**89, 90**) as the major products and the 6-oxo-(4S,8S,11S)-triol and its (8R)-epimer (**87, 88**) as the minor products. All four compounds have most recently been isolated from an extract of flowers of tobacco (**97**).

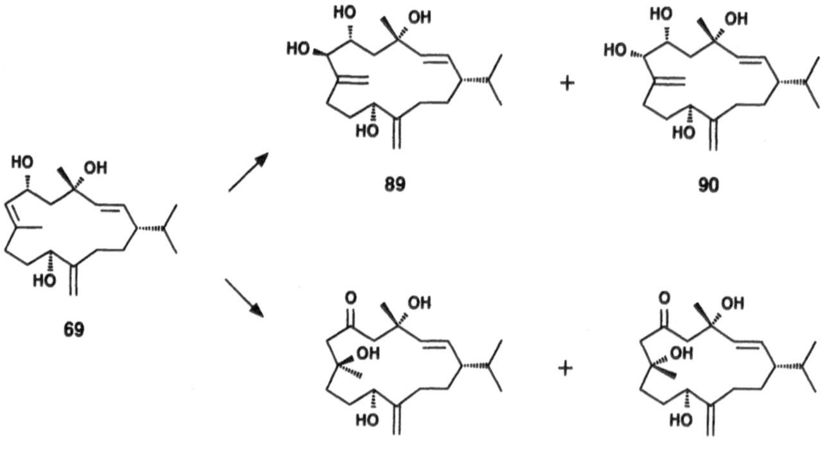

Scheme 5. Proposed biogenesis of compounds **87–90** from the (4S,6R,11S)-triol **69**

The 4,6-diols **35** and **36** are also converted into 11,12-epoxides, this constituting the second principal route for oxidation of the 11,12 double bond (Scheme 6). While the (11S,12S)-epoxides **47** and **48** are abundant tobacco constituents (**74, 98**), the (11R,12R)-epoxides **49** and **50** have only most recently been isolated in small amount from tobacco (**99**), indicating that the oxidation process taking place in tobacco is highly stereoselective. This was to be expected, since model experiments using m-chloroper-oxybenzoic acid as the oxidizing agent have shown that the formation of the (11S,12S)-epoxides (**47, 48**) is highly favoured over that of the (11R,12R)-epoxides (**49, 50**) (**98**).

The formulation of the (11S,12S)-epoxides **47** and **48** as precursors of the (8R,11S,12R)-8,11-epoxy-bridged 4,12-diols **56** and **57** (and their dehydration products **51–55**) and of the (4S,8R,11S,12R)-8,12-epoxy-bridged 4,11-diol **59** is also supported by biomimetic experiments, weak

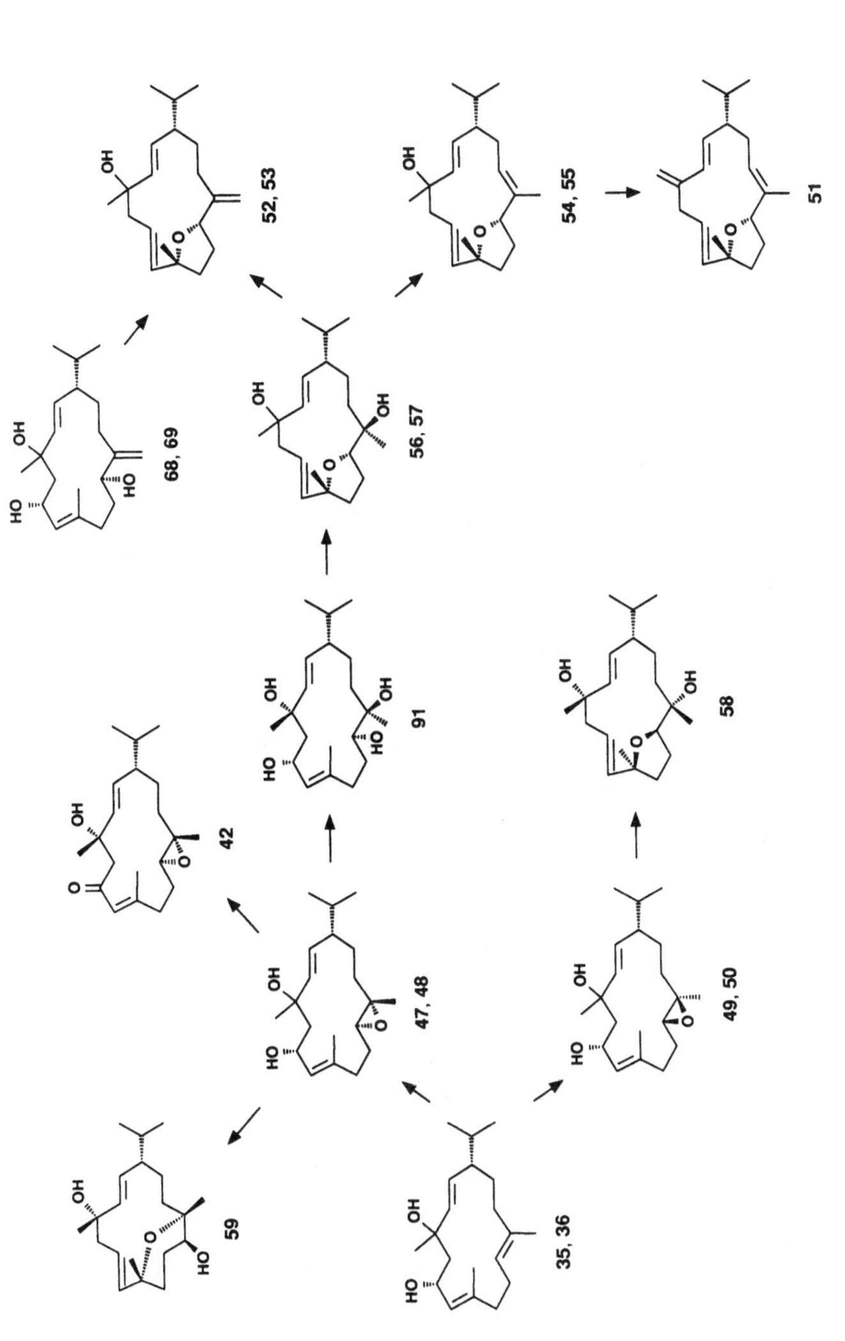

Scheme 6. Proposed biogenesis of compounds **42**, **47–59** and **91** from the 4,6-diols **35** and **36** and of compounds **52** and **53** from the (4,6R,11S)-triols **68** and **69**

acid now being used as the reagent. It has been suggested that these products arise *via anti*-addition of water to the epoxide group, elimination of the hydroxyl group at C-6, double bond migration and attack of the newly formed 11- or 12-hydroxy group on C-8 (*98*). Consistent with this, the presumed intermediate of the (4*S*)-group, the (4*S*,6*R*,11*S*,12*R*)-tetraol **91**, which has recently been found in tobacco, gives, in fact, rise to the (8*R*,11*S*)-epoxide **57** when treated with weak acid (*97*).

By analogous processes, the (11*R*,12*R*)-epoxide **50** is converted into **58**, the only (8*S*,11*R*)-epoxy-bridged compound so far encountered in tobacco (see also Scheme 9) (*100*), and the (4,6*R*,11*S*)-triols **68** and **69** into the (8*R*,11*S*)-epoxides **52** and **53** (*90*).

2. Oxidation of the 7,8 Double Bond

The (7*S*,8*S*)- and (7*R*,8*R*)-epoxides **43** and **44** are the only tobacco constituents encountered so far in which the 7,8 but not the 11,12 double bond is oxidized (*101*). They are obtained synthetically in the ratio 76:24 (**43**:**44**) by treatment of the 4*S*,6*R*-diol **36** with t-butyl hydroperoxide in the presence of a catalytic amount of VO(acac)$_2$ (Scheme 7).

The major epoxide (**43**) is an experimentally verified precursor of the (8*R*,11*S*)-epoxide **92** and the (8*R*,11*S*)-epidioxide **93** (*102*). Thus, sensitized photooxygenation of **43** gives as one of the two major products the (11*S*)-hydroperoxide **321**. Reduction of the hydroperoxide group by using triethyl phosphite results in spontaneous cyclization with the formation of **92**. The reaction probably takes place by attack of the newly formed 11-hydroxy group on C-8, thereby causing opening of the epoxide group. An analogous type of mechanism accounts for the formation of **92** *via* epoxidation of the 7,8 double bond in the (4*S*,6*R*,11*S*)-triol **69** and also for the formation of isoincensole oxide (**12**) *via* epoxidation of the 7,8 double bond in incensole (**10**) (see Scheme 2).

The (8*R*,11*S*)-epidioxide **93**, whose structure has been unambiguously settled by X-ray analysis (*102*), can be prepared by treatment of the (11*S*)-hydroperoxide **321** (obtained from **43** or **63**) with aqueous HCl in CHCl$_3$ (see Scheme 7). Rather unexpectedly, the major product of this reaction was identified as the (8*R*,11*S*)-epoxide **92**. It has been suggested that its formation, which formally involves loss of hydroxyl proceeds with participation of a peroxonium ion. Support for this view was provided by the observation that when methyl phenyl sulfoxide is present in the reaction mixture this is oxidized to the corresponding sulfone (*102*).

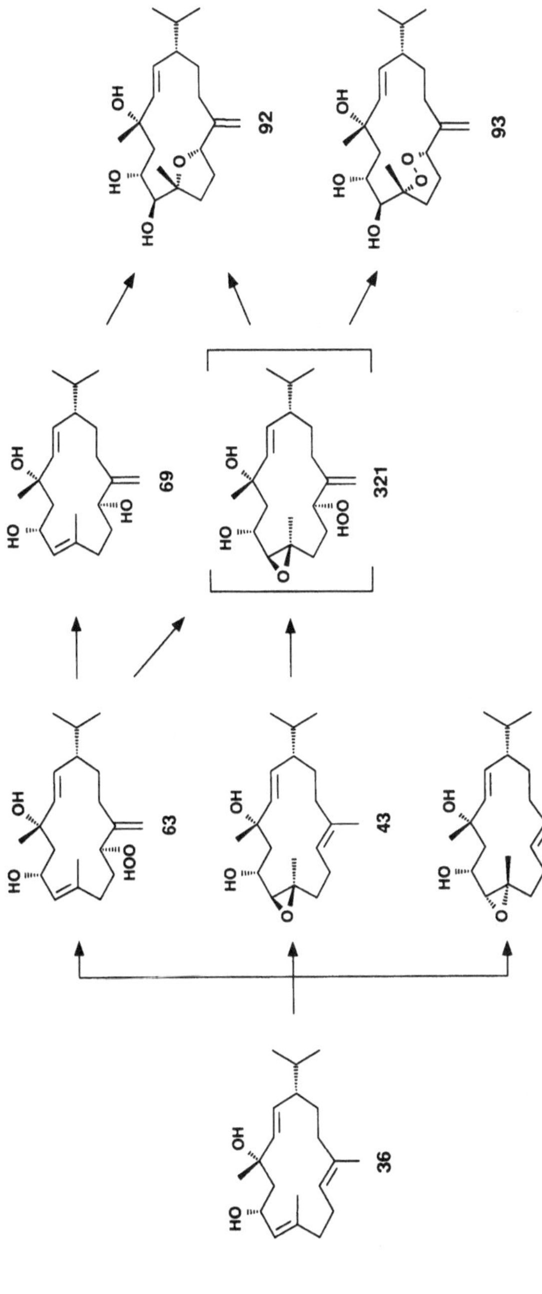

Scheme 7. Proposed biogenesis of compounds **43**, **44**, **92** and **93** from the (4*S*,6*R*)-diol **36**

3. Oxidation at Positions Allylic to the 11,12 Double Bond

It has been demonstrated by the recent isolation of several isomers of 4,6,10-triol (**67**) (*103*), of the (4*R*,6*R*,13*R*)-triol **75** (*104*) and of the (4*S*,6*R*,20)-triol **76** (*104*) that the 4,6-diols (**35**, **36**) are also susceptible to oxidation at all three positions allylic to the 11,12-double bond. Preliminary chemical studies have shown that treatment of the 6-acetates of **35** and **36** with selenium dioxide and t-butyl hydroperoxide leads to the formation of products oxygenated at C-13 and C-20 but not to products oxygenated at C-10 (*104*). There are, however, patents on the preparation of (a) 4,6,10-triol(s) (**67**) and also of (a) 4,6,20-triol(s) (**76**) from (a) 4,6-diol(s) (**35**, **36**) with the aid of bacteria (*105*, *106*).

67   75   76

4. Oxidation of the Hydroxy Group at C-6

Oxidation of the hydroxy group at C-6 in the two 4,6-diols (**35**, **36**) yields the ketols **33** and **34**, respectively (Scheme 8). Both of these are experimentally verified precursors of the *seco*-diketone **95**, to which they are converted by a retroaldol type of reaction. All three have long been known as constituents of dark-fired tobacco (*107*), while the presence of the remaining 6-oxo cembranoids (**42**, **61**, and **84–86**) was disclosed recently (*95*). As suggested in Scheme 8, the (4*S*)-ketol **34** may serve as an alternative precursor of the (11*S*,12*S*)-epoxide **42** and the (4*S*,12*S*)-diol **61** (see also Schemes 4 and 6). Like the (11*S*, 12*S*)-epoxides **47** and **48**, epoxide **42** undergoes an acid-induced rearrangement thus accounting for the biogenesis of the (8*R*,11*S*,12*R*)-8,11-epoxy bridged 4,12-diol **86** and its dehydration products **84** and **85**. The reaction of **42** with acid does not, however, take place with the high stereoselectivity that was shown by the reaction of **47** and **48** with acid (*95*).

The route to the (4*S*,12*S*)-diol **61** has also been explored experimentally. Sensitized photooxygenation of ketol **34** and subsequent reduction of the hydroperoxides generated yielded **61** as the major product. The (4*S*,11*S*)-diol **322**, hitherto not encountered in tobacco, is the second major product. It gives rise to the (8*R*,11*S*)-epoxide **84** on treatment with

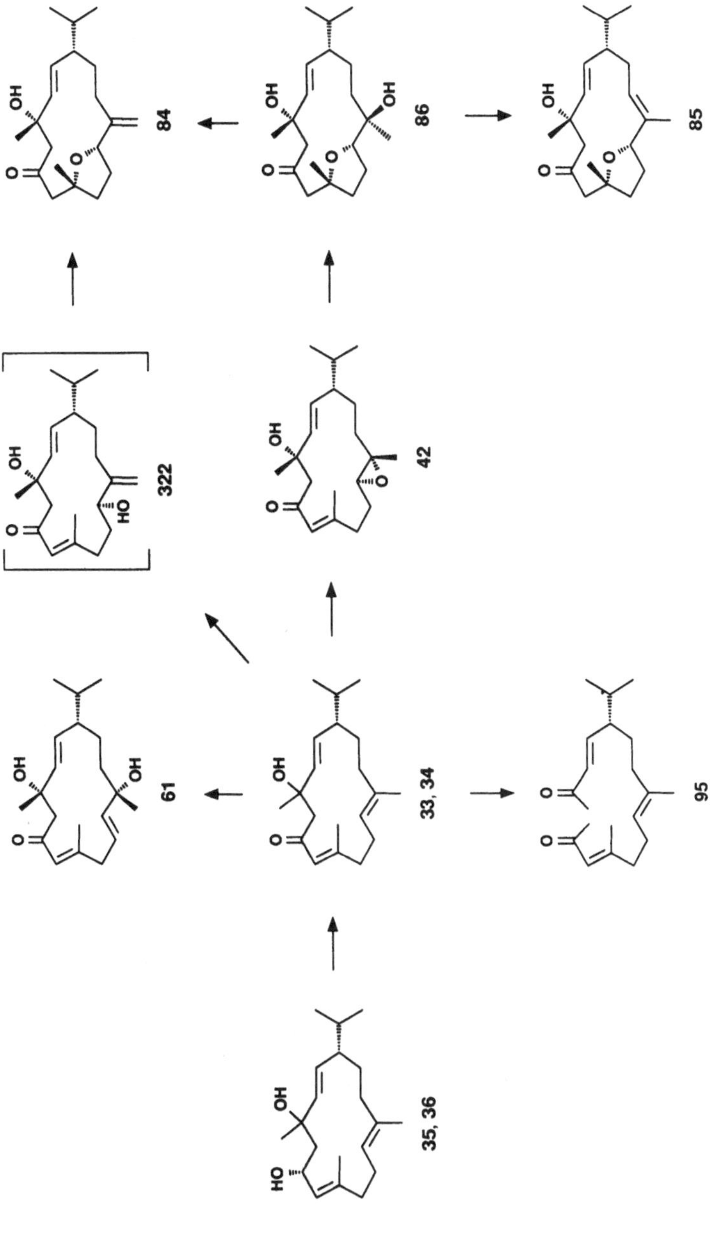

Scheme 8. Proposed biogenesis of compounds **33**, **34**, **42**, **61**, **84-86** and **95** from the 4,6-diols **35** and **36**

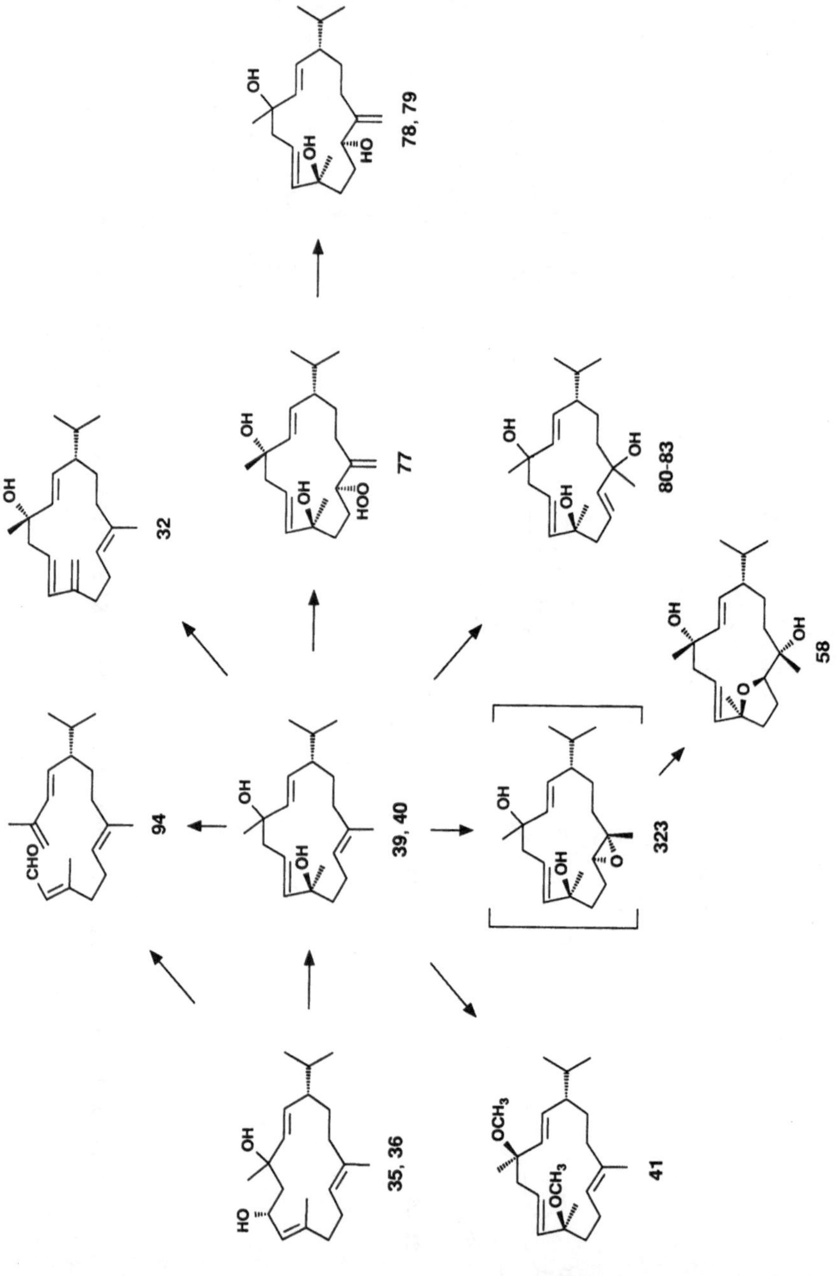

Scheme 9. Proposed biogenesis of compounds **32**, **39–41**, **58**, **77–83** and **94** from the 4,6-diols **35** and **36**

Table 3. *Cembrenes from Insects*

Structure no.	Compound Name	Source		Synthesis		Biological activity	
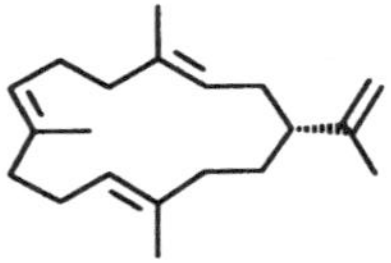 \n\n**2**	(−)-Cembrene-A (Neocembrene) (Neocembrene-A) (1R,3E,7E,11E)-Cembra-3,7,11,15-tetraene	*Cubitermes glebae* *Cubitermes umbratus* *Monomorium pharaonis* *Nasutitermes exitiosus* *Trinervitermes bettonianus*	*(133)* *(132)* *(131)* *(128)* *(130)*	(±)-**2** *(51–54)* (−)-**2** *(50, 55, 56)* (+)-**2** *(55)*		Queen-recognition pheromone Scent trail pheromone Scent trail pheromone	*(131)* *(128)* *(130)*
\n\n**105**	(3Z)-Cembrene-A (3Z,7E,11E)-Cembra-3,7,11,15-tetraene	*Cubitermes umbratus* *Cubitermes ugandensis* *Cubitermes* sp. N.D.	*(132, 133)* *(133)* *(133)*	(±)-**105** *(134, 135)* (−)-**105** *(135)*			

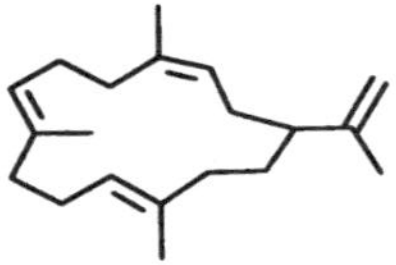

acid thus providing an alternative pathway for the biogenesis of this compound (95).

5. Acid-Induced Reactions

As outlined in Scheme 9, the 4,6-diols **35** and **36** give rise to the *seco*-aldehyde **94** (*108*) by fragmentation and to the (4,8S)-diols **39** and **40** (*109, 110*) by allylic rearrangement. An insight into these transformations, which are acid-induced, was provided by studies of the reaction of the 4,6-diols **35** and **36** with dilute sulphuric acid in dioxane/water. The results confirmed that under weakly acidic conditions a 4,6-diol (**35, 36**) undergoes competing allylic rearrangement and fragmentation reactions but also that epimerization reactions at C-4 take place (*110*).

The subsequent metabolism of the 4,8-diols (**39, 40**) involves dehydration to form **32** (*111*) and methylation to form **41** (*112*). Oxidation of the 11,12 double bond provides alternative routes for the biogenesis of the (11S)-hydroperoxide **77** (*83*), the 4,8,11- and 4,8,12-triols (**78–83**) (*90, 94*) and the (8S,11R)-epoxide **58** (*100*). Consistent with this, the triols **78, 80,** and **82** can be prepared *via* sensitized photo-oxygenation of the (4R,8S)-diol **39** and subsequent reduction, while the triols **79, 81,** and **83** were obtained from the (4S,8S)-diol **40** (*94*).

The route from the (4S,8S)-diol **40** to the (8S,11R)-epoxide **58** has also been explored experimentally. Epoxidation of **40** with the use of *m*-chloroperoxybenzoic acid gives the (11S,12S)-epoxide **323**. This undergoes a facile S_N2 type opening of the epoxide group at C-11 by the hydroxy group at C-8 to give **58** when treated with a trace of hydrochloric acid in chloroform (*100*).

III. Cembrenes from Insects

Cembrene-A (**2**) was first discovered as a trail-following pheromone of the termite *Nasutitermes exitiosus* (*128*). Later studies have shown that *Nasutitermes* termites also elaborate diterpenoids such as the bicyclic *seco*-trinervitanes, tricyclic trinervitanes and tetracyclic kempanes, and that cembrene-A (**2**) is a plausible intermediate in the biosynthesis of these diterpenoids (*129*).

Cembrene-A (**2**) has also been identified as a biologically active compound in other insect species. It is a trail pheromone of the grass feeding termite *Trinervitermes bettonianus* (*130*) and may serve as a queen-recognition pheromone for the ant *Monomorium pharaonis* (*131*).

Furthermore, cembrene-A (**2**) and/or its (3*Z*)-isomer (**105**) are present in defense secretions of soldiers of various *Cubitermes* termites (*132, 133*).

IV. Cembranoids from Marine Invertebrates

Cembranoids are also well-established constituents of marine invertebrates, sea fans and sea whips of the order *Gorgonacea* and soft corals of the order *Alcyonacea* being the principal sources. These animals dwell on tropical reefs; the gorgonians are most abundant in the Caribbean region and the alcyonaceans in the Indo-Pacific region. They have long been noted for their chemical defence systems, by which they are left relatively free of predators and microbial growth. This was an impetus to the initial investigations of the chemical composition of various gorgonians and alcyonaceans as were the facts that these animals do contain substantial amounts of extractable organic matter and grow in shallow water. They are hence easily collected. The studies carried out since the 1960's have been most rewarding. They have revealed that in soft corals diterpenoids are the most frequently encountered secondary metabolites and that the cembrane group is predominant. Gorgonians also elaborate a wide array of diterpenoids, the cembranoids constituting one of the principal groups. In all, of the cembranoids reported to-date some 200 are derived from marine invertebrates. These are listed in Table 4. Of considerable interest are the findings that a remarkably large number has been shown to exhibit pharmacologically important effects in mammal systems.

Most gorgonians and alcyonaceans live in symbiosis with intracellular dinoflagellate algae (zooxanthellae). By photosynthesis the algae produce both organic carbon and energy. It is believed that this energy is utilized for the production of secondary metabolites. The contention is now that the biosynthesis of cembranoids takes place in the coral polyps (*136*) and not, as previously suggested, in the symbiotic algae (*137*).

The early suggestions that the secondary metabolites in gorgonians and alcyonaceans take part in the chemical defence, either as toxins or feeding deterrents, are supported by experimental results. Thus among the cembranic metabolites, sarcoglaucol (**197**) (*138*), sarcophine (**209**) (*139*), denticulatolide (**228**) (*140*), crassolide (**230**) (*141*), and lobophytolide (**239**) (*142*) have all demonstrated toxicity towards large predators such as fish. Recent studies carried out by COLL et al. (*143*) have suggested that the role of the terpenoid metabolites may, in fact, be more extensive. They were able to show that eggs spawned by the soft coral *Lobophytum compactum* contain a considerable quantity of thunbergol (**5**). This compound is not present in pre- or postspawning colonies and appears to be biosynthesized specifically for egg release. It may serve as a release factor for ovulation, a function reminiscent of that of the prostaglandins also present in soft corals (*143*). Epoxypukalide (**282**) and/or pukalide (**281**), present solely in eggs and probably formed during the month prior to spawning, have been assigned a similar function in *Sinularia* species (*144*).

The literature on cembranoids derived from marine invertebrates or special aspects thereof has been reviewed by FAULKNER (*145–149*), TURSCH *et al.* (*142, 150*), FENICAL (*151*), KASHMAN (*152*) and KREBS (*153*). These duplicate some of the material in the present review.

A. Gorgonacea

Research on the chemistry of gorgonians was initiated in the 1960's by the group at the University of Oklahoma. It has been disclosed by their work and subsequent studies by others that the sea fans and sea whips produce structurally fairly complex cembranoids.

1. Eunicea Species

A number of cembran-16,14-olides (**238, 240, 241, 243, 244,** and **248–253**) have been isolated from various *Eunicea* species. All but two have a *cis*-fused lactone ring, the exceptions being epipeunicin (**241**) and 13,14-bisepijeunicin (**253**). The configuration at C-1 is *S* in all compounds for which the absolute configuration has been determined, *i.e.* they belong to the β-series. It is evident from other structural and stereochemical features that there exists a close biogenetic relationship within this group, and a hypothetical pathway for their formation is shown in Scheme 10.

Euniolide (**238**) (*154*) may be a precursor of 12,13-bisepieupalmerin (**243**) (*154, 155*), which is converted to eunicin (**248**) (*154–157*) and jeunicin (**252**) (*155, 159*) *via* S_N2 type opening of the epoxide group at C-3 or C-4, respectively, by attack of the hydroxy group at C-13. The validity of this route is borne out by chemical results (*154*) and by the fact that both eunicin (**248**) and jeunicin (**252**) were found along with the presumed precursor (**243**) in an extract of *E. succinea* (*155*).

Cueunicin (**250**) (*160*), the (13*R*)-epimer of eunicin (**248**), would then arise by a similar process occurring with the (13*R*)-isomer of **243**, *i.e.* **324**, while the diastereomeric (12*S*,13*R*)-derivative **325** is the immediate precursor of the naturally occurring eupalmerin acetate (**244**) (*154, 161, 162*).

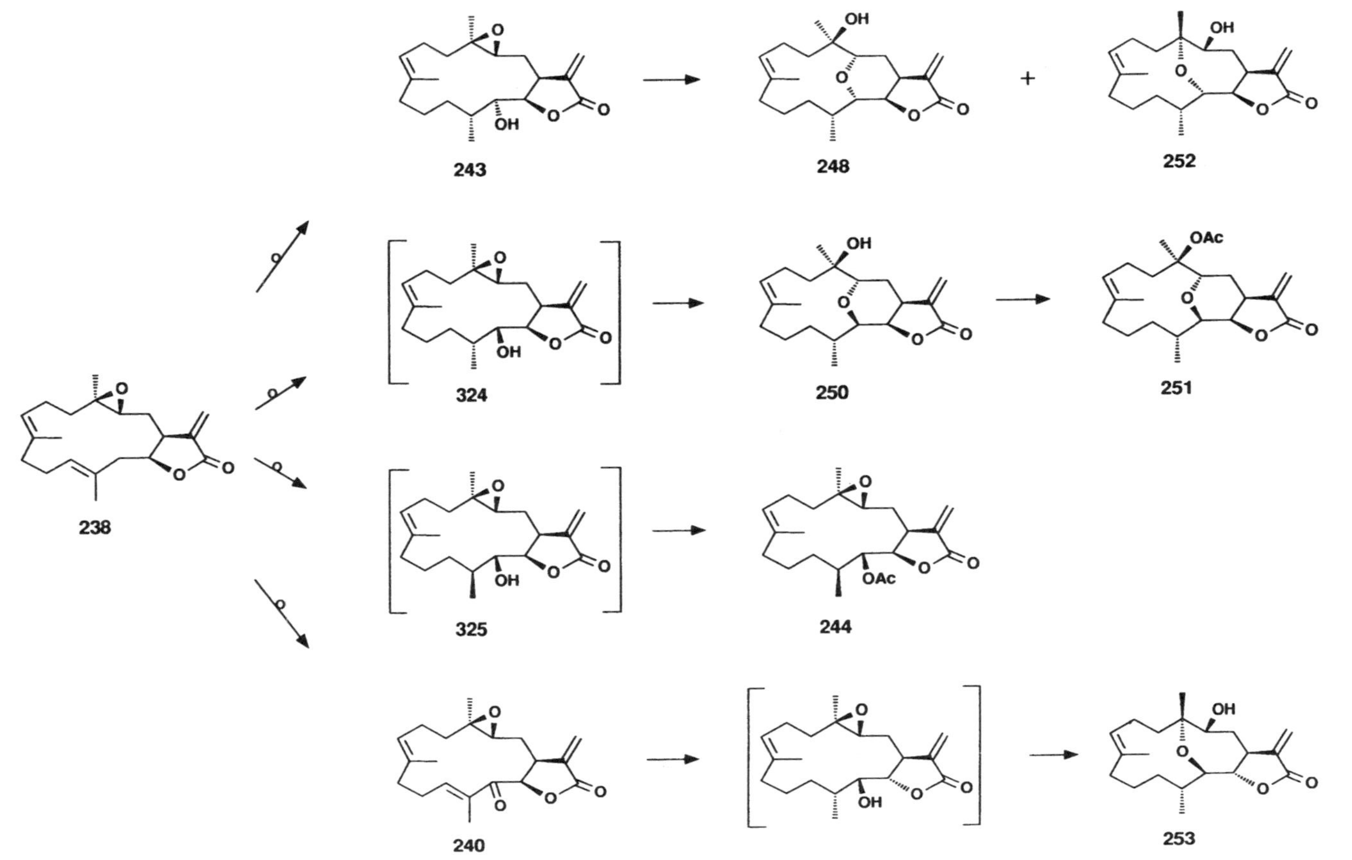

Scheme 10. Proposed biogenesis of peunicin (**240**), 12,13-bisepieupalmerin (**243**), eupalmerin acetate (**244**), eunicin (**248**), cueunicin (**250**), cueunicin acetate (**251**), jeunicin (**252**) and 13,14-bisepijeunicin (**253**) from euniolide (**238**)

Allylic oxidation of euniolide (**238**) is a proposed route to peunicin (**240**) (*163, 164*). Epimerization at C-14 followed by reduction and rearrangement may account for the formation of 13,14-bisepijeunicin (**253**) (*4*).

Asperdiol (**193**), a non-lactonic member of the β-series, has been isolated from *E. asperula* and *E. tourneforti* (*165*). It displays anticancer activity.

2. *Lophogorgia and Leptogorgia Species*

Lophodione (**264**), isolophodione (**265**), and epoxylophodione (**271**) are three 3,6-dioxocembren-20,10-olides isolated from *Lophogorgia alba*, a Pacific sea whip that does not contain zooxanthellae (*166*). Also present in this as well as in other *Lophogorgia* species is the potent neurotoxin lophotoxin (**280**) (*166, 167*). Pukalide (**281**) and the related **283** have been found in *L. rigida* (*167*) as well as in certain soft corals (see Table 4), while 11,12-epoxypukalide (**282**) is a constituent of *Leptogorgia setacea* (*168*) and of a *Placogorgia* species (*169*).

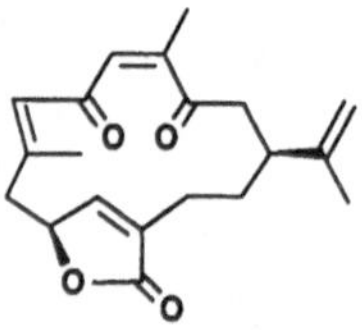

264

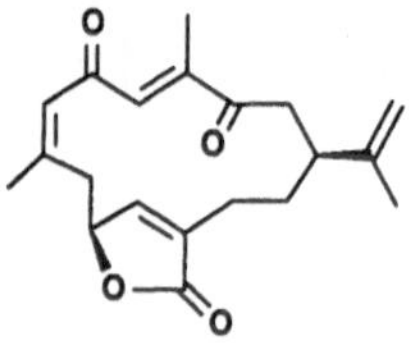

265

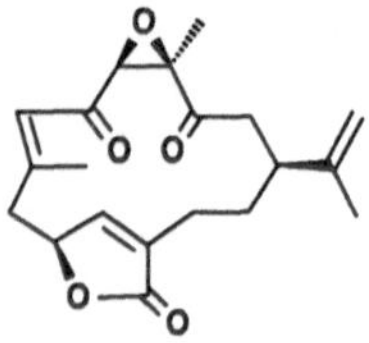

271

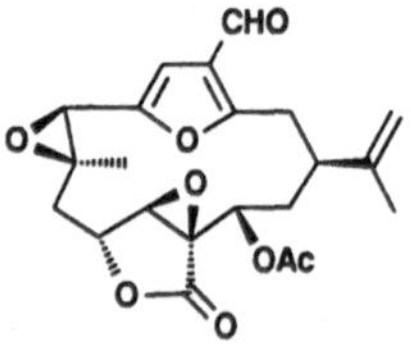

280

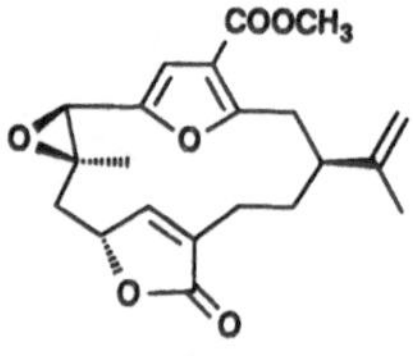

281

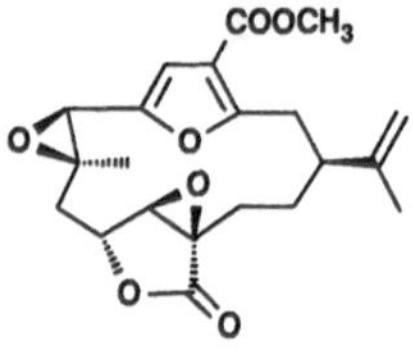

282

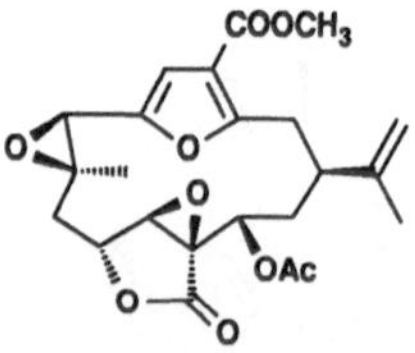

283

Table 4. *Cembranoids from Marine Invertebrates*

Compound Structure no. / Name	Source	X-ray and spectral studies	Synthesis	Biological activity
(−)-Cembrene-A (Neocembrene) (Neocembrene-A) (1R,3E,7E,11E)-Cembra-3,7,11,15-tetraene **2**	*Nephthea brassica* *Nephthea* species *Sarcophyton* species *Sinularia conferta* *Sinularia flexibilis*	(227) (228) (229) (219) (213,216)	(±)-**2** (*51–54*) (−)-**2** (*50, 55, 56*) (+)-**2** (*55*)	
Thunbergol (1S,2E,4R,7E,11E)-Cembra-2,7,11-trien-4-ol **5**	*Sarcophyton decaryi* Eggs spawned by *Lobophytum compactum*	(206) (143)	(±)-**5** (*61, 62*)	
Cembrene-C (Isoneocembrene A) (1E,3E,7E,11E)-Cembra-1,3,7,11-tetraene **106**	*Alcyonium flaccidum* *Alcyonium utinomii* *Armina maculata* *Lobophytum* species *Nephthea brassica* *Nephthea chabrolii* *Nephthea* species *Sarcophyton* cf. *birklandi* *Sarcophyton ehrenbergi* *Sarcophyton* species *Sinularia mayi*	(186) (187) (230) (231) (227) (232) (228, 233) (200) (199) (313) (207)	(±)-**106** (*314*)	

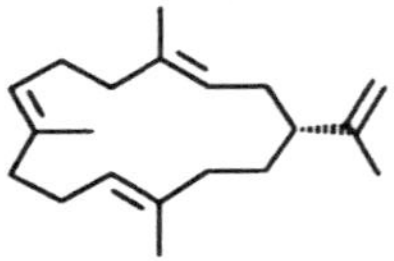

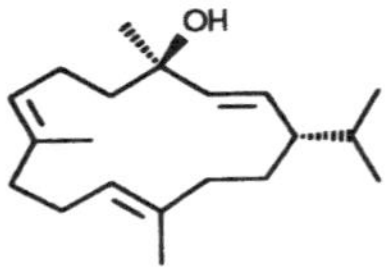

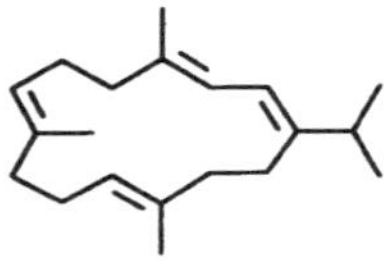

Table 4 (continued)

Structure no.	Compound Name	Source	X-ray and spectral studies	Synthesis	Biological activity
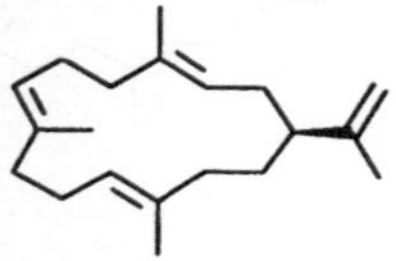 **107**	(+)-Cembrene-A (1S,3E,7E,11E)-Cembra-3,7,11,15-tetraene	*Clavularia* species	(234)		
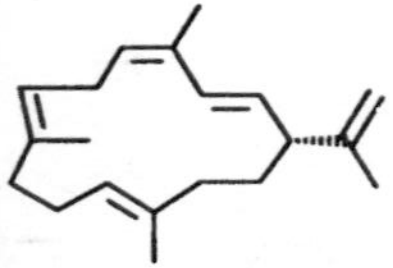 **108**	Cembrenene (1R,2E,4Z,7E,11E)-Cembra-2,4,7,11,15-pentaene	*Sinularia mayi*	(207)	(±)-**108** (134) (+)-**108** (50, 235)	
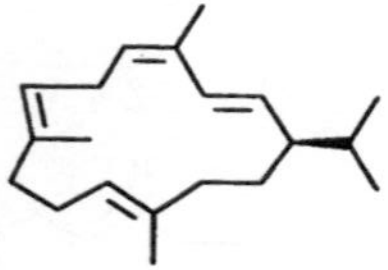 **109**	(−)-Cembrene (1R,2E,4Z,7E,11E)-Cembra-2,4,7,11-tetraene	*Sinularia mayi*	(207)	(±)-**109** (47–49) (+)-**109** (50)	
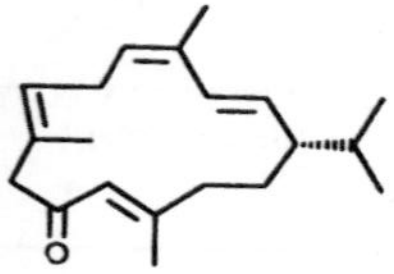 **110**	10-Oxocembrene (1S,2E,4Z,7E,11E)-Cembra-2,4,7,11-tetraen-10-one	*Sarcophyton elegans*	(236)		

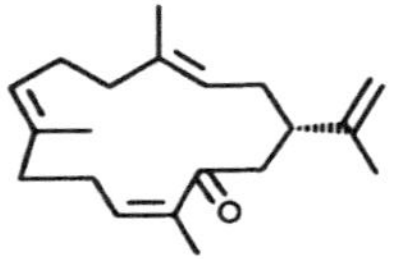

(±)-**114** (64)

(229)

(229)

(208)

(191)

Sarcophyton species

Sarcophyton species

Sinularia mayi

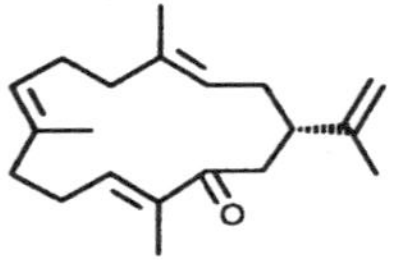

Sarcophyton glaucum

(1S*,3E,7E,11Z)-
Cembra-3,7,11,15-
tetraen-13-one

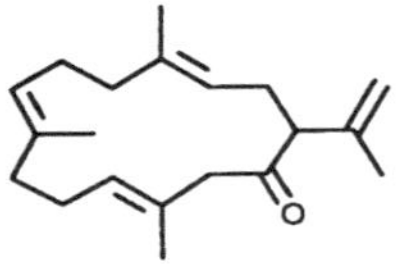

(1S*,3E,7E,11E)-
Cembra-3,7,11,15-
tetraen-13-one

Sinularone A
(3E,7E,11E)-Cembra-
3,7,11,15-tetraen-
14-one

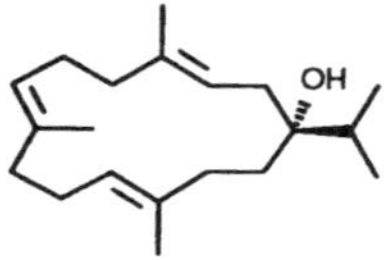

Sarcophytol M
(1R,3E,7E,11E)-Cembra-
3,7,11-trien-1-ol

111

112

113

114

Table 4 (continued)

Structure no.	Compound Name	Source	X-ray and spectral studies	Synthesis	Biological activity
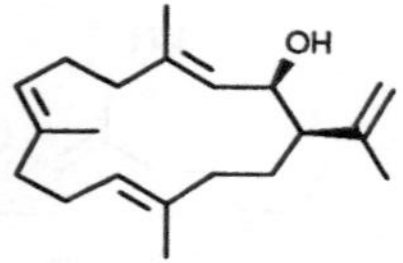 115	Mayol (1R,2S,3E,7E,11E)-Cembra-3,7,11,15-tetraen-2-ol	Simularia mayi Lobophytum denticulatum	(207, 208) (237)	(+)-115 (235)	
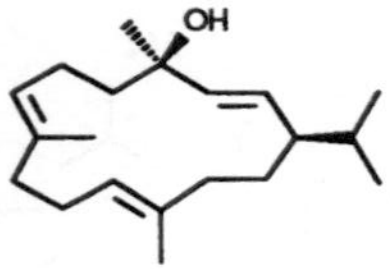 116	(1R,2E,4R,7E,11E)-Cembra-2,7,11-trien-4-ol	Briareum species Lobophytum species Simularia facile Simularia polydactyla	(238) (231) (239) (240)		
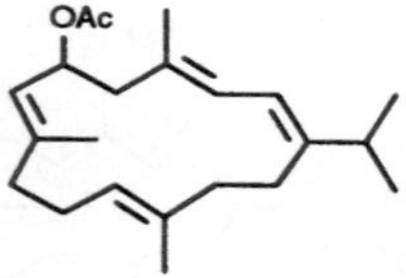 117	Preverecynarmin (1E,3E,7E,11E)-Cembra-1,3,7,11-tetraen-6-oyl acetate	Armina maculata Vereillum cynomorium	(230) (230)		
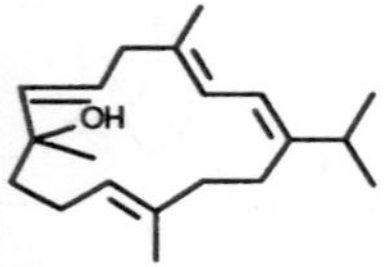 118	Alcyonol-B (1E,3E,6E,11E)-Cembra-1,3,6,11-tetraen-8-ol	Alcyonium uinomii	(187)		

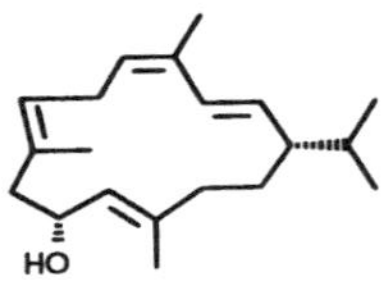

10-Hydroxycembrene
(1S,2E,4Z,7E,10R,11E)-
Cembra-2,4,7,11-
tetraen-10-ol

Sarcophyton elegans
Sarcophyton glaucum

(236)
(152, 196)

X-ray analysis
of the *p*-bro-
mobenzoyl
ester (absolute
configuration)
(236)

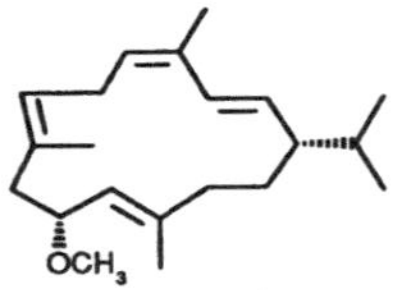

10-Methoxycembrene
(1S,2E,4Z,7E,10R,11E)-
Cembra-2,4,7,11-tetraene-
10-methoxide

Sarcophyton elegans

(236)

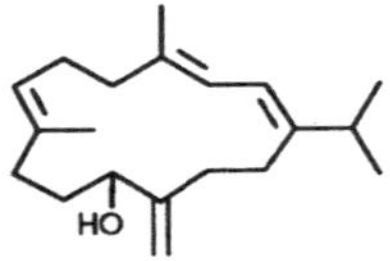

Alcyonol-C
(1E,3E,7E)-Cembra-
1,3,7,12(20)-tetraen-
11-ol

Alcyonium utinomii

(187)

Alcyonol-A
(1E,3E,7E,10E)-
Cembra-1,3,7,10-
tetraen-12-ol

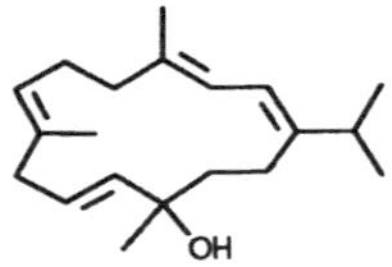

Alcyonium utinomii

(187)

119

120

121

122

Table 4 (continued)

Structure no.	Compound Name	Source	X-ray and spectral studies	Synthesis	Biological activity
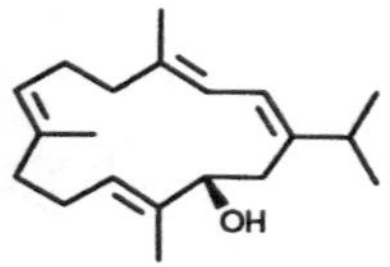 123	13-Hydroxycembrene-C (1E,3E,7E,11E,13S)-Cembra-1,3,7,11-tetraen-13-ol	Nephthea brassica Nephthea species	(227) (233)		
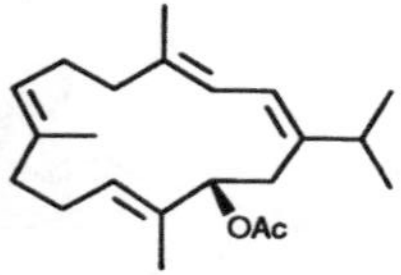 124	(1E,3E,7E,11E,13S)-Cembra-1,3,7,11-tetraen-13-oyl acetate	Nephthea brassica Nephthea species	(227) (233)		
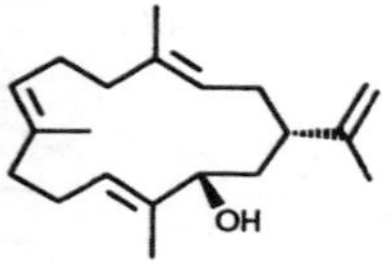 125	(13S)-Hydroxy-(−)-neocembrene (1S,3E,7E,11E,13S)-Cembra-3,7,11,15-tetraen-13-ol	Sarcophyton trochelio-phorum	(241)	X-ray analysis of the p-nitro-benzoate of 125 (relative configuration) (241)	Cytostatic effect on Ehrlich ascites tumor cells (242)

Structure	Name	Source	Ref.			
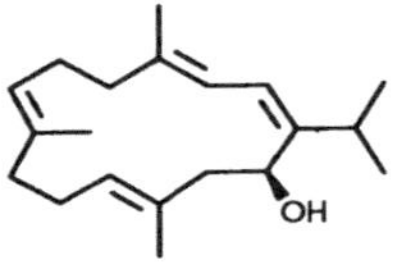 **126**	Sarcophytol A (1Z,3E,7E,11E,14S)-Cembra-1,3,7,11-tetraen-14-ol	Lobophytum pauciflorum Sarcophyton glaucum Sarcophyton species	(200) (190, 243) (227)	X-ray analysis of the α-methoxy-α-tri-fluoromethyl-phenyl acetate of **126** (absolute configuration) (244) MM Calculations (244)	(+)-**126** (245)	Inhibits the activity of the tumor-promoter teleocidin (192)
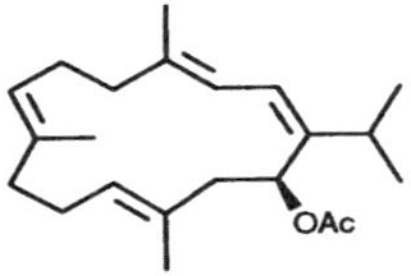 **127**	Sarcophytol A acetate (1Z,3E,7E,11E,14S)-Cembra-1,3,7,11-tetraen-14-oyl acetate	Sarcophyton glaucum Sarcophyton species	(190) (227)			
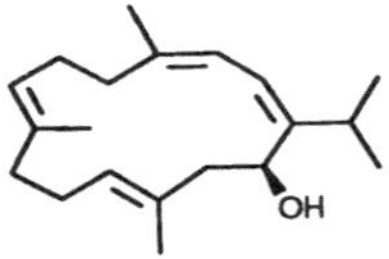 **128**	Sarcophytol N (1Z,3Z,7E,11E,14S)-Cembra-1,3,7,11-tetraen-14-ol	Sarcophyton glaucum	(191)			
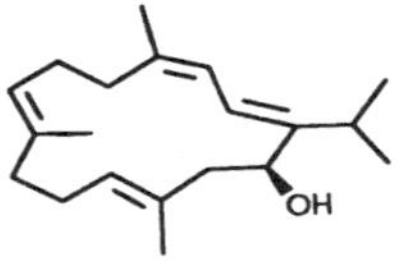 **129**	Sarcophytol T (1E,3Z,7E,11E,14S)-Cembra-1,3,7,11-tetraen-14-ol	Sarcophyton glaucum	(193)			

Table 4 (continued)

Structure no.	Compound Name	Source	X-ray and spectral studies	Synthesis	Biological activity
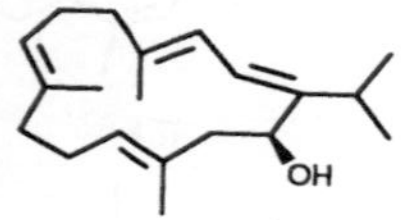 **130**	Sarcophytol F (1E,3E,7E,11E,14S)-Cembra-1,3,7,11-tetraen-14-ol	Sarcophyton glaucum	(194)		
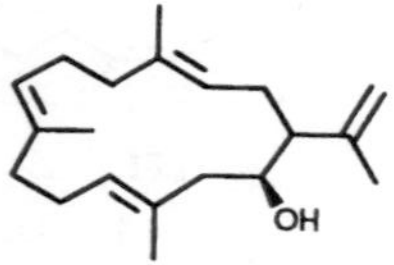 **131**	Sinulariol C (3E,7E,11E,14S)-Cembra-3,7,11,15-tetraen-14-ol	Simularia mayi	(208)		
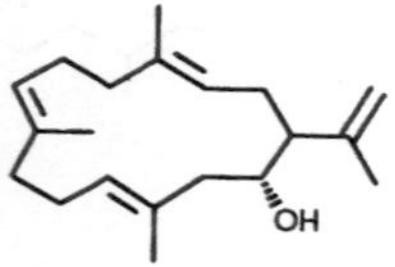 **132**	14-Hydroxycembrene-A (3E,7E,11E,14R)-Cembra-3,7,11,15-tetraen-14-ol	Lobophytum denticulatum Simularia mayi	(237) (208)		
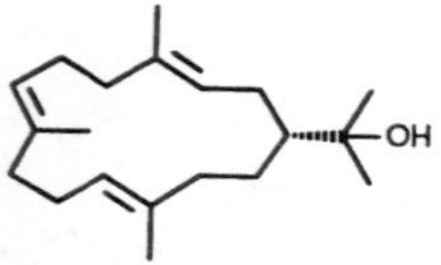 **133**	(−)-Nephthenol (1R,3E,7E,11E)-Cembra-3,7,11-trien-15-ol	Litophyton arboreum Litophyton viridis Lobophytum pauciflorum Nephthea species Sarcophyton decaryi Sarcophyton glaucum	(246) (247) (187, 246) (248) (206, 249) (191)	(±)-133 (51) (−)-133 (50, 56)	

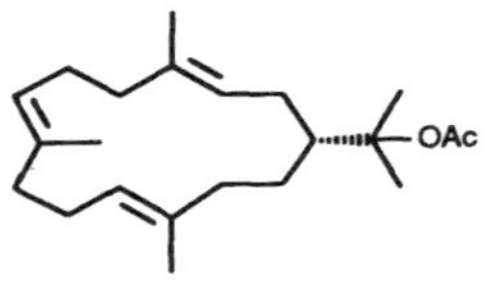

(±)-**24** (58)

(227)

Nephthea brassica

Nephthenol acetate
(1R,3E,7E,11E)-
Cembra-3,7,11-trien-
15-oyl acetate

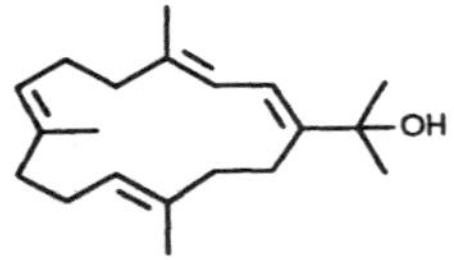

134

(200)

(1E,3E,7E,11E)-Cembra- *Lobophytum pauciflorum*
1,3,7,11-tetraen-15-ol

135

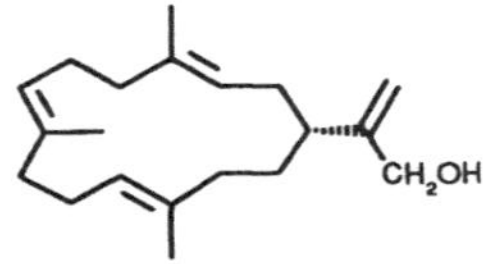

(191)
(208)

Sinulariol D
(1R,3E,7E,11E)-
Cembra-3,7,11,15(17)-
tetraen-16-ol

Sarcophyton glaucum
Sinularia mayi

136

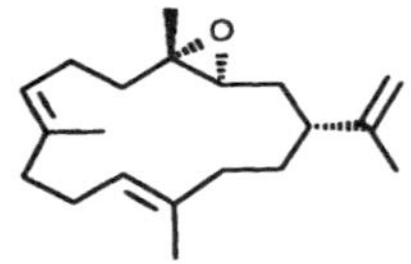

(227)
(239)

(1R,3S,4S,7E,11E)-
Cembra-7,11,15-
trien-3,4-epoxide

Nephthea brassica
Sinularia facile

137

Table 4 (continued)

Compound Structure no. / Name	Source	X-ray and spectral studies	Synthesis	Biological activity
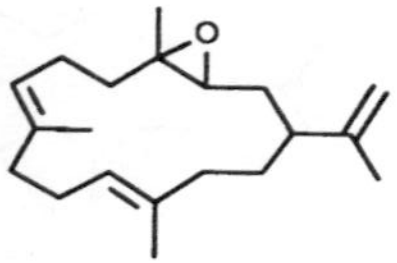 138 (7E,11E)-Cembra-7,11,15-trien-3,4-epoxide	Sarcophyton species	(229)		
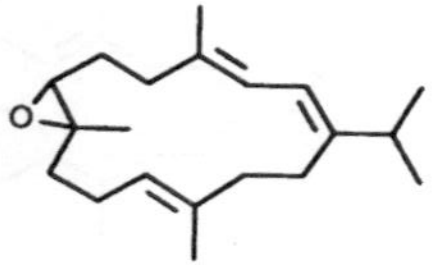 139 (1E,3E,11E)-Cembra-1,3,11-trien-7,8-epoxide	Lobophytum species Sarcophyton crassocaule	(143) (198)		
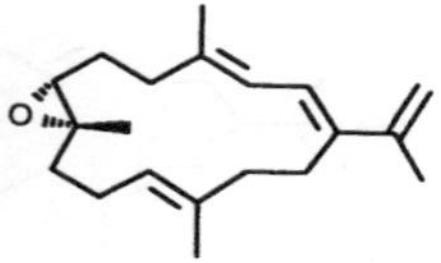 140 (1E,3E,7R*,8R*,11E)-Cembra-1,3,11,15-tetraen-7,8-epoxide	Lobophytum species Sarcophyton crassocaule	(143) (198)		
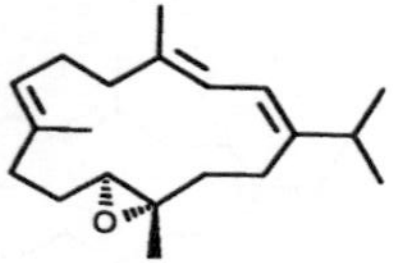 141 (1E,3E,7E,11S,12S)-Cembra-1,3,7-trien-11,12-epoxide	Lobophytum microlobulatum Lobophytum species Nephthea species Sarcophyton cf. birklandi Sinularia grayi	(143) (231) (233) (200) (199)		

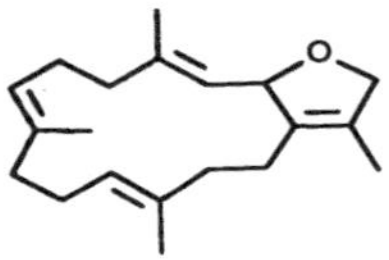

Sarcophytonin-A
(3E,7E,11E)-
Cembra-1(15),3,7,11-
tetraen-2,16-epoxide

Sarcophyton glaucum (2R)   (190)
Sarcophyton species (2S)   (313)
Ovula ovum (2S)   (315)

Sinularial A
(1R,3E,7E,11E)-Cembra-
3,7,11,15-tetraen-17-al

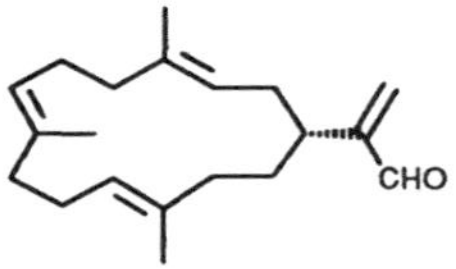

Sinularia mayi   (208)

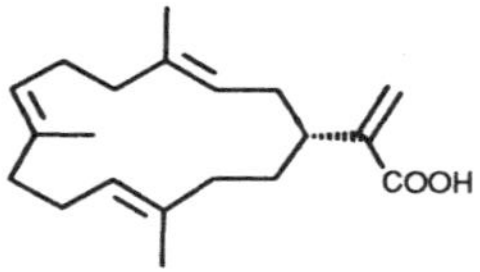

Sinularic acid
(1R,3E,7E,11E)-Cembra-
3,7,11,15-tetraen-
17-oic acid

Sinularia mayi   (208)

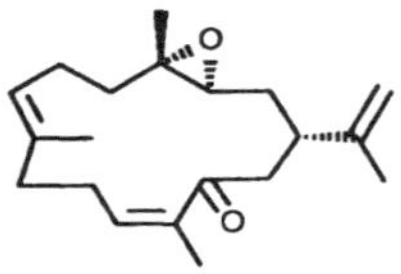

(1S*,3S*,4S*,7E,11Z)-
3,4-Epoxycembra-
7,11,15-trien-13-one

Sarcophyton species   (229)

142

143

144

145

Table 4 (*continued*)

Structure no.	Compound Name	Source	X-ray and spectral studies	Synthesis	Biological activity
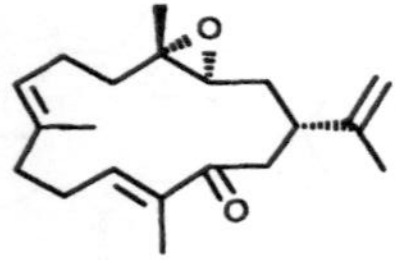 146	(1S*,3S*,4S*,7E,11E)-3,4-Epoxycembra-7,11,15-trien-13-one	*Sarcophyton* species	(229)		
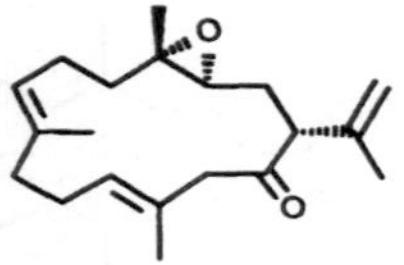 147	(1R*,3S*,4S*,7E,11E)-3,4-Epoxycembra-7,11,15-trien-14-one	*Sarcophyton* species	(229)		
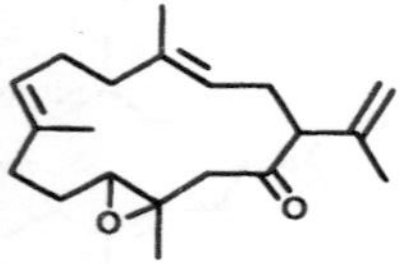 148	Sinularone B (3E,7E)-11,12-Epoxycembra-3,7,15-trien-14-one	*Sinularia mayi*	(208)		
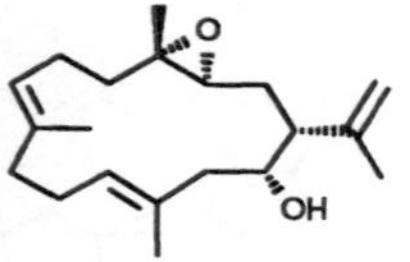 149	(1R*,3S*,4S*,7E,11E,14R*)-3,4-Epoxy-cembra-7,11,15-trien-14-ol	*Sarcophyton* species	(229)		

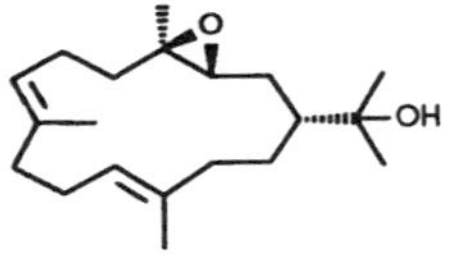

150

3,4-Epoxynephthenol
(1R,3R,4R,7E,11E)-3,4-Epoxycembra-7,11-dien-15-ol

Sarcophyton decaryi (206)
Eggs spawned by Lobophytum microlobulatum (143)

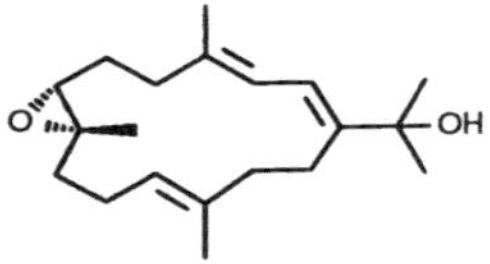

151

(1E,3E,7R*,8R*,11E)-7,8-Epoxycembra-1,3,11-trien-15-ol

Sarcophyton species (250)
Sarcophyton crassocaule (198)
Sarcophyton trocheliophorum (251)

X-ray analysis of **151** (relative configuration) (250)

(±)-**151** (64)

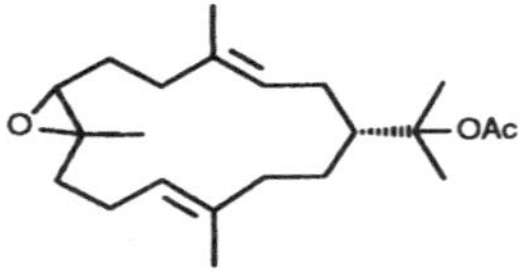

152

Epoxynephthenol acetate (1R,3E,11E)-15-Acetoxy-cembra-3,11-dien-7,8-epoxide

Nephthea brassica (227)
Nephthea species (248)

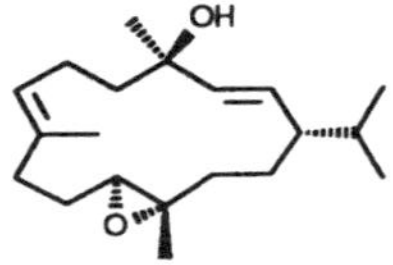

153

Trocheliophorol
(1S,2E,4R,7E,11S,12S)-11,12-Epoxycembra-2,7-dien-4-ol

Lobophytum species (143)
Sarcophyton decaryi (206)
Sarcophyton trocheliophorum (142, 249, 251)

Table 4 (continued)

Compound Structure no.	Name	Source	X-ray and spectral studies	Synthesis	Biological activity
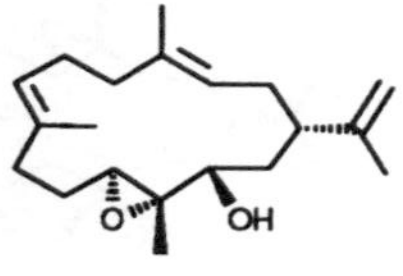 154	(13S)-Hydroxy-11,12-epoxy-(−)-neocembrene (1S,3E,7E,11S,12R,13S)-11,12-Epoxycembra-3,7,15-trien-13-ol	Sarcophyton trochelio-phorum	(241)		Cytostatic effect on Ehrlich ascites tumor cells (242)
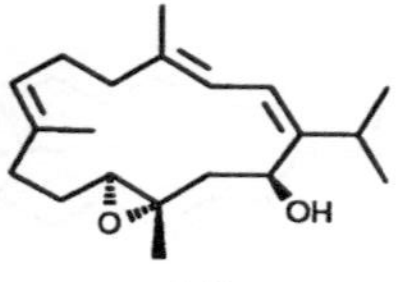 155	(1Z,3E,7E,11S,12S,14S)-11,12-Epoxycembra-1,3,7-trien-14-ol	Lobophytum species	(231)		
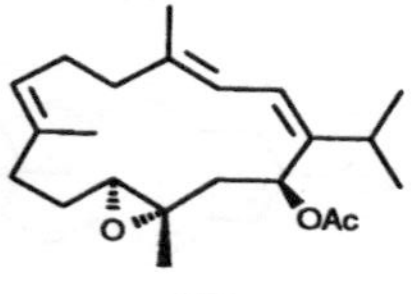 156	(1Z,3E,7E,11S,12S,14S)-14-Acetoxycembra-1,3,7-trien-11,12-epoxide	Lobophytum species	(231)		
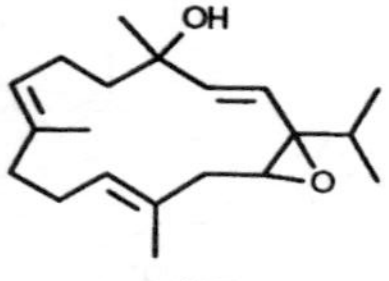 157	Sarcophytol-C (2E,7E,11E)-1,14-Epoxycembra-2,7,11-trien-4-ol	Sarcophyton glaucum	(195)		

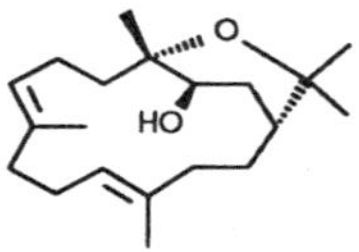

158	Decaryiol (1R,3R,4S,7E,11E)-4,15-Epoxycembra-7,11-dien-3-ol	*Sarcophyton decaryi* *Lobophytum microlobulatum* and eggs spawned thereof	(206) (143)
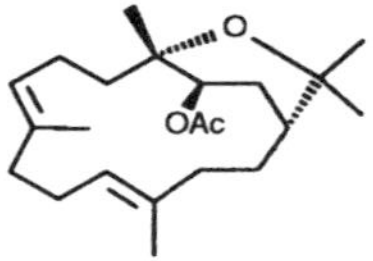 **159**	Decaryiol acetate (1R,3R,4S,7E,11E)-4,15-Epoxycembra-7,11-dien-3-oyl acetate	*Lobophytum microlobulatum* and eggs spawned thereof	(143)
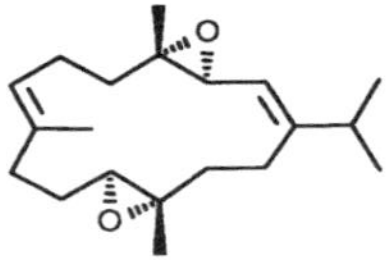 **160**	(1E,3S,4S,7E,11S,12S)-Cembra-1,7-diene-3,4:11,12-diepoxide	*Lobophytum species*	(231)
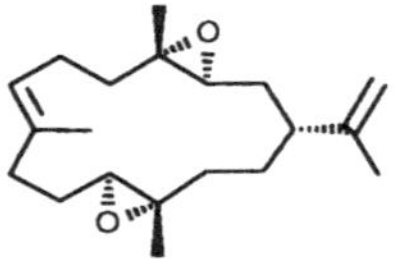 **161**	(1R,3S,4S,7E,11S,12S)-Cembra-7,15-diene-3,4:11,12-diepoxide	*Sinularia flexibilis*	(216)

Table 4 (continued)

Structure no.	Compound Name	Source	X-ray and spectral studies	Synthesis	Biological activity
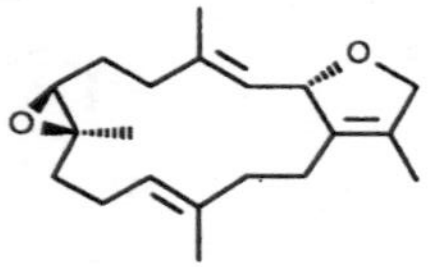 162	Deoxosarcophine (+)-Sarcophytoxide (2S,3E,7S,8S,11E)-Cembra-1(15),3,11-triene-2,16:7,8-diepoxide	Lobophytum species (246) Sarcophyton glaucum (196) Sarcophyton species (184, 252, 313)	X-ray analysis of 162 (relative configuration) (252)		Facilitates neuromuscular transmission in a rat diaphragm (184, 253) Calcium antagonist activity on the isolated rabbit aorta (252)
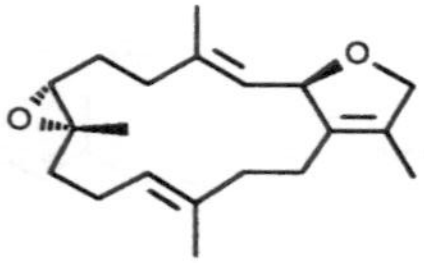 163	(−)-Sarcophytoxide (2R,3E,7R,8R,11E)-Cembra-1(15),3,11-triene-2,16:7,8-diepoxide	Sarcophyton cf. birklandi (200, 254) Sarcophyton crassocaule (198) Sarcophyton ehrenbergi (199) Sarcophyton gemmatum (249) Sarcophyton pauciplicatum (249) Sarcophyton trocheliophorum (142)			Ichthyotoxic to the fish Gambusia affinis (315)
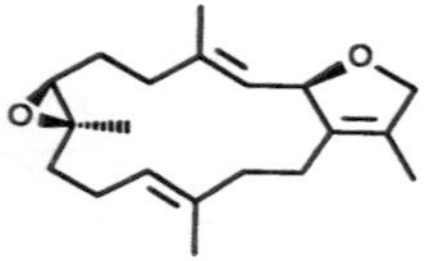 164	(2R,3E,7S,8S,11E)-Cembra-1(15),3,11-triene-2,16:7,8-diepoxide	Lobophytum pauciflorum (200) Sarcophyton glaucum (196)			

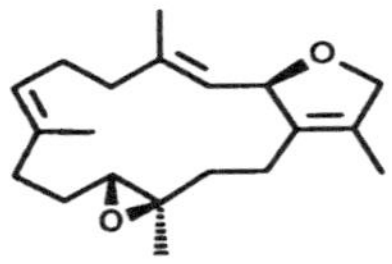

Structure	Name	Source	
165	Isosarcophytoxide (2R,3E,7E,11R,12R)-Cembra-1(15),3,7-triene-2,16:11,12-diepoxide	*Sarcophyton* cf. *birklandi* (200, 254) *Sarcophyton glaucum* (200) *Sarcophyton* species (255) *Sarcophyton trochelio-phorum* (142)	
166	(2S,3E,7E,11R,12R)-Cembra-1(15),3,7-triene-2,16:11,12-diepoxide	*Sarcophyton* cf. *birklandi* (200, 254) *Sarcophyton glaucum* (200) *Sinularia mayi* (316)	X-ray analysis of **166** (relative configuration) (200)
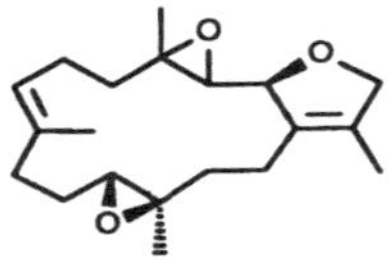 **167**	(2S,7E,11R,12R)-Cembra-1(15),7-diene-2,16:3,4:11,12-triepoxide	*Sarcophyton* species (255)	
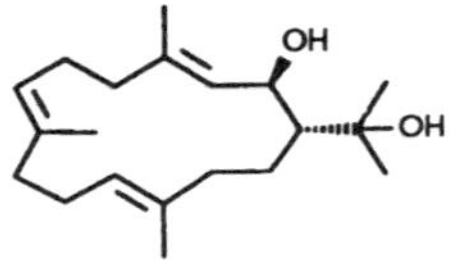 **168**	2-Hydroxynephthenol (1R*,2R*,3E,7E,11E)-Cembra-3,7,11-triene-2,15-diol	*Litophyton viridis* (247)	(±)-**168** (64)

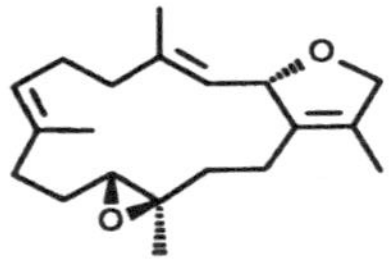

Table 4 (continued)

Compound Structure no.	Name	Source	X-ray and spectral studies	Synthesis	Biological activity
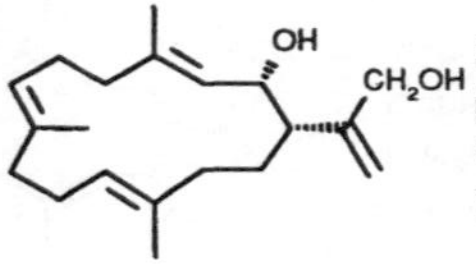 169	Sinulariol A (1S,2S,3E,7E,11E)-Cembra-3,7,11,15-tetraene-2,17-diol	Sinularia mayi	(180)		
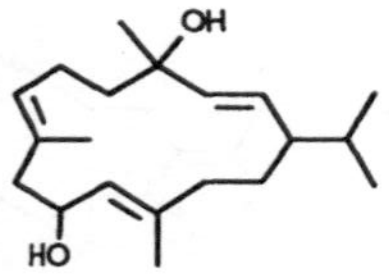 170	(2E,7E,11E)-Cembra-2,7,11-triene-4,10-diol	Sarcophyton glaucum	(196)		
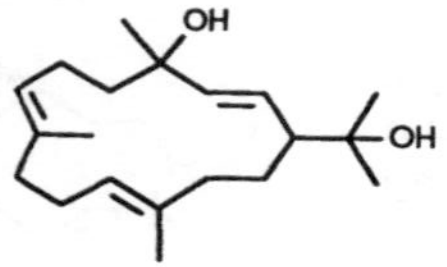 171	Pauciflorol-A (2E,7E,11E)-Cembra-2,7,11-triene-4,15-diol	Lobophytum pauciflorum	(187)	(±)-171 (256)	
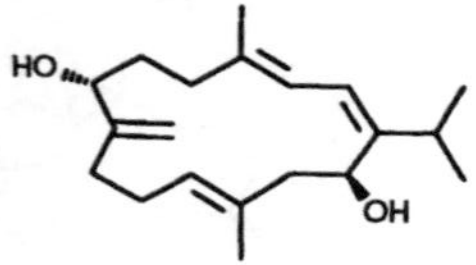 172	Sarcophytol-H (1Z,3E,7R,11E,14S)-Cembra-1,3,8(19),11-tetraene-7,14-diol	Sarcophyton glaucum	(191)		

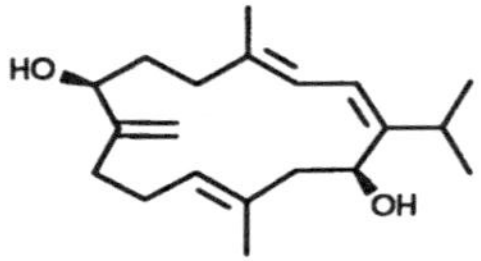 173	Sarcophytol-O (1*Z*,3*E*,7*S*,11*E*,14*S*)-Cembra-1,3,8(19),11-tetraene-7,14-diol	*Sarcophyton glaucum*	(*191*)
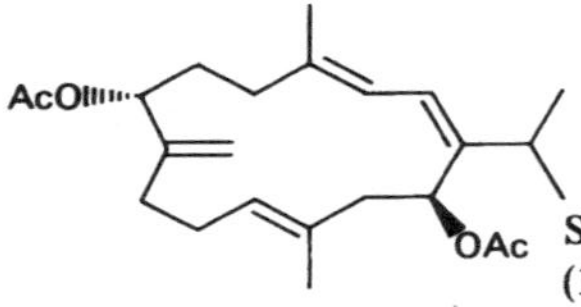 174	Sarcophytol-H diacetate (1*Z*,3*E*,7*R*,11*E*,14*S*)-Cembra-1,3,8(19),11-tetraene-7,14-dioyl diacetate	*Sarcophyton glaucum*	(*191*)
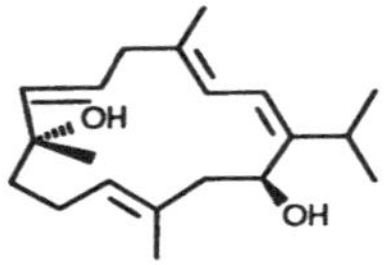 175	Sarcophytol R (1*Z*,3*E*,6*E*,8*R*,11*E*,14*S*)-Cembra-1,3,6,11-tetraene-8,14-diol	*Sarcophyton glaucum*	(*194*)
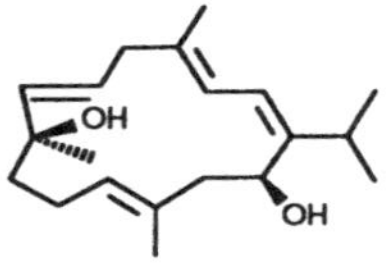 176	Sarcophytol S (1*Z*,3*E*,6*E*,8*S*,11*E*,14*S*)-Cembra-1,3,6,11-tetraene-8,14-diol	*Sarcophyton glaucum*	(*194*)

Table 4 (continued)

Structure no.	Compound Name	Source	X-ray and spec-tral studies	Synthesis	Biological activity
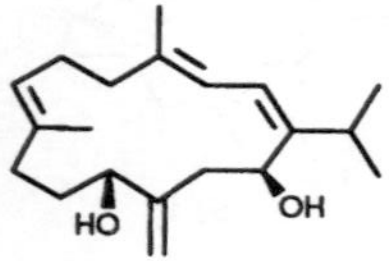 177	Sarcophytol-I (1Z,3E,7E,11R,14S)-Cembra-1,3,7,12(20)-tetraene-11,14-diol	Sarcophyton glaucum	(191)		
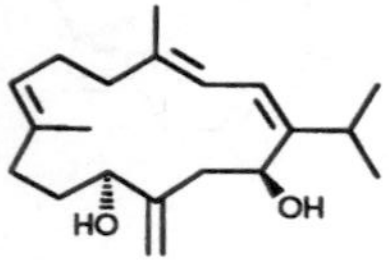 178	Sarcophytol-E (1Z,3E,7E,11S,14S)-Cembra-1,3,7,12(20)-tetraene-11,14-diol	Sarcophyton glaucum	(191, 195)		
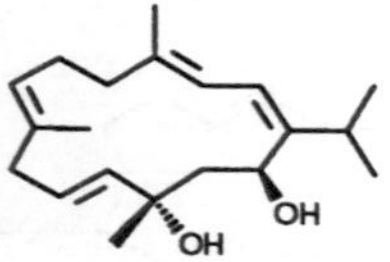 179	Sarcophytol-G (1Z,3E,7E,10E,12S,14S)-Cembra-1,3,7,10-tetraene-12,14-diol	Sarcophyton glaucum	(191)		
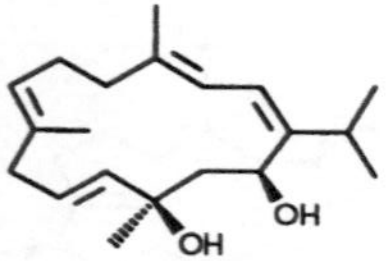 180	Sarcophytol-D (1Z,3E,7E,10E,12R,14S)-Cembra-1,3,7,10-tetraene-12,14-diol	Sarcophyton glaucum	(191, 195)		

Structure	Name	Source	Ref.	Comments
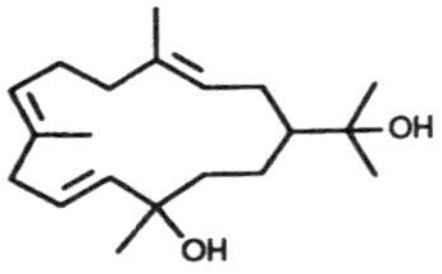 **181**	Pauciflorol-B (3E,7E,10E)-Cembra-3,7,10-triene-12,15-diol	*Lobophytum pauciflorum*	(187)	
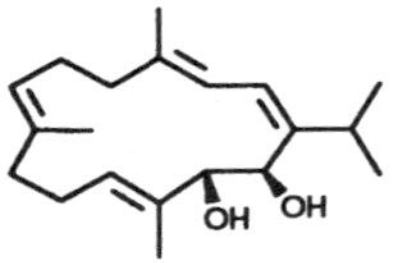 **182**	Sarcophytol-B (1Z,3E,7E,11E,13R,14R)-Cembra-1,3,7,11-tetraene-13,14-diol	*Alcyonium flaccidum* *Sarcophyton glaucum*	(186) (190, 191, 194)	X-ray analysis of the *p*-bromophenyl-boronate derivative of (±)-**182** (201) (±)-**182** (201) Inhibits the activity of the tumor-promoter teleo-cidin (192)
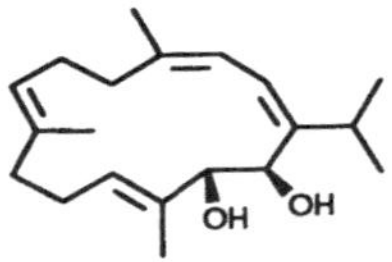 **183**	Sarcophytol-J (1Z,3Z,7E,11E,13R,-14R)-Cembra-1,3,7,11-tetraene-13,14-diol	*Sarcophyton glaucum*	(191, 194)	
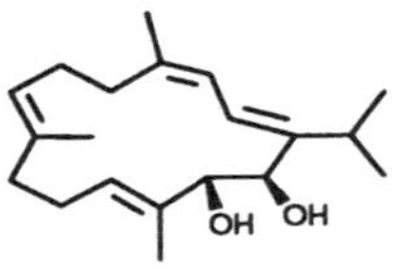 **184**	Sarcophytol K (1E,3Z,7E,11E,13R,14R)-Cembra-1,3,7,11-tetraene-13,14-diol	*Sarcophyton glaucum*	(194)	

Table 4 (continued)

Compound Structure no.	Name	Source	X-ray and spectral studies	Synthesis	Biological activity
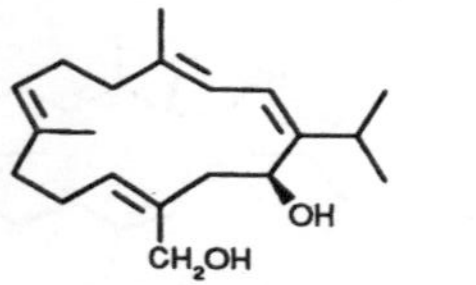 185	Sarcophytol P (1Z,3E,7E,11Z,14S)-Cembra-1,3,7,11-tetraene-14,20-diol	Sarcophyton glaucum	(194)		
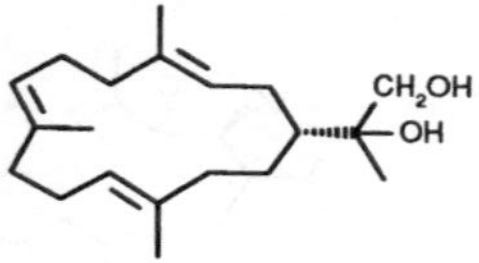 186	Sinulariol B (1R,3E,7E,11E)-Cembra-3,7,11-triene-15,16-diol	Sinularia mayi	(180)		
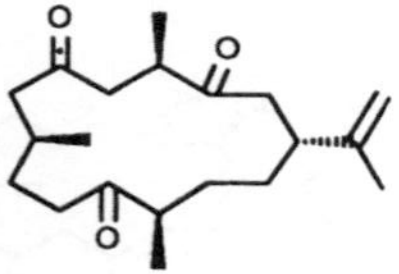 187	Dehydroplexaurone (1R,4R,8S,12R)-Cembr-15-ene-3,6,11-trione	Plexaura A	(171)		
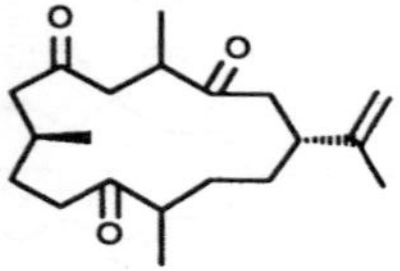 188	(1R,8S)-Cembr-15-ene-3,6,11-trione	Plexaura A	(171)		

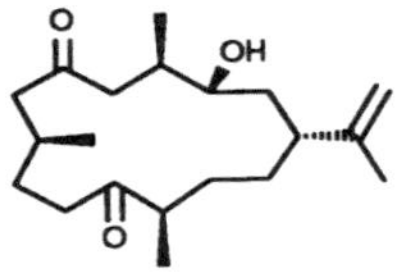

Plexaurolone
(1R,3R,4R,8S,12R)-3-
Hydroxycembr-15-
ene-6,11-dione

Plexaura A

(170) X-ray analysis
of the acetate of
189 (absolute
configuration)
(170)

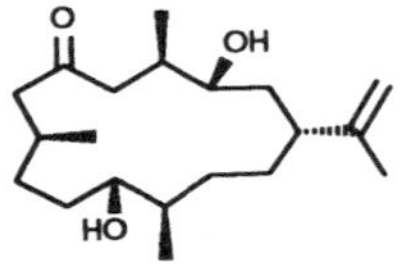

Dihydroplexaurolone
(1R,3R,4R,8S,11R,12R)-
3,11-Dihydroxycembr-
15-en-6-one

Plexaura A

(171) X-ray analysis
of **190** (relative
configuration)
(171)

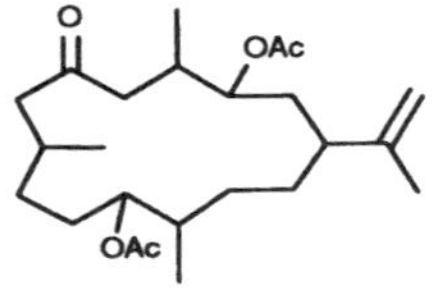

3,11-Diacetoxycembr-
15-en-6-one

Planaxis sulcatus

(257)

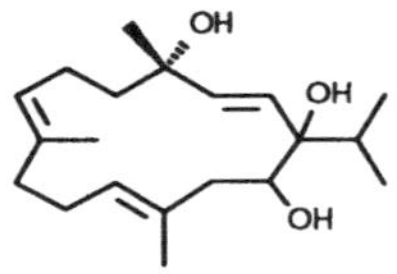

Sarcophytol Q
(2E,4S,7E,11E)-
Cembra-2,7,11-triene-
1,4,14-triol

Sarcophyton glaucum

(194)

189

190

191

192

Table 4 (continued)

Compound		Source	X-ray and spectral studies	Synthesis	Biological activity
Structure no.	Name				
193	Asperdiol (1R,2R,3E,7E,7R,8R,11E)-7,8-Epoxycembra-3,11,15-triene-2,18-diol	Eunicea asperula Eunicea tourneforti	(165) (165) ^{13}C NMR study (258) X-ray analysis of 193 (absolute configuration) (165)	(±)-193 (259–261) (–)-193 (262)	Antitumor activity. ED$_{50}$ of the in vitro KB, PE and LE cell lines are 24, 6 and 6 µg/ml resp. (165)
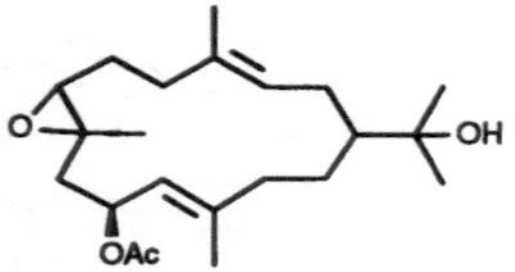 194	(3E,10S,11E)-10-Acetoxy-7,8-epoxycembra-3,11-dien-15-ol	Nephthea brassica	(227)		
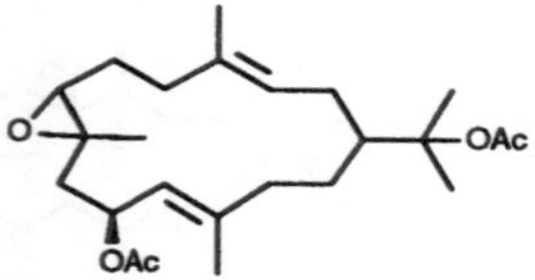 195	(3E,10S,11E)-7,8-Epoxycembra-3,11-diene-10,15-dioyl diacetate	Nephthea brassica	(227)		
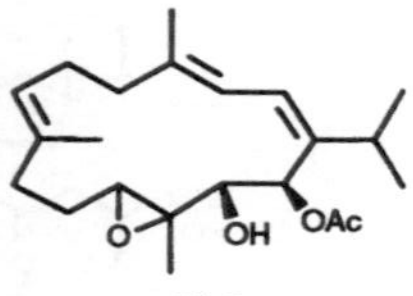 196	Flaccidoxide (1Z,3E,7E,13S,14R)-14-Acetoxy-11,12-epoxycembra-1,3,7-trien-13-ol	Alcyonium flaccidum	(186)		

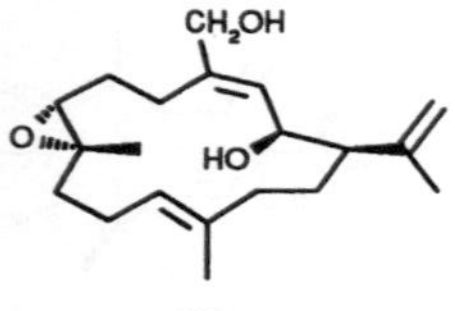

No.	Name	Species	Ref.	Analysis	Bioactivity
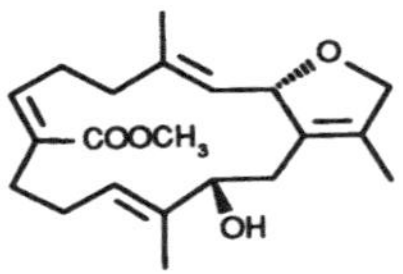 197	Sarcoglaucol Methyl (2S*,3E,7Z,11E,13S*)-2,16-Epoxy-13-hydroxycembra-1(15),3,7,11-tetraen-19-oate	Sarcophyton auritum Sarcophyton glaucum Sarcophyton species	(186) (138) (202)	X-ray analysis of 197 (relative configuration) (138)	Ichtyotoxic: Lethal to the fish Lebistes reticulatus at 15 mg/l (138)
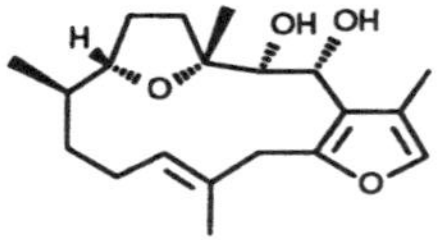 198	Pachyclavulariadiol (2R*,3S*,4R*,7S*,8R*,11E)-4,7:14,16-Diepoxycembra-1(14),11,15-triene-2,3-diol	Pachyclavularia violacea	(225)	X-ray analysis of 198 (relative configuration) (225)	
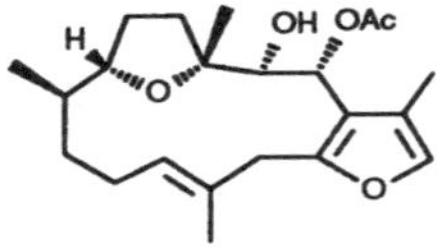 199	Pachyclavulariadiol 2-acetate (2R*,3S*,4R*,7S*,8R*,11E)-2-Acetoxy-4,7:14,16-diepoxycembra-1(14),11,15-trien-3-ol	Pachyclavularia violacea	(225)		
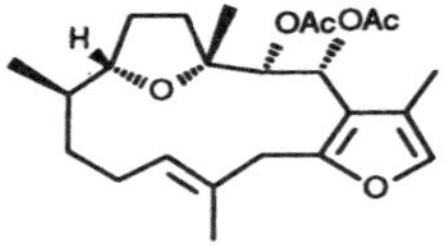 200	Pachyclavulariadiol 2,3-diacetate (2R*,3S*,4R*,7S*,8R*,11E)-4,7:14,16-Diepoxycembra-1(14),11,15-trien-2,3-dioyl diacetate	Pachyclavularia violacea	(225)		

Table 4 (continued)

Compound Structure no.	Name	Source	X-ray and spectral studies	Synthesis	Biological activity
201	(1S,2S,3E,7E,11E)-Cembra-3,7,11,15(17)-tetraen-16,2-olide	Lobophytum michaelae Sinularia mayi	(189, 263) (208, 209)	(±)-201 (264–266) (+)-201 (267)	
202	(1S,2R,3E,7E,11E)-Cembra-3,7,11,15(17)-tetraen-16,2-olide	Sinularia mayi	(208, 209)	(±)-202 (268)	
203	Sarcophinone (2S*,3E,8S*,11E)-7-Oxocembra-1(15),3,11-trien-16,2-olide	Sarcophyton decaryi	(269) X-ray analysis of 203 (relative configuration) (269)		
204	Kericembrenolide A (1R,2S,3E,6S,7E,11E)-6-Acetoxycembra-3,7,11,15(17)-tetraen-16,2-olide	Clavularia koellikeri	(224)		Growth inhibitory effect against B-16 Melanoma cells $IC_{50} = 3.8$ μg/ml (224)

Compound	Name	Source	Ref.	Notes
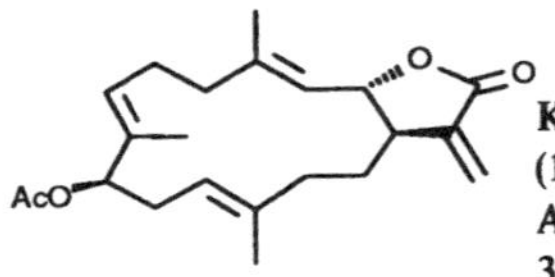 205	Kericembrenolide B (1R,2S,3E,7E,9R,11E)-9-Acetoxycembra-3,7,11,15(17)-tetraen-16,2-olide	Clavularia koellikeri	(224)	Growth inhibitory effect against B-16 Melanoma cells IC$_{50}$ = 2.5 μg/ml (224)
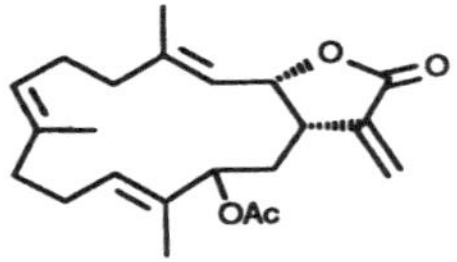 206	(1S,2S,3E,7E,11E)-13-Acetoxycembra-3,7,11,15(17)-tetraen-16,2-olide	Lobophytum denticulatum Sinularia mayi	(180) (180)	
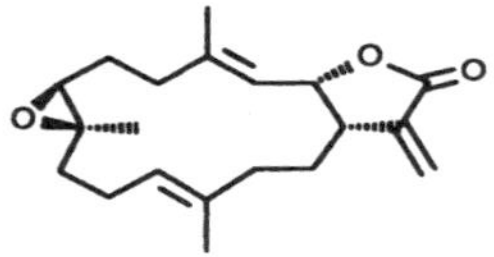 207	(1S,2S,3E,7S,8S,11E)-7,8-Epoxycembra-3,11,15(17)-trien-16,2-olide	Efflatounaria species	(270)	X-ray analysis of 207 (relative configuration) (270)
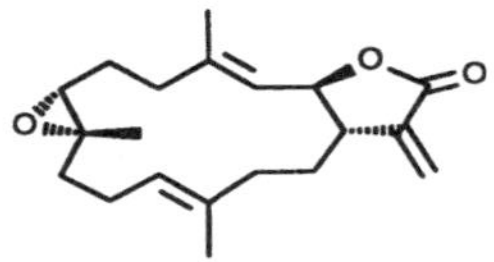 208	(1S,2R,3E,7R,8R,11E)-7,8-Epoxycembra-3,11,15(17)-trien-16,2-olide	Efflatounaria variabilis	(270)	

Table 4 (continued)

Compound Structure no.	Name	Source	X-ray and spectral studies	Synthesis	Biological activity
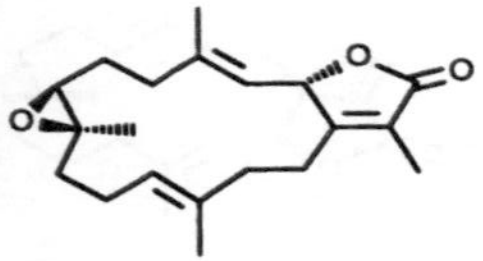 209	(+)-Sarcophine (2S,3E,7S,8S,11E)-7,8-Epoxycembra-1(15),3,11-trien-16,2-olide	Sarcophyton glaucum (139, 196, 197)	X-ray analysis of 209 (relative configuration) (197)		Inhibition of phosphofructokinase (271) Competitive inhibition of cholinesterase in vitro (139) Strong anti-acetylcholine action on isolated guinea pig ileum (139) Ichthyotoxic: Lethal to the fish Gambusia affinis: LD$_{50}$: 3 mg/l after 3 h (139) Toxic to mice, rats and guinea pigs (139)
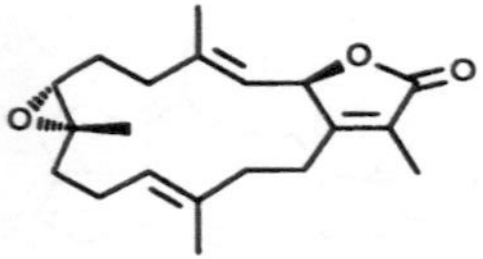 210	(−)-Sarcophine (2R,3E,7R,8R,11E)-7,8-Epoxycembra-1(15), 3,11-trien-16,2-olide	Sarcophyton crassocaule (198, 270) Sarcophyton gemmatum (249)			

Structure	Name	Source	References	Notes	
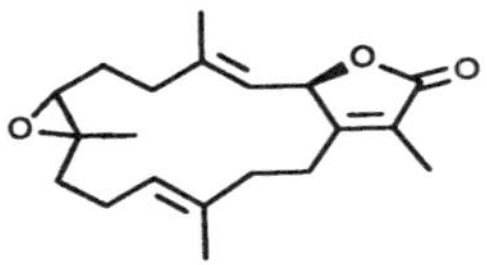 211	2-Episarcophine (2R,3E,11E)-7,8-Epoxy-cembra-1(15),3,11-trien-16,2-olide	*Sarcophyton glaucum*	(196)		
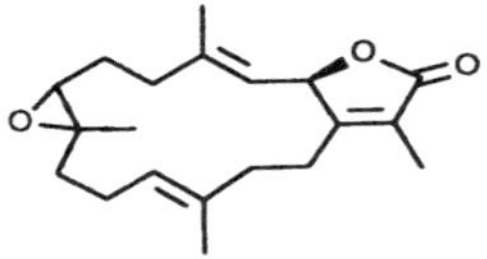 212	2,8-Bisepisarcophine (2R,3E,11E)-7,8-Epoxy-cembra-1(15),3,11-trien-16,2-olide	*Sarcophyton glaucum*	(196)		
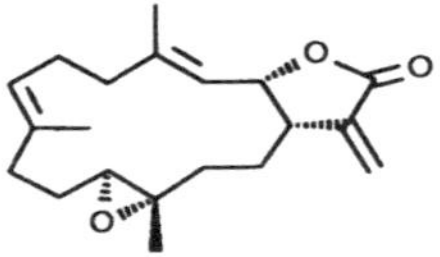 213	(1S,2S,3E,7E,11S,12S)-11,12-Epoxycembra-3,7,15(17)-trien-16,2-olide	*Lobophytum pauciflorum* *Sinularia mayi*	(272) (180, 211)		
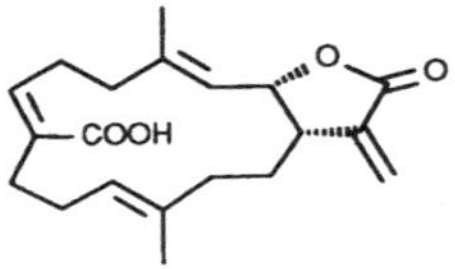 214	Lobohedleolide (1S,2S,3E,7Z,11E)-16,2-Olidylcembra-3,7,11,15(17)-tetraen-19-oic acid	*Lobophytum hedleyi*	(182)	X-ray analysis of the *p*-bromophenacyl ester of **214** (absolute configuration) (182)	Growth inhibitory effect against Hella cells at 5 µg/ml (182)

Table 4 (continued)

Structure no.	Compound Name	Source	X-ray and spectral studies	Synthesis	Biological activity
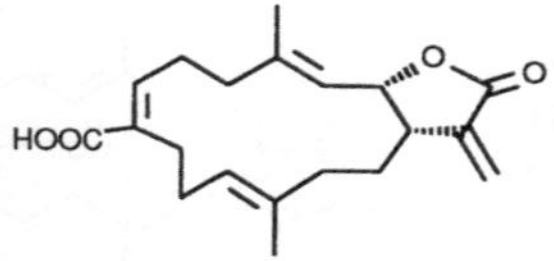 215	(7Z)-Lobohedleolide (1S,2S,3E,7E,11E)-16,2-Olidylcembra-3,7,11,15(17)-tetraen-19-oic acid	Lobophytum hedleyi	(182)		
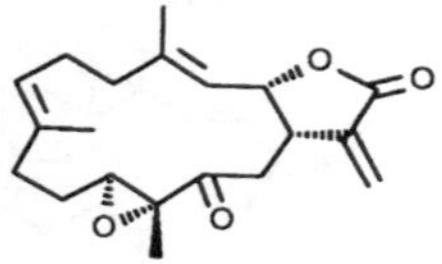 216	(1S,2S,3E,7E,11S,12R)-11,12-Epoxy-13-oxo-cembra-3,7,15(17)-trien-16,2-olide	Lobophytum species Sinularia mayi	(273) (180, 181)		Cytotoxic effect against B16 mouse melanoma cells at 2.1 ppm (181)
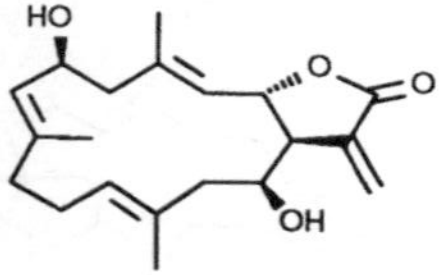 217	Kericembrenolide E (1S,2S,3E,6S,7E,11E, 14S)-6,14-Dihydroxy-cembra-3,7,11,15(17)-tetraen-16,2-olide	Clavularia koellikeri	(224)		Growth inhibitory effect against B16 Melanoma cells: IC_{50} 1.8 μg/ml (224)
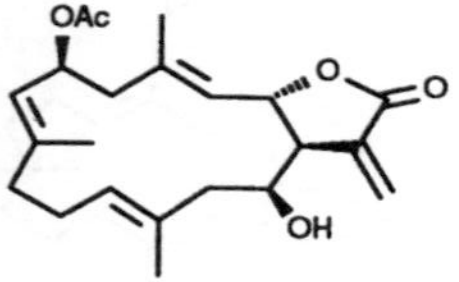 218	Kericembrenolide D (1S,2S,3E,6S,7E,11E, 14S)-6-Acetoxy-14-hydroxycembra-3,7,11,15(17)-tetraen-16,2-olide	Clavularia koellikeri	(224)		Growth inhibitory effect against B16 Melanoma cells: IC_{50} 1.2 μg/ml (224)

Structure	Name	Source	Ref.	Notes
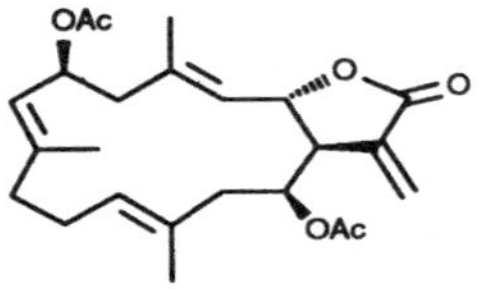 **219**	Kericembrenolide C (1S,2S,3E,6S,7E,11E,14S)-6,14-Diacetoxycembra-3,7,11,15(17)-tetraen-16,2-olide	*Clavularia koellikeri*	(224)	Growth inhibitory effect against **B16 melanoma cells:** IC_{50} 1.3 µg/ml (224)
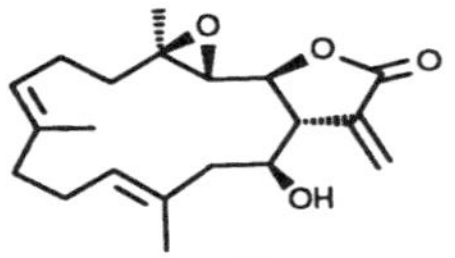 **220**	(1R,2S,3R,4R,7E,11E,14S)-3,4-Epoxy-14-hydroxycembra-7,11,15(17)-trien-16,2-olide	*Lobophytum cristigalli*	(274)	
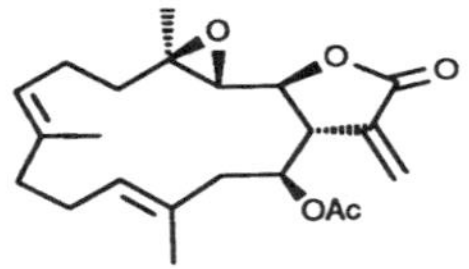 **221**	(1R,2S,3R,4R,7E,11E,14S)-14-Acetoxy-3,4-epoxycembra-7,11,15(17)-trien-16,2-olide	*Lobophytum cristigalli*	(274)	X-ray analysis of **221** (relative configuration) (274)
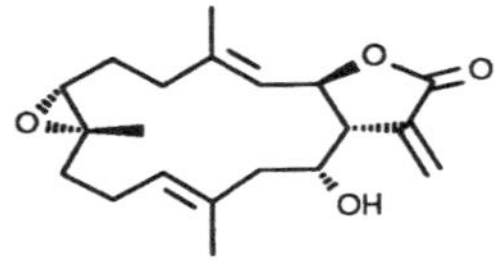 **222**	(1R,2R,3E,7R,8R,11E,14R)-7,8-Epoxy-14-hydroxycembra-3,11,15(17)-trien-16,2-olide	*Cespitularia subviridis*	(275)	X-ray analysis of **222** (relative configuration) (275)

Table 4 (continued)

Structure no.	Compound Name	Source	X-ray and spectral studies	Synthesis	Biological activity
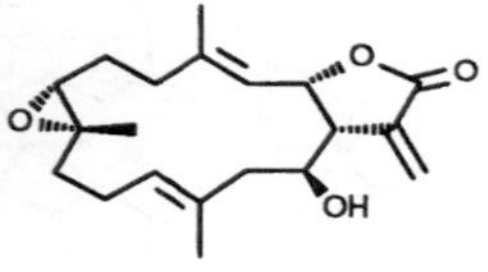 223	(1R,2S,3E,7R,8R,11E,14S)-7,8-Epoxy-14-hydroxycembra-3,11,15(17)-trien-16,2-olide	Cespitularia species	(275)		
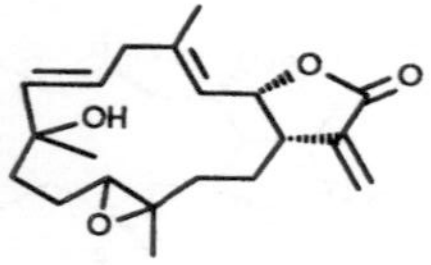 224	Mayolide B (1S,2S,3E,6E)-11,12-Epoxy-8-hydroxycembra-3,6,15(17)-trien-16,2-olide	Simularia mayi	(210)		
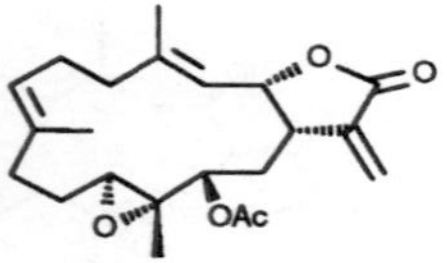 225	(1S,2S,3E,7E,11S,12R,13S)-13-Acetoxy-11,12-epoxycembra-3,7,15(17)-trien-16,2-olide	Lobophytum pauciflorum Lobophytum species Simularia mayi	(272) (276) (180, 181) X-ray analysis of 225 (absolute configuration) (181)		Cytotoxic effect against B16 melanoma at 8.4 ppm (181)
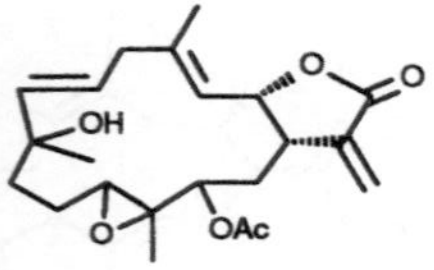 226	Mayolide C (1S,2S,3E,6E)-13-Acetoxy-11,12-epoxy-8-hydroxycembra-3,6,15(17)-trien-16,2-olide	Simularia mayi	(210)		

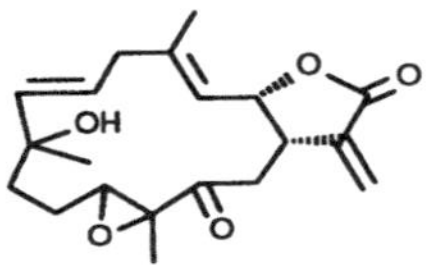

Mayolide D
(1S,2S,3E,6E)-11,12-Epoxy-8-hydroxy-13-oxocembra-3,6,15(17)-trien-16,2-olide

Sinularia mayi

(210)

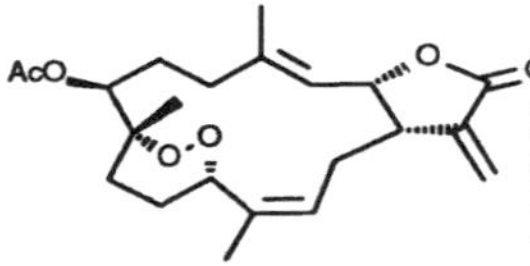

227

Denticulatolide
(1S,2S,3E,7S,8R,11S,12Z)-7-Acetoxy-8,11-epidioxycembra-3,12,15(17)-trien-16,2-olide

Lobophytum denticulatum
Sinularia mayi

(140)
(180, 181)

X-ray analyses of the methanol adduct of **228** (relative configuration) (140) and of the *p*-bromo-benzoate of **228** (absolute configuration) (181) MM Calculations (277)

Ichtyotoxic to Medaka (*Oryzias latipes*) at 5 μg/ml (140)
Cytotoxic effect against B16 melanoma cells at 3.6 ppm (181)

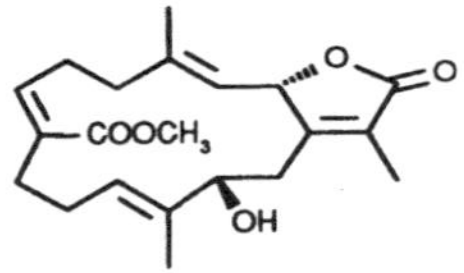

228

Methyl (2S,3E,7Z,11E,13S)-13-Hydroxy-16,2-olidylcembra-3,7,11,15-tetraen-19-oate

Sarcophyton species

(202)

Transient convulsions in mice and potentiated barbiturate induced sleeping (202)

229

Table 4 (continued)

Structure no.	Compound Name	Source	X-ray and spectral studies	Synthesis	Biological activity
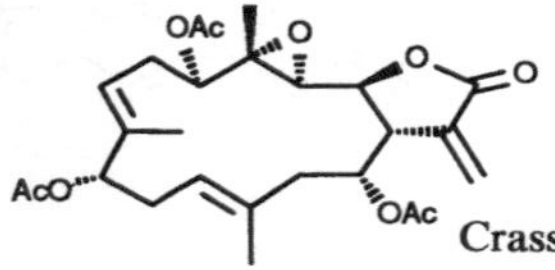 230	Crassolide (1R*,2S*,3S*,4S*,5R*,7E,9S*,11E,14R*)-3,4-Epoxy-5,9,14-triacetoxy-cembra-7,11,15(17)-trien-16,2-olide	Lobophytum crassum	(141, 142) X-ray analysis of a triol obtained by base methanolysis. (relative configuration) (141)		Ichthyotoxic to the fish Lebistes reticulatus: LD_{50}: 7 mg/l (141)
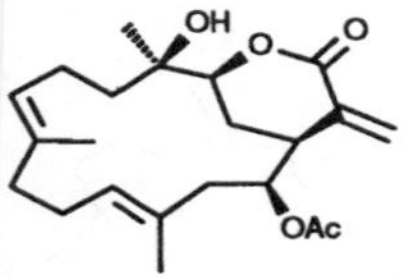 231	Crassin Acetate (1S,3S,4R,7E,11E,14S)-14-Acetoxy-4-hydroxycembra-7,11,15(17)-trien-16,3-olide	Eunicea calyculata Pseudoplexaura crucis Pseudoplexaura flagellosa Pseudoplexaura porosa Pseudoplexaura wagenaari	(278) (172) (157, 172) (157, 172) (157, 172) X-ray analyses of the benzenethiol 1,4 adduct of crassin (relative configuration) (279) and of the p-iodobenzoate of crassin (absolute configuration) (280) ^{13}C NMR study (281)	(±)Crassin (282, 283) Methyl ether of (±)-231 (284)	Toxic to the rotifer Brachionus plicatilis and the amphipod Parhyale hawaiensis (285) Affects the motility of marine flagellates (286) Antibiotic effect (157) Antineoplastic activity (172)

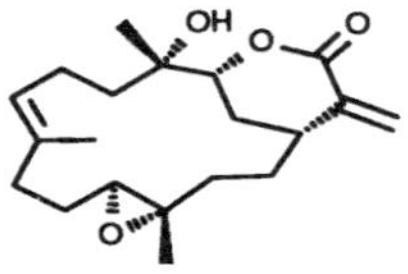 **232**	Sinularin (Flexibilide) (1*R*,3*R*,4*S*,7*E*,11*S*,12*S*)-11,12-Epoxy-4-hydroxycembra-7,15(17)-dien-16,3-olide	*Sinularia flexibilis*	(*212*, *213*, *217*)	X-ray analysis of **232** (absolute configuration) (*212*, *213*, *287*) Conformational and ^{13}C NMR studies (*288*, *289*)	Inhibitory effect against KB and PS cell lines ED_{50} (KB): 0.3 μg/ml ED_{50} (PS): 0.3 μg/ml (*212*) Anti-inflammatory and anti-arthritic activity in the rat (*289*)
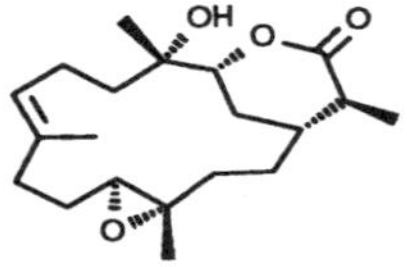 **233**	Dihydrosinularin (Dihydroflexibilide) (1*R*,3*R*,4*S*,7*E*,11*S*,12*S*,15*S*)-11,12-Epoxy-4-hydroxycembra-7-en-16,3-olide	*Sinularia flexibilis* *Planaxis sulcatus*	(*212*, *213*, *217*) (*290*)	X-ray analysis of **233** (absolute configuration) (*212*, *287*)	Inhibitory effect against KB and PS cell lines ED_{50} (KB): 16 μg/ml ED_{50} (PS): 1.1 μg/ml (*212*)
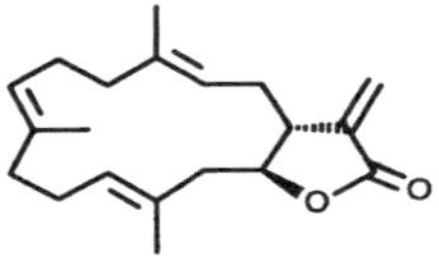 **234**	(1*R*,3*E*,7*E*,11*E*,14*S*)-Cembra-3,7,11,15(17)-tetraen-16,14-olide	*Lobophytum crassospiculatum*	(*189*)		
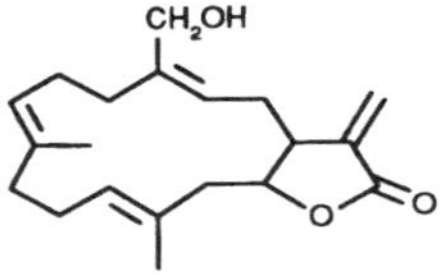 **235**	(3*Z*,7*E*,11*E*)-18-Hydroxycembra-3,7,11,15(17)-tetraen-16,14-olide	*Lobophytum crassum*	(*187*)		

Table 4 (continued)

Structure no.	Compound Name	Source	X-ray and spectral studies	Synthesis	Biological activity
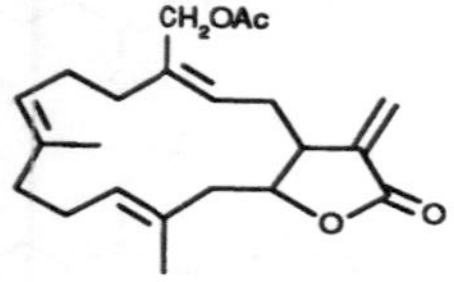 236	(3Z,7E,11E)-18-Acetoxycembra-3,7,11,15(17)-tetraen-16,14-olide	Lobophytum crassum	(187)		
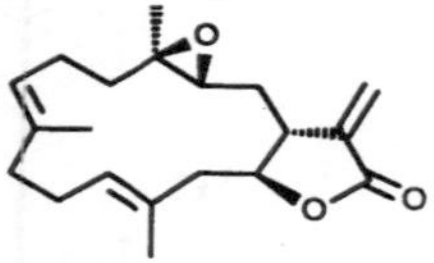 237	Isolobophytolide (1R*,3R*,4R*,7E,11E,14S*)-3,4-Epoxycembra-7,11,15(17)-trien-16,14-olide	Lobophytum crassum Lobophytum crassospiculatum	X-ray analysis of 237 (relative configuration) (191) (188) (189)	(±)-237 (283, 291, 292)	Effect against murine P-338 lymphocytic leukemia (283)
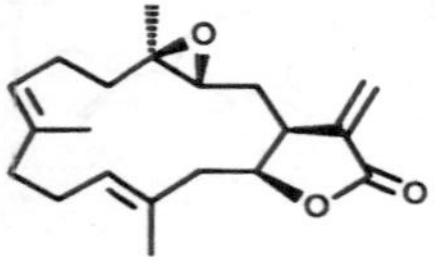 238	Euniolide (1-Epiisolobophytolide) (1S*,3R*,4R*,7E,11E,14S*)-3,4-Epoxycembra-7,11,15(17)-trien-16,14-olide	Eunicea mammosa Eunicea succinea	(154) (154)	(1S)-238 (173)	Effect against CHO-K1 cells: ED$_{50}$: 3.80 µg/ml (154) Active against P388 lymphocytic leukemia in mice (173)
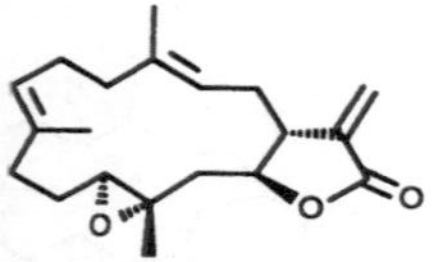 239	Lobophytolide (1R,3E,7E,11S,12S,14S)-11,12-Epoxycembra-3,7,15(17)-trien-16,14-olide	Lobophytum cristagalli	X-ray analysis of 239 (absolute configuration) (293) (183)		Ichthyotoxic to the fish Lebistes reticulatus: LD$_{50}$: 12 mg/l (142)

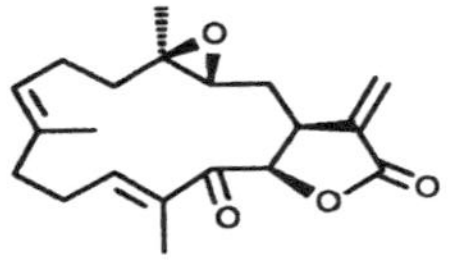

Structure	Name	Source	References	Activity
240	Peunicin (1S,3R,4R,7E,11E,14R)-3,4-Epoxy-13-oxo-cembra-7,11,15(17)-trien-16,14-olide	Eunicea succinea	(163) X-ray analysis of **240** (absolute configuration) (164)	Inhibitory effect against KB and PS cell lines ED$_{50}$ (KB): 30 µg/ml ED$_{50}$ (PS): 32 µg/ml (164)
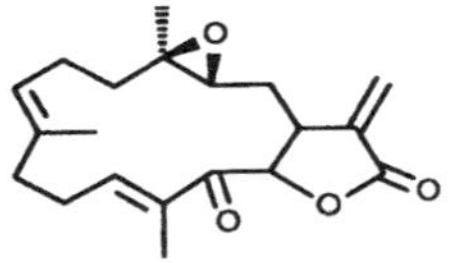 **241**	Epipeunicin (3R*,4R*,7E,11E)-3,4-Epoxy-13-oxocembra-7,11,15(17)-trien-16,14-olide	Eunicea succinea	(163)	
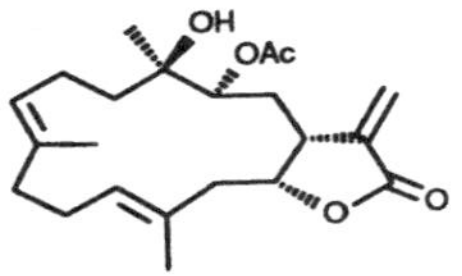 **242**	Styeolide (1R*,3S*,4R*,7E,11E,14R*)-3-Acetoxy-4-hydroxycembra-7,11,15(17)-trien-16,14-olide	Styela plicata	(294)	
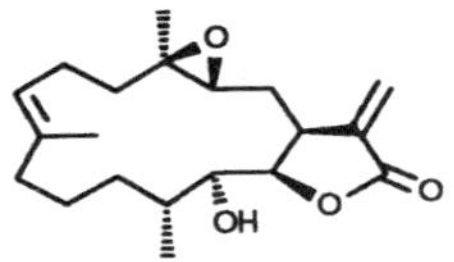 **243**	12,13-Bisepieupalmerin (1S,3R,4R,7E,12R,13S,14R)-3,4-Epoxy-13-hydroxycembra-7,15(17)-dien-16,14-olide	Eunicea succinea	(154, 155) X-ray analysis of **243** (absolute configuration) (155)	Effect against CHO-K1 cells: ED$_{50}$: 4.80 µg/ml (154)

Table 4 (continued)

Structure no.	Compound Name	Source	X-ray and spectral studies	Synthesis	Biological activity
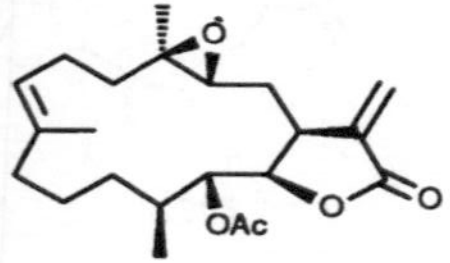 244	Eupalmerin acetate (1S,3R,4R,7E,12S,13R,14R)-13-Acetoxy-3,4-epoxycembra-7,15(17)-dien-16,14-olide	Eunicea mammosa Eunicea succinea Eunicea palmeri	2D NMR study (295) X-ray analyses of 244 (296) and of the dibromide of 244 (absolute configuration) (296, 297) (154, 162) (162) (161)		Immobilizes dinoflagellates at a concentration of 5 ppm (286) Effect against CHO-K1 cells: ED$_{50}$: 9.30 µg/ml (154)
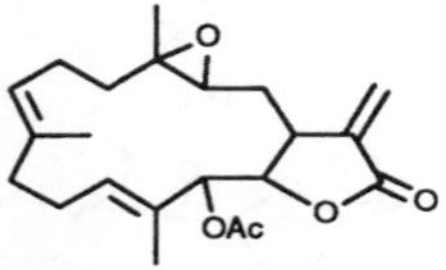 245	(7E,11E)-13-Acetoxy-3,4-epoxycembra-7,11,15(17)-trien-16,14-olide	Lobophytum crassospiculatum	(189)		
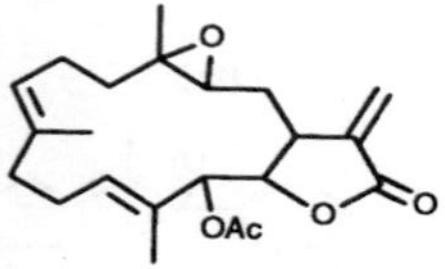 246	(7E,11E)-13-Acetoxy-3,4-epoxycembra-7,11,15(17)-trien-16,14-olide	Lobophytum crassospiculatum	(189)		

Structure	Name	Organism	References	Studies	Activity
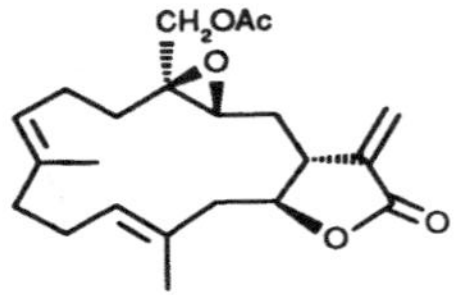 **247**	Lobolide (1R*,3R*,4S*,7E,11E,14S*)-18-Acetoxy-3,4-epoxycembra-7,11,15(17)-trien-16,14-olide	Lobophytum crassum	(185–187, 246)	X-ray analysis of (±)-**247** (152)	Ichthyotoxic effect (185)
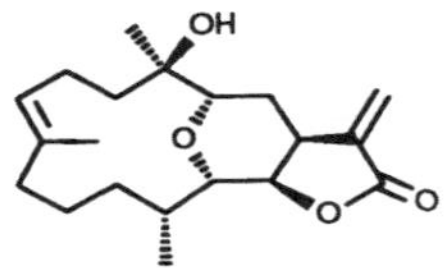 **248**	Eunicin (1S,3S,4R,7E,12R,13S,14R)-3,13-Epoxy-4-hydroxycembra-7,15(17)-dien-16,14-olide	Eunicea mammosa Eunicea succinea	(156, 157) (154, 155)	X-ray analysis of the iodo-acetate of **248** (absolute configuration) (298) ^{13}C NMR study (299)	Immobilizes dinoflagellates at a concentration of 5 ppm (286) Effect against CHO-K1 cells: ED$_{50}$: 4.14 μg/ml (154) Antibiotic effect (156) Toxic to the ciliate Tetrahymena pyriformis (159)
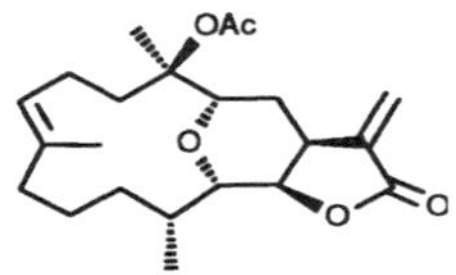 **249**	Eunicin acetate (1S,3S,4R,7E,12R,13S,14R)-4-Acetoxy-3,13-epoxycembra-7,15(17)-dien-16,14-olide	Eunicea succinea	(155)		
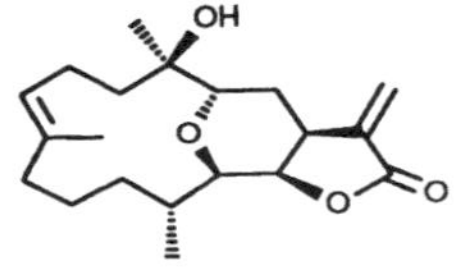 **250**	Cueunicin (1S,3S,4R,7E,12R,13R,14R)-3,13-Epoxy-4-hydroxycembra-7,15(17)-dien-16,14-olide	Eunicea mammosa	(160)		

Table 4 (continued)

Structure no.	Compound Name	Source	X-ray and spectral studies	Synthesis	Biological activity
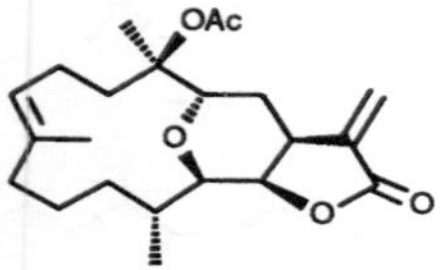 251	Cueunicin acetate (1S,3S,4R,7E,12R,13R,14R)-4-Acetoxy-3,13-epoxycembra-7,15(17)-dien-16,14-olide	Eunicea mammosa	X-ray analysis of 251 (relative configuration) (300)	(160)	
252	Jeunicin (1S,3R,4S,7E,12R,13S,14R)-4,13-Epoxy-3-hydroxycembra-7,15(17)-dien-16,14-olide	Eunicea mammosa Eunicea succinea Planaxis sulcatus	2D NMR study (158) X-ray analysis of the iodobenzoate of 252 (absolute configuration) (159)	(159) (155) (158)	Toxic to the ciliate Tetrahymena pyriformis (159) Cytotoxic to the KB cell line (159)
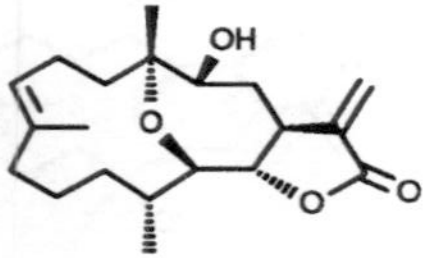 253	13,14-Bisepijeunicin (1S,3R,4S,7E,12R,13R,14S)-4,13-Epoxy-3-hydroxycembra-7,15(17)-dien-16,14-olide	Eunicea mammosa	(4)		
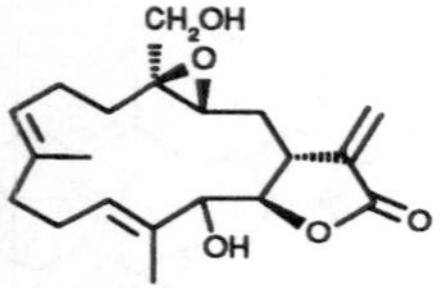 254	(1R*,3R*,4S*,7E,11E,14R*)-13,18-Dihydroxy-3,4-epoxycembra-7,11,15(17)-trien-16,14-olide	Lobophytum crassum	(187)		

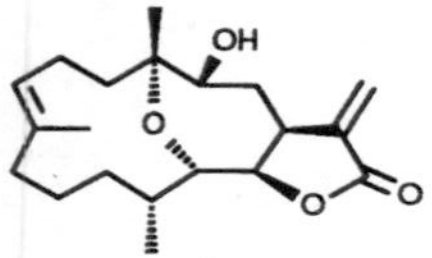

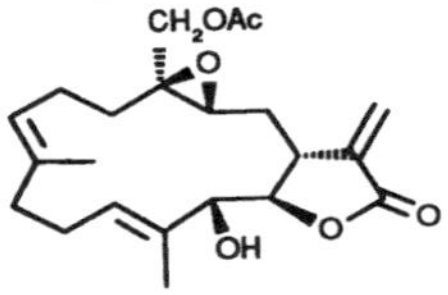

13-Hydroxylobolide
(1R*,3R*,4S*,7E,11E,13R*,14R*)-18-Acetoxy-3,4-epoxy-13-hydroxy-cembra-7,11,15(17)-trien-16,14-olide

Labophytum crassum

(152, 186)

13-Hydroxylobolide
(1R*,3R*,4S*,7E,11E,13S*,14R*)-18-Acetoxy-3,4-epoxy-13-hydroxy-cembra-7,11,15(17)-trien-16,14-olide

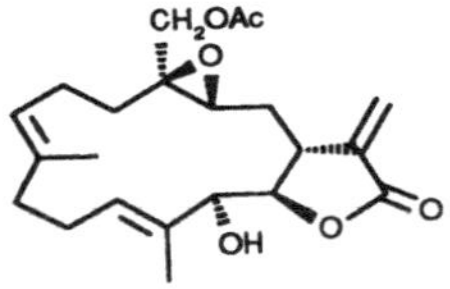

Lobophytum crassum

(152, 186)

Pachyclavulariolide
(2R*,3S*,4R*,7S*,8R*,11E,14S*)-2,3-Dihydroxy-4,7-epoxycembra-1(15),11-dien-16,14-olide

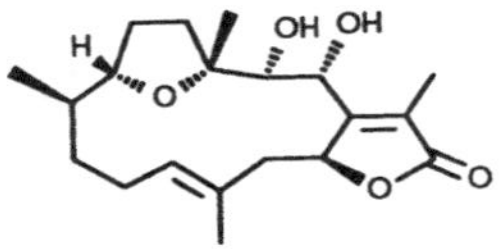

Pachyclavularia violacea

(226)   MM Calculations (226)

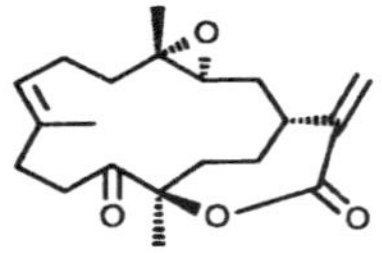

11-Dehydrosinulariolide
(1R,3S,4S,7E,12R)-3,4-Epoxy-11-oxocembra-7,15(17)-dien-16,12-olide

Simularia flexibilis
Simularia notanda

(213, 216, 217)
(246)

Table 4 (continued)

Structure no.	Compound Name	Source	X-ray and spectral studies	Synthesis	Biological activity
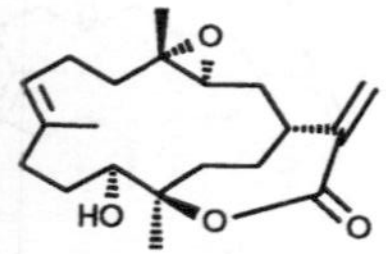 259	Sinulariolide (1R,3S,4S,7E,11S,12R)-3,4-Epoxy-11-hydroxycembra-7,15(17)-dien-16,12-olide	Sinularia flexibilis	(212–214, 216, 217, 246) X-ray analysis of 259 (absolute configuration) (215)		Inhibitory effect against KB and PS cell lines ED$_{50}$ (KB): 20 µg/ml ED$_{50}$ (PS): 7.0 µg/ml (212)
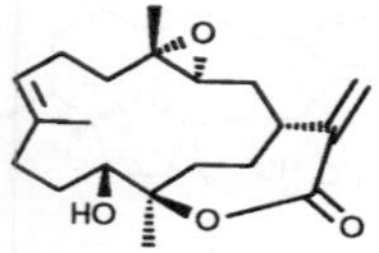 260	11-Episinulariolide (1R,3S,4S,7E,11R,12R)-3,4-Epoxy-11-hydroxycembra-7,15(17)-dien-16,12-olide	Sinularia flexibilis Planaxis sulcatus	(217) (290)		
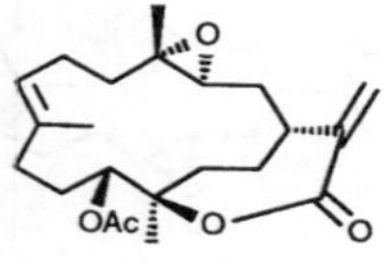 261	11-Episinulariolide acetate (1R,3S,4S,7E,11R,12R)-11-Acetoxy-3,4-epoxycembra-7,15(17)-dien-16,12-olide	Sinularia flexibilis Sinularia notanda Sinularia querciformis	(4, 216, 217) (246) (246) X-ray analysis of 261 (relative configuration) (4)		
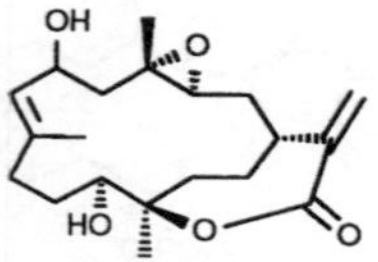 262	6-Hydroxysinulariolide (1R,3S,4S,7E,11S,12R)-6,11-Dihydroxy-3,4-epoxycembra-7,15(17)-dien-16,12-olide	Sinularia flexibilis	(216)		

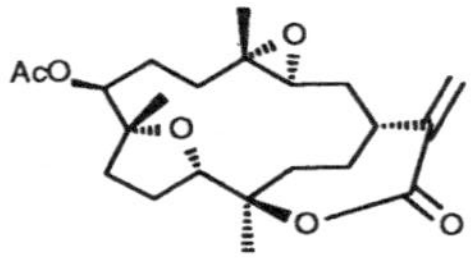

Structure	Name	Source	Reference / Notes
263	(1R,3S,4S,7S,8R,11S, 12R)-7-Acetoxy-3,4:8,11-diepoxy-cembr-15(17)-en-16,12-olide	*Sinularia flexibilis*	(217) Cytotoxic effect against DBA 1 MC fibrosarcoma at 100 μg/ml (217)
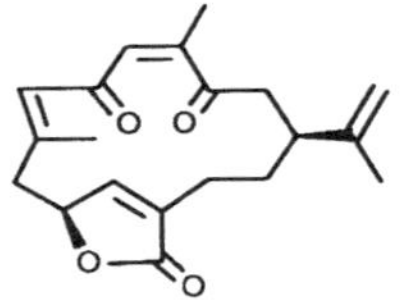 **264**	Lophodione (1S*,4Z,7E,10S*)-3,6-Dioxocembra-4,7,11,15-tetraen-20,10-olide	*Lophogorgia alba*	(166) X-ray analysis of **264** (relative configuration) (166)
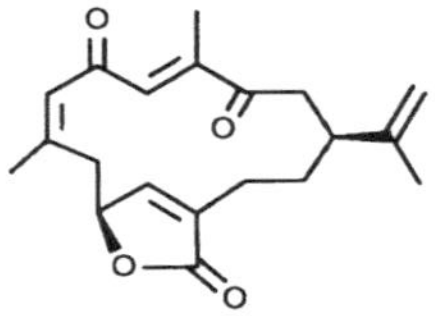 **265**	Isolophodione (1S*,4E,7Z,10S*)-3,6-Dioxocembra-4,7,11,15-tetraen-20,10-olide	*Lophogorgia alba*	(166)
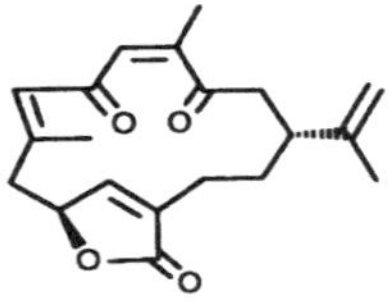 **266**	Epilophodione (1R*,4Z,7E,10S*)-3,6-Dioxocembra-4,7,11,15-tetraen-20,10-olide	*Gersemia rubiformis*	(178)

Table 4 (continued)

Compound Structure no.	Name	Source	X-ray and spectral studies	Synthesis	Biological activity
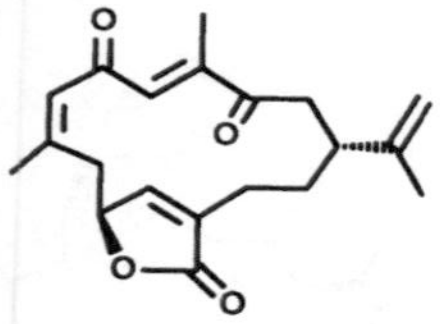 267	Isoepilophodione A (1R*,4E,7Z,10S*)-3,6-Dioxocembra-4,7,11,15-tetraen-20,10-olide	Gersemia rubiformis	(179)		
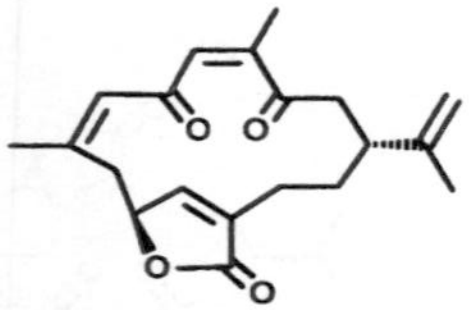 268	Isoepilophodione B (1R*,4Z,7Z,10S*)-3,6-Dioxocembra-4,7,11,15-tetraen-20,10-olide	Gersemia rubiformis	(179)		
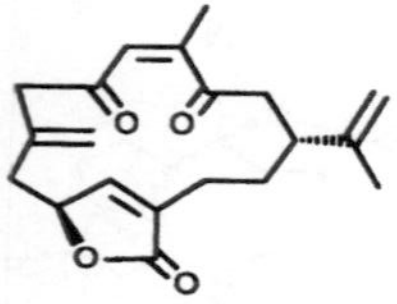 269	Isoepilophodione C (1R*,4Z,10S*)-3,6-Dioxocembra-4,8(19),11,15-tetraen-20,10-olide	Gersemia rubiformis	(179)		

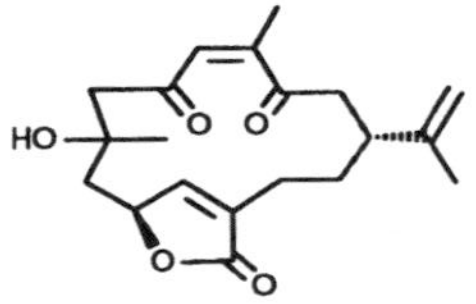 **270**	Rubifol (1*R**,4*Z*,10*S**)-3,6-Dioxo-8-hydroxy-cembra-4,11,15-trien-20,10-olide	*Gersemia rubiformis*	(*179*)
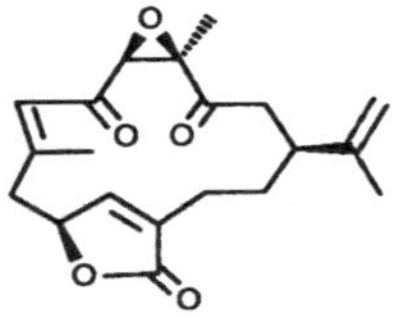 **271**	Epoxylophodione (1*S**,4*R**,5*S**,7*E*,10*S**)-3,6-Dioxo-4,5-epoxycembra-7,11,15-trien-20,10-olide	*Lophogorgia alba*	(*166*)
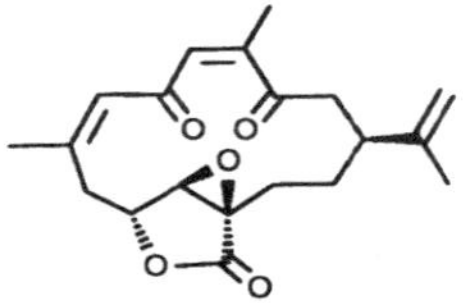 **272**	Coralloidolide E (1*S**,4*Z*,7*Z*,10*R**,11*R**,12*R**)-3,6-Dioxo-11,12-epoxycembra-4,7,15-trien-20,10-olide	*Alcyonium coralloides*	(*177*)
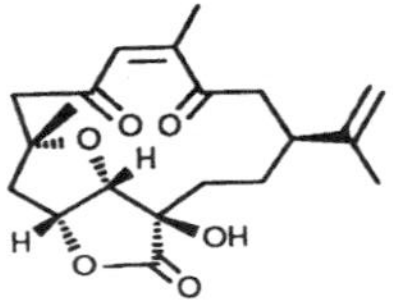 **273**	Coralloidolide D (1*S**,4*Z*,8*R**,10*R**,11*S**,12*R**)-3,6-Dioxo-8,11-epoxy-12-hydroxy-cembra-4,15-dien-20,10-olide	*Alcyonium coralloides*	(*177*)

Table 4 (continued)

Structure no.	Compound Name	Source	X-ray and spectral studies	Synthesis	Biological activity
274	(1S,2R,3S,4Z,7Z,10R,11R, 12S,13S)-2,3-Dihydroxy-6-oxo-3,13:11,12:15,16-triepoxycembra-4,7-dien-20,10-olide	Pseudopterogorgia bipinnata	(174)		Anti-inflammatory effect (174)
275	Methyl (1R,2R,3S,4Z,7Z,10R, 11R,12S,13S)-2-Acetoxy-3,13:11,12-diepoxy-3-hydroxy-20,10-olidyl-6-oxocembra-4,7,15(17)-trien-16-oate	Pseudopterogorgia bipinnata	(174)		Anti-inflammatory effect (174)
276	Coralloidolide B (1S*,3S*,6S*,7Z,10R*, 11S*,12R*)-3,6:6,11-Diepoxy-3,12-dihydroxy-cembra-4,7,15-trien-20,10-olide	Alcyonium coralloides	(176)		

Structure	Name	Organism	Ref.	Activity
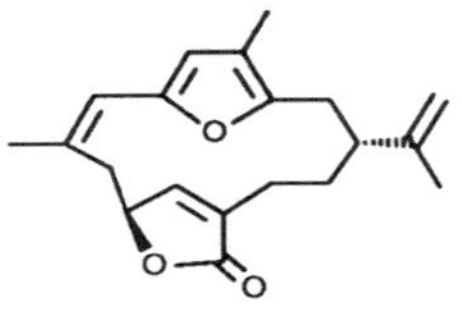	Rubifolide (1R*,7Z,10S*)-3,6-Epoxycembra-3,5,7,11,15-pentaen-20,10-olide	*Gersemia rubiformis* *Tochuina tetraquetra*	(178) (301)	
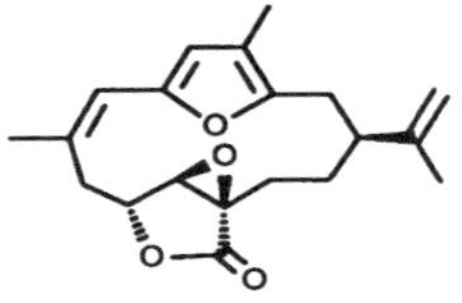	Coralloidolide A (1S*,7Z,10R*,11R*,12R*)-3,6:11,12-Diepoxycembra-3,5,7,15-tetraen-20,10-olide	*Alcyonium coralloides*	(176)	
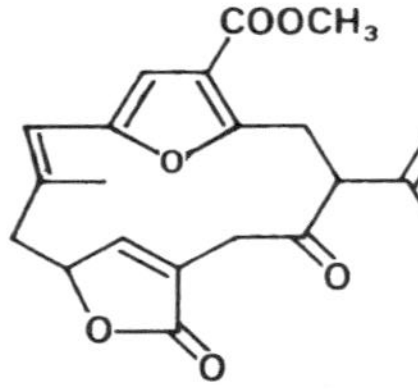	Acerosolide Methyl 3,6-Epoxy-20,10-olidyl-14-oxo-cembra-3,5,7,11,15-pentaen-18-oate	*Pseudopterogorgia acerosa*	(302, 317)	
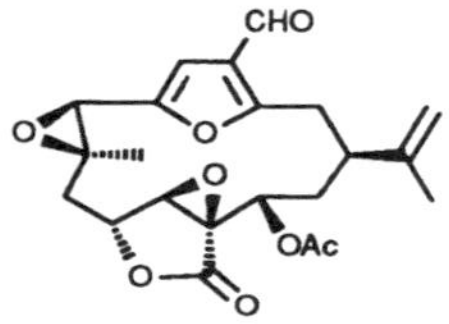	Lophotoxin (1S*,7R*,8S*,10R*,11R*,12R*,13S*)-13-Acetoxy-3,6:7,8:11,12-triepoxy-18-oxocembra-3,5,15-trien-20,10-olide	*Lophogorgia alba* *Lophogorgia chilensis* *Lophogorgia cuspidata* *Lophogorgia rigida*	(166, 167) (167) (167) (167)	Neuromuscular toxin (167) Lethal to mice on subcutaneous injection LD$_{50}$: 8.0 µg/g (167) Inactivates the nicotinic acetylcholine receptor (303)

Table 4 *(continued)*

Compound Structure no.	Name	Source	X-ray and spectral studies	Synthesis	Biological activity
281	Pukalide Methyl (1S*,7R*,8S*,10R*)-3,6:7,8-Diepoxy-20,10-olidylcembra-3,5,11,15-tetraen-18-oate	*Lophogorgia rigida* *Sinularia abrupta* *Sinularia polydactyla* *Tochuina tetraquetra* Eggs spawned by various *Sinularia* species	*(167)* *(304)* *(240)* *(301)* *(144)*	X-ray analysis of **281** (relative configuration) *(144, 167)*	
282	11β,12β-Epoxypukalide Methyl (1S*,7R*,8S*,10R*,11R*,12R*)-20,10-Olidyl-3,6:7,8:11,12-triepoxycembra-3,5,15-trien-18-oate	*Leptogorgia setacea* *Placogorgia* species Eggs spawned by various *Sinularia* species	*(168)* *(169)* *(144)*	X-ray analysis of **282** (relative configuration) *(144)*	
283	13α-Acetoxy-11β,12β-epoxypukalide Methyl (1S*,7R*,8S*,10R*,11R*,12R*,13S*)-13-Acetoxy-20,10-olidyl-3,6:7,8:11,12-triepoxycembra-3,5,15-trien-18-oate	*Lophogorgia rigida* *Sinularia polydactyla*	*(167)* *(240)*		

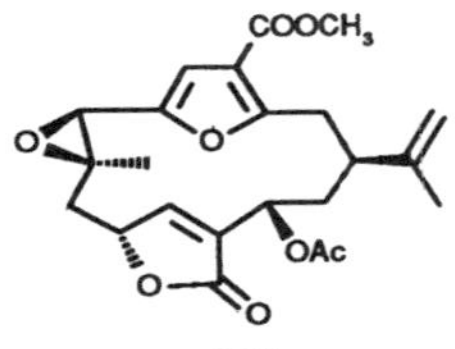

Effect against
the P388 murine
cell line with
IC$_{50}$: 0.9 μg/ml
(175)

(240)

(240)

(240)

(175)

Sinularia polydactyla

Sinularia polydactyla

Sinularia polydactyla

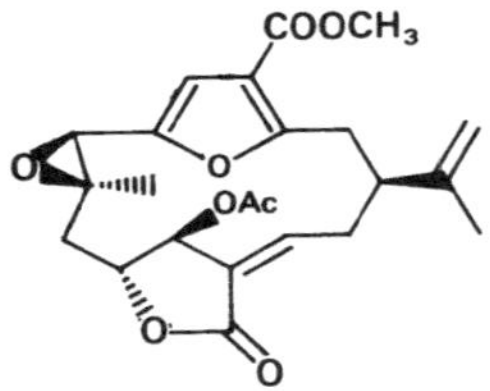

*Pseudopterogorgia
bipinnata*

13α-Acetoxypukalide
Methyl (1S*,7R*,8S*,
10R*,13S*)-13-Acetoxy-
3,6:7,8-diepoxy-20,10-
olidylcembra-3,5,11,15-
tetraen-18-oate

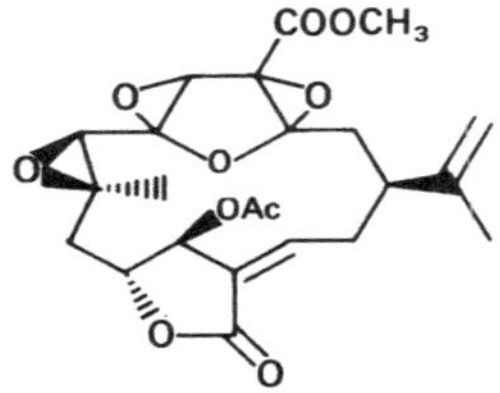

Methyl (1S*,7R*,8S*,
10R*,11S*,12Z)-11-
Acetoxy-3,6:7,8-diepoxy-
20,10-olidylcembra-3,5,
12,15-tetraen-18-oate

Methyl (1S*,7R*,8S*,
10R*,11S*,12Z)-11-
Acetoxy-20,10-olidyl-
3,4:3,6:5,6:7,8-tetra-
epoxycembra-12,15-
dien-18-oate

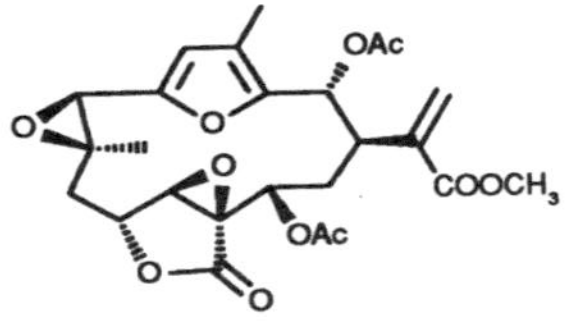

Bipinnatin a
Methyl (1R*,2R*,7R*,
8S*,10R*, 11R*,12R*,
13S*)-2,13-Diacetoxy-
20,10-olidyl-3,6:7,8:11,12-
triepoxycembra-
3,5,15(17)-trien-16-oate

284

285

286

287

Table 4 (*continued*)

Compound Structure no.	Name	Source	X-ray and spectral studies	Synthesis	Biological activity
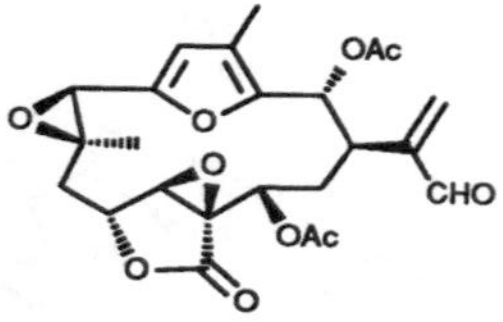 288	Bipinnatin b (1R*,2R*,7R*,8S*,10R*, 11R*,12R*,13S*)-2,13-Diacetoxy-16-oxo-3,6:7,8:11,12-triepoxy-cembra-3,5,15(17)-trien-20,10-olide	*Pseudopterogorgia bipinnata*	(175)		Effect against the P388 murine cell line with IC$_{50}$: 3.2 μg/ml (175)
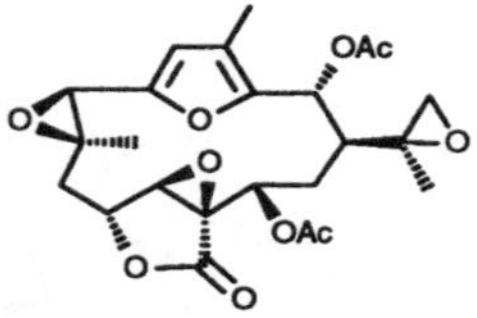 289	Bipinnatin c (1S*,2R*,7R*,8S*,10R*, 11R*,12R*,13S*,15S*)-2,13-Diacetoxy-3,6:7,8:11,12:15,16-tetraepoxycembra-3,5-dien-20,10-olide	*Pseudopterogorgia bipinnata*	(175) X-ray analysis of **289** (relative configuration) (175)		Effect against the P388 murine cell line with IC$_{50}$: 46.6 μg/ml (175)
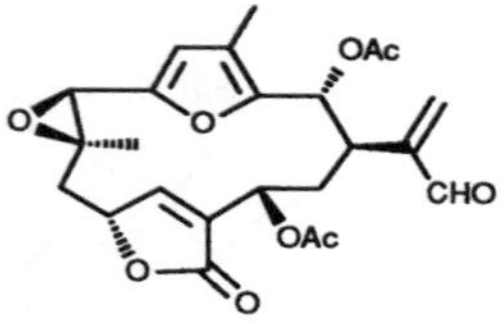 290	Bipinnatin d (1R*,2R*,7R*,8S*,10R*, 13S*)-2,13-Diacetoxy-16-oxo-3,6:7,8-diepoxy-cembra-3,5,11,15(17)-tetraen-20,10-olide	*Pseudopterogorgia bipinnata*	(175)		Effect against the P388 murine cell line with IC$_{50}$: 1.5 μg/ml (175)

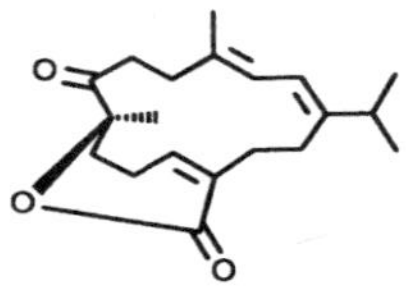

Sarcophytolide (1E,3E,8S,11Z)-7-Oxocembra-1,3,11-trien-20,8-olide	*Sarcophyton elegans* *Sarcophyton glaucum*	(205) (203)

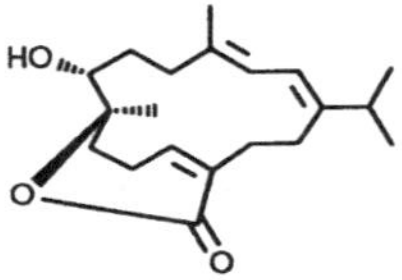

291

(1E,3E,7R,8S,11Z)-7-Hydroxycembra-1,3,11-trien-20,8-olide	*Sarcophyton glaucum*	(203)

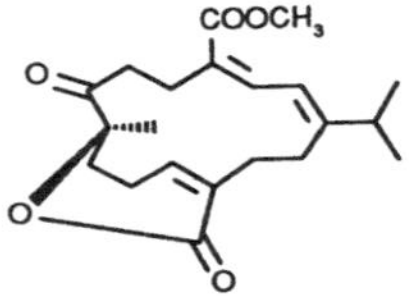

292

Ketoemblide Methyl (1E,3Z,8S*,11Z)-20,8-Olidyl-7-oxo-cembra-1,3,11-trien-18-oate	*Sarcophyton elegans*	(205)

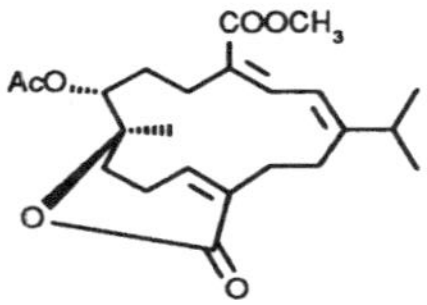

293

Emblide Methyl (1E,3Z,7R*,8S*,11Z)-7-Acetoxy-20,8-olidylcembra-1,3,11-trien-18-oate	*Sarcophyton glaucum*	(204) X-ray analysis of **294** (relative configuration) (204)

294

Table 4 (continued)

Compound Structure no.	Name	Source	X-ray and spectral studies	Synthesis	Biological activity
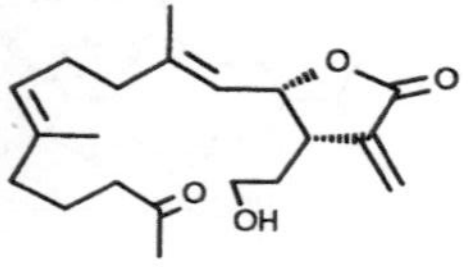 **295**	Mayolide A (3S,4S,5E,9E)-6,10-Dimethyl-3-(1-hydroxy-2-ethyl)-2-methylen-14-oxopentadeca-5,9-dien-4-olide	Sinularia mayi	(210)	(+)-**295** (305)	
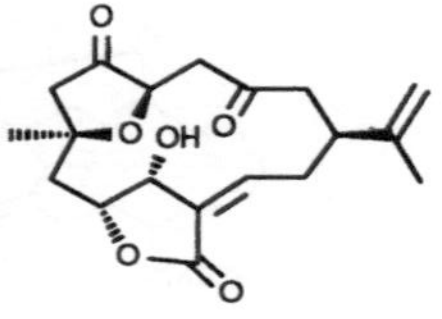 **296**	(1S*,5R*,8S*,10R*,11R*,12Z)-18-Nor-3,6-dioxo-5,8-epoxy-11-hydroxycembra-12,15-dien-20,10-olide	Sinularia foeta Sinularia gyrosa Sinularia leptocladus Sinularia numerosa Sinularia quersiformis	(223) (222) (221, 222) (220) (220)	X-ray analysis of **296** (relative configuration) (221)	
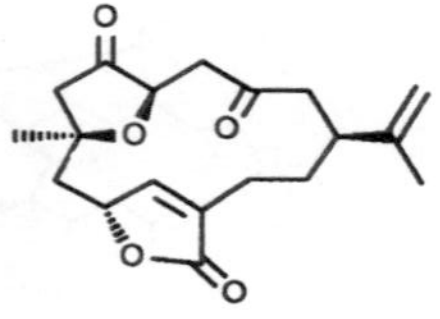 **297**	(1S*,5R*,8S*,10R*)-18-Nor-3,6-dioxo-5,8-epoxycembra-11,15-dien-20,10-olide	Sinularia numerosa Sinularia species	(220) (220)	X-ray analysis of **297** (relative configuration) (220)	
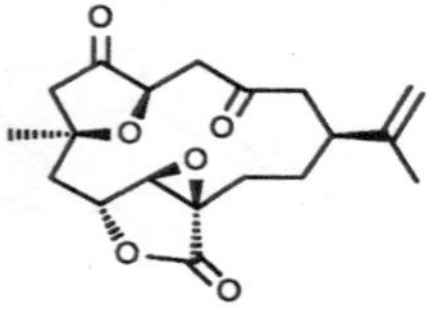 **298**	(1S*,5R*,8S*,10R*,11R*-12R*)-18-Nor-5,8:11,12-diepoxy-3,6-dioxocembr-15-en-20,10-olide	Sinularia inelegans	(220)	X-ray analysis of **298** (relative configuration) (220)	

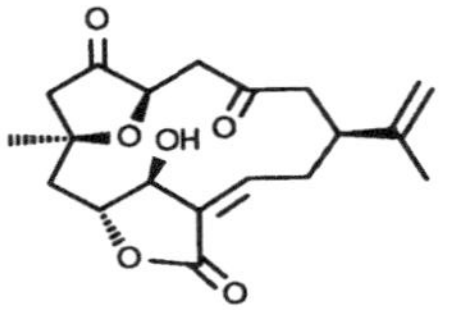

(1S*,5R*,8S*,10R*,11S*,12Z)-18-Nor-3,6-dioxo-5,8-epoxy-11-hydroxy-cembra-12,15-dien-20,10-olide	*Simularia inelegans*	(220)
299		
(5R*,8S*,10R*,11R*,12Z)-18-Nor-3,6-dioxo-5,8-epoxy-11-hydroxy-cembra-1(15),12-dien-20,10-olide	*Simularia leptocladus*	(222)
300		
Methyl (1S*,5R*,8S*,12Z)-18-Nor-5,8-epoxy-3,6,10-trioxocembra-12,15-dien-20-oate	*Simularia quersiformis*	(220)
301		
Flexibilene 1,1,5,9,13-Pentamethyl-cyclopentadeca-2E,5E,9E,13E-tetraene	*Simularia flexibilis*	(213, 217, 218)
	Simularia conferta	(219)
302		(±)-**302** (306, 312)

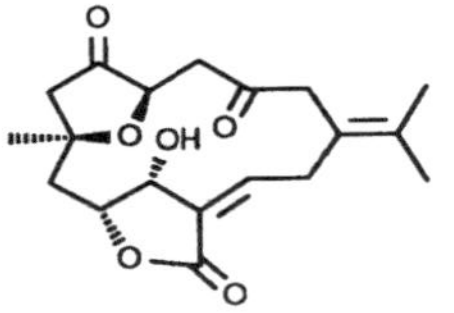

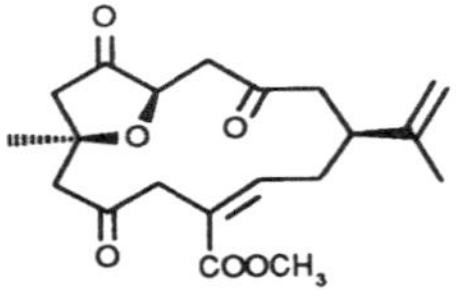

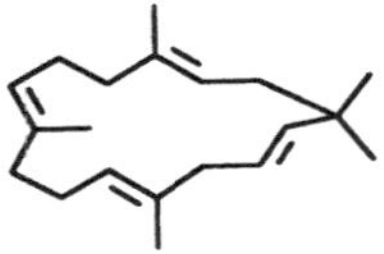

Table 4 (continued)

Structure no.	Compound Name	Source	X-ray and spectral studies	Synthesis	Biological activity
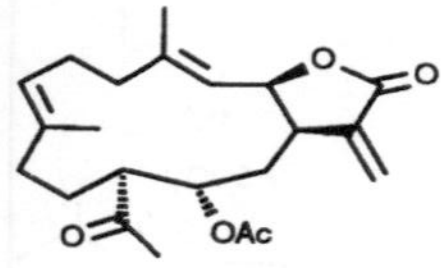 303	(1R,2E,6E,10S,11S,13R)-11-Acetoxy-10-acetyl-3,7-dimethyl-14-methylen-16-oxabicyclo [11.3.0] hexadeca-2,6-dien-15-one	Lobophytum pauciflorum Lobophytum species Simularia mayi	(307, 308) (273) (180)		
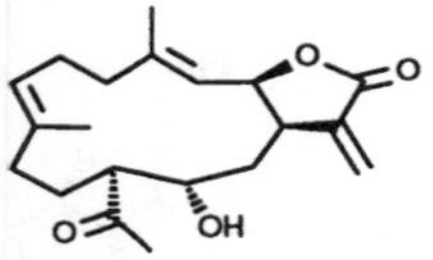 304	(1R,2E,6E,10S,11S,13R)-10-Acetyl-3,7-dimethyl-11-hydroxy-14-methylen-16-oxabicyclo [11.3.0] hexadeca-2,6-dien-15-one	Lobophytum pauciflorum	(307, 308)		
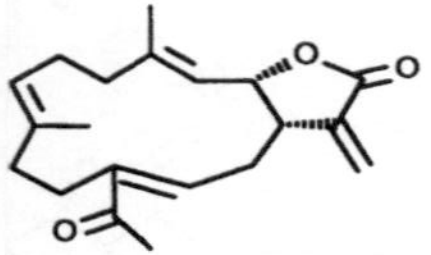 305	(1S*,2E,6E,10E,13S*)-10-Acetyl-3,7-dimethyl-14-methylen-16-oxabicyclo [11.3.0] hexadeca-2,6,10-trien-15-one	Simularia mayi	(180)		

It has been suggested that the 1,4-diketones **264**, **265**, and **271** may be precursors of the furan-containing lophotoxin (**280**). It should be noted, however, that lophodione (**264**) has (1S*,10S*)-configuration and lophotoxin (**280**) (1S*,10R*)-configuration.

3. Plexaura Species

Plexaurolone (**189**) (*170*) was the first of the four structurally related compounds (**187–190**) (*171*) isolated from gorgonians of the genus *Plexaura*. Although its structure and absolute configuration have been firmly determined by X-ray analysis of the corresponding monoacetate, the classification of plexaurolone (**189**) as a member of the α-series is not unambiguous. The reason is that **189** lacks olefinic unsaturation in the macrocyclic ring, rendering the configuration convention not directly applicable. It was argued by EALICK *et al.* (*170*), however, that if (−)-cembrene-A (**2**) and a 3,4-epoxide thereof are precursors, oxygenation is to be expected at C-3 and/or C-4. Plexaurolone, as formulated in **189**, fulfills this requirement. The oxo group at C-6 may then be introduced by allylic oxidation and that at C-11 *via* epoxidation of the 11,12 double bond.

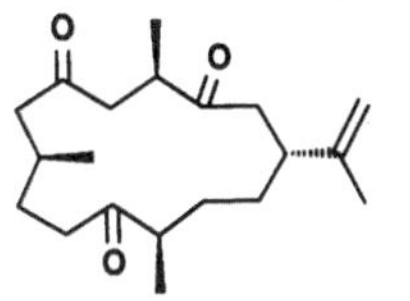

187 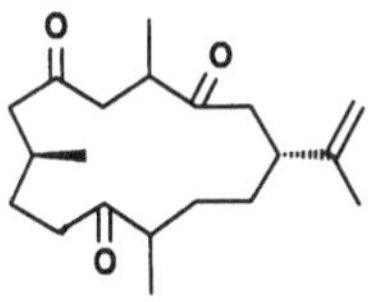   188 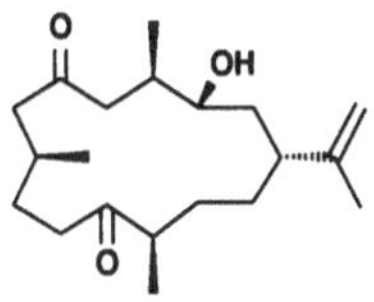   189 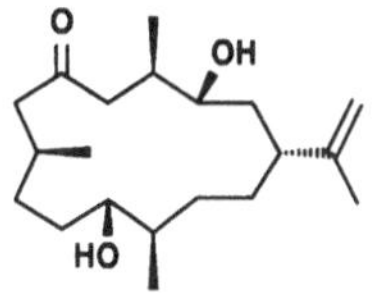   190

Crassin acetate (**231**), an antineoplastic and cytotoxic 16,3-olide, has been found in four *Plexaura* species (*157, 172*). In their early work WEINHEIMER *et al.* (*4*) reported that selective hydrogenation of the acetate **231** followed by saponification and acidification resulted in trans-lacton-

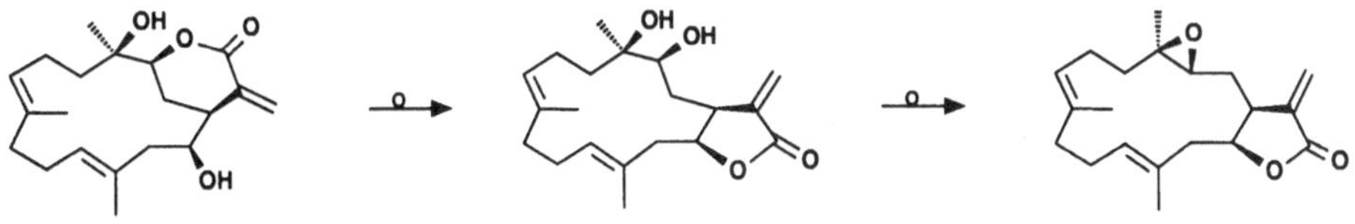

326   327   238

Scheme 11. Conversion of crassin alcohol (**326**) into euniolide (**238**) via isocrassin alcohol (**327**)

ization with the formation of a 16,14-olide. MARSHALL *et al.* (*173*) later pursued this route and developed a synthetic sequence by which crassin alcohol (**326**) is converted to euniolide (**238**) *via* isocrassin alcohol (**327**) (Scheme 11).

4. *Pseudopterogorgia Species*

Although there is an indication in the literature that more than fifteen cembranoids have been encountered in *Pseudopterogorgia bipinnata* (*174*), the structures of only six (**274**, **275**, **287–290**) have been published to-date (*174, 175*). These compounds are highly oxygenated and structurally closely related. They are all 20,10-olides and are oxygenated at C-2, C-3, C-6, C-13 and on the isopropyl group. Three of the four bipinnatins (**287**, **288**, and **290**) exhibit potent cytotoxic activity (*175*), while the hemiacetals **274** and **275** are anti-inflammatory agents (*174*). The latter two have been suggested to be formed by hydrolysis of a furanoid precursor (*174*).

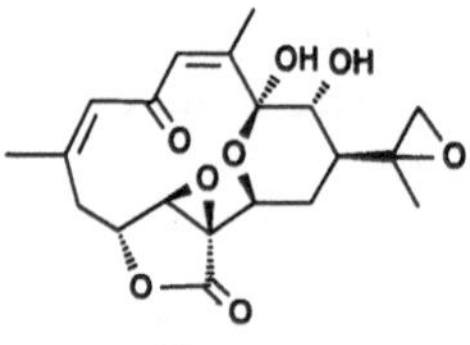

274

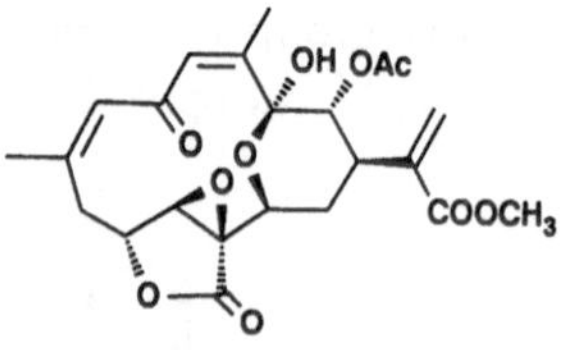

275

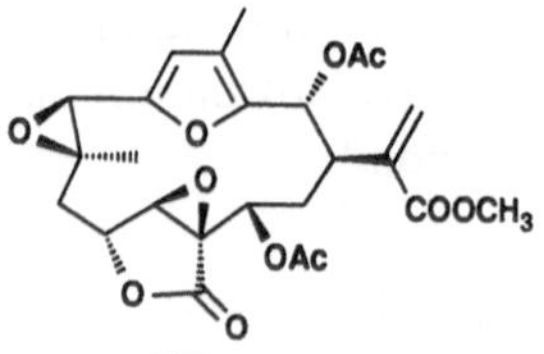

287

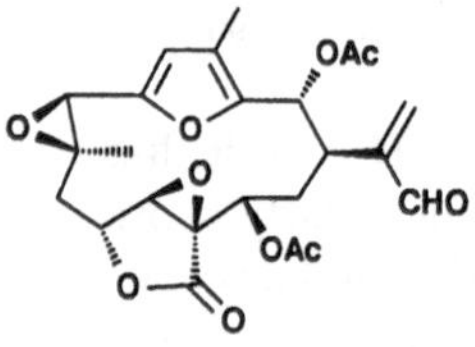

288

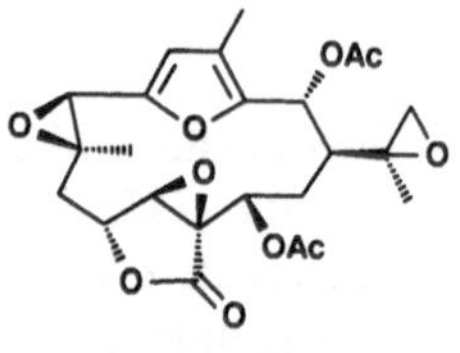

289

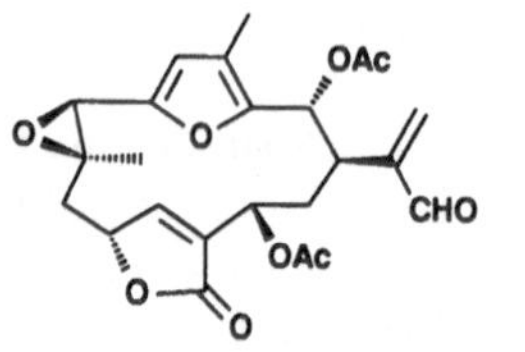

290

B. Alcyonacea

Soft corals of the order *Alcyonacea* have been the subject of extensive chemical studies during the past two decades. The results show that a wide array of structurally different cembranoids are found within this order.

1. Alcyonium Species

Four structurally complex cembran-20,10-olides, the coralloidolides A, B, D, and E (**278, 276, 273, 272**), have been reported as constituents of *Alcyonium coralloides*, a Mediterranean soft coral (*176, 177*). PIETRA *et al.* have suggested (*177*) that coralloidolide A (**278**) may be a precursor of the other three coralloidolides (**276, 273, 272**). As illustrated in Scheme 12, the furan group in **278** would then undergo oxygenation to form an epidioxide intermediate (**328**), which is converted to the coralloidolides B and E (**276, 272**). There may also be a route leading from **272** to **276** which involves the rearrangement shown. Hydration of the 7,8 double bond in coralloidolide E (**272**) and attack of the newly formed hydroxy group on C-11 explains the generation of coralloidolide D (**273**).

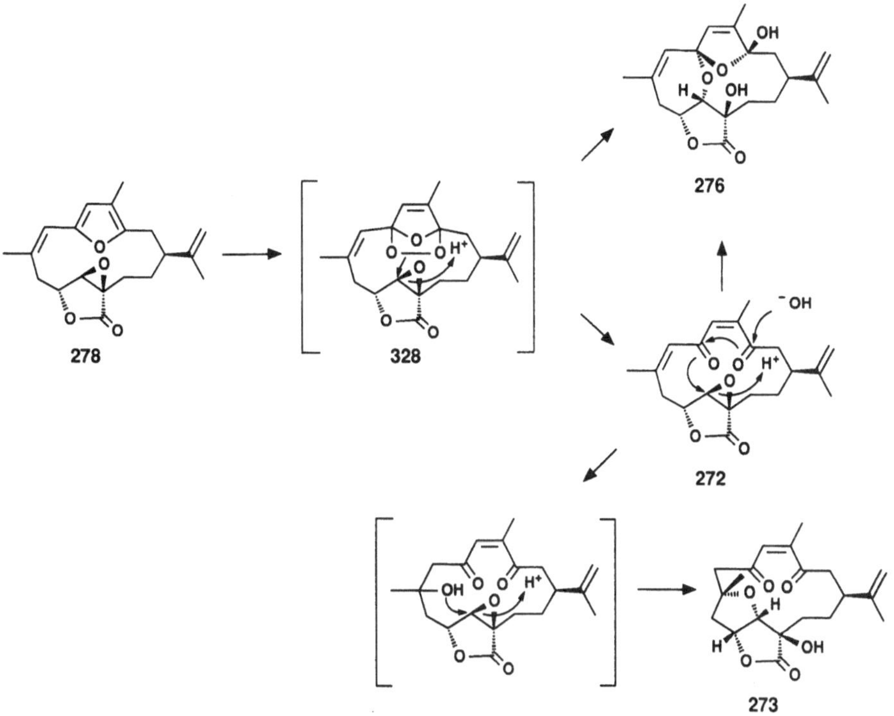

Scheme 12. Proposed biogenesis of the coralloidolides B (**276**), D (**273**) and E (**272**) from coralloidolide A (**278**)

2. Gersemia Species

Recent studies have revealed that *Gersemia rubiformis*, a soft coral living in cold Canadian waters, produces cembranoids similar to those found in tropical corals but in lower concentrations. The compounds reported so far, **266–270**, and **277**, are all 20,10-olides possessing a 3,6-dioxo-4-ene system or a furan group (*178, 179*). It is noteworthy that epilophodione (**266**) of *G. rubiformis* differs from the *Lophogorgia* metabolite lophodione (**264**) solely with respect to the stereochemistry at C-1.

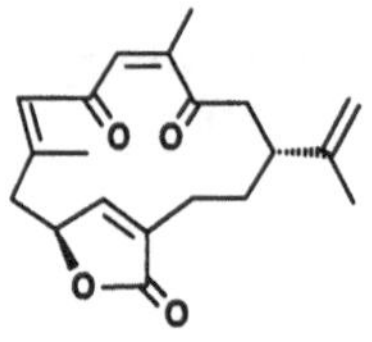

266 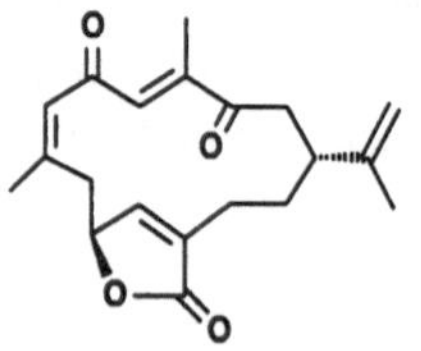   **267** 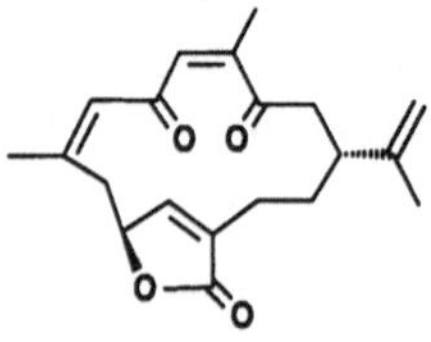   **268**

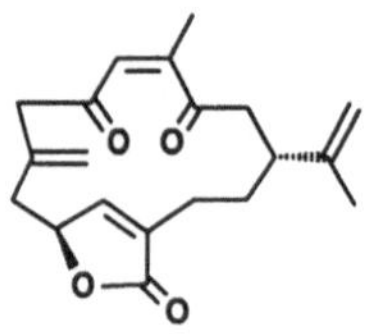

269 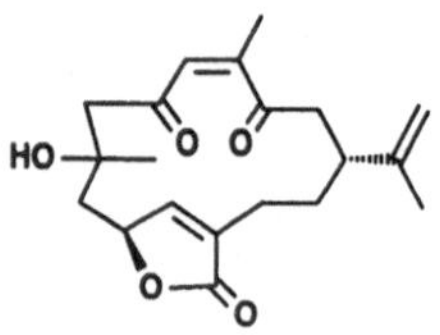   **270** 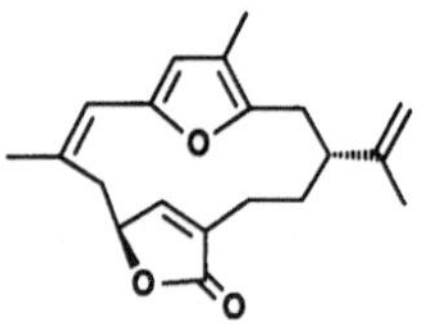   **277**

3. Lobophytum Species

In addition to structurally simple alcohols and ethers, species of the genus *Lobophytum* have been found to elaborate cembran-16,2- and -16,14-olides. Several of these are of interest from the structural and biological point of view. Denticulatolide (**228**), obtained from *L. denticulatum* and *Sinularia mayi* (*140, 180, 181*), merits comments. It is a 16,2-olide of the α-series possessing an 8,11-epidioxide group. Denticulatolide (**228**) is hence one of the two cembranic epidioxides so far encountered in nature, the other being compound **93** isolated from tobacco. Denticulatolide (**228**) is ichthyotoxic (*140*) and cytotoxic (*181*).

L. *hedleyi*, collected from Japanese coastal waters, contains the cytotoxic 16,2-olide lobohedleolide (**214**) and its (7*E*)-isomer **215** (*182*). It is noteworthy that lobohedleolide (**214**) is the (2*S*)-isomer of anisomelic acid (**27**), which is of plant origin (*29*).

Lobophytolide (**239**), an ichthyotoxic 16,14-olide, and the two 16,2-olides **220** and **221** have been found in *L. cristigalli* (*183, 274*). All three possess transfused lactone rings and belong to the α-series.

Crassolide (**230**), a highly oxygenated 16,2-olide that is toxic to fish, has been reported as the major cembranoid in *L. crassum* collected from Indonesian waters (*141, 142*). Corals from the Red Sea have given the fish toxin lobolide (**247**) and *i.a.* the 13-hydroxylobolides **255** and **256** (*185–187*), whereas isolobophytolide (**237**) has been found as the predominant terpenoid in a specimen of *L. crassum* from the Great Barrier Reef (*188*). Isolobophytolide (**237**) is also the major cembranoid in *L. crassospiculatum* (*189*).

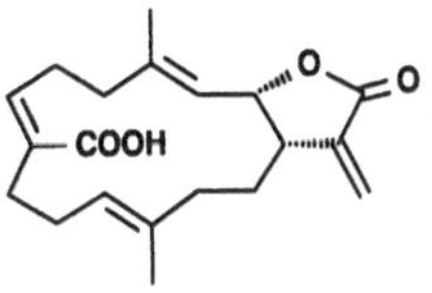

214

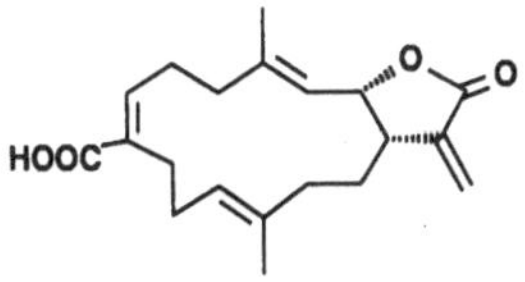

215

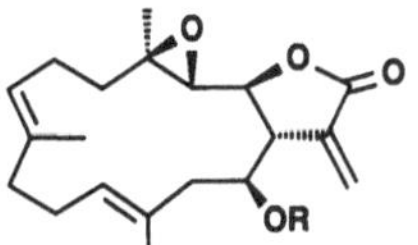

220 R = H
221 R = Ac

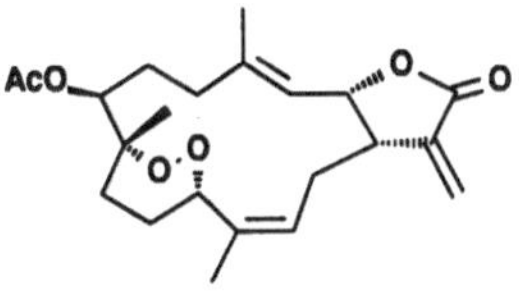

228

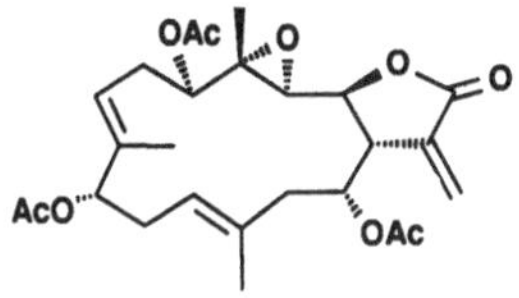

230

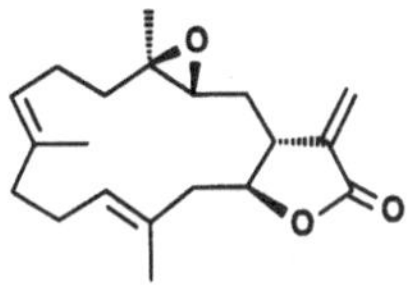

237

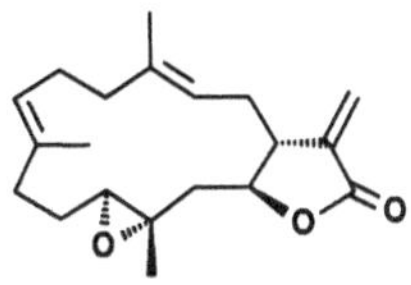

239

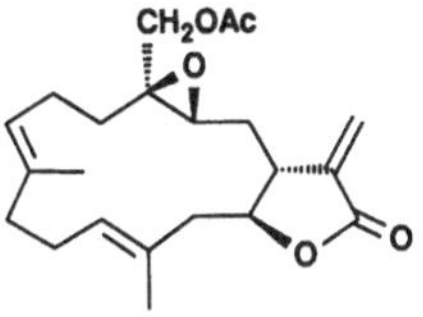

247

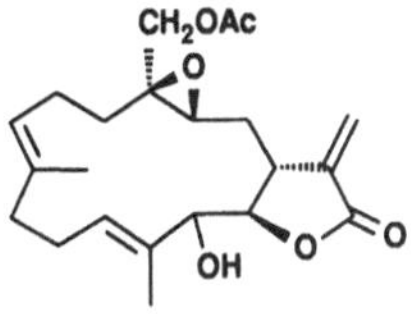

255 13R
256 13S

4. Sarcophyton Species

Soft corals of the genus *Sarcophyton* are common in the reefs in the Indo-Pacific coastal waters, *S. glaucum* being the species studied most extensively. These studies have shown that the content of cembranoids varies considerably depending upon the collection locality and the period of the year.

Sarcophytol-A (**126**), the (14*S*)-hydroxy derivative of cembrene-C (**106**), is the predominant cembranoid in Okinawaian *S. glaucum*, where it constitutes about one third of the total lipids (*190*). Both sarcophytol-A (**126**) and sarcophytol-B (**182**), a (13*R*,14*R*)-diol* also present in this species (*190, 191*), are noted for their potent anti-tumor promoting effects (*192*). Of the remaining cembranoids, five (**128, 129, 130, 183, 184**) (*191, 193, 194*) are geometrical isomers of the sarcophytols-A and -B (**126, 182**) and several are apparently derived from sarcophytol-A (**126**) by oxidation.

Thus, as proposed in Scheme 13, allylic oxidation of the methyl group at C-12 gives sarcophytol-P (**185**) (*194*), while oxygenation of the 11,12 double bond, either *via* sensitized photooxygenation or through the assistance of an oxygenase, explains the biogenesis of the sarcophytols-I, -E, -G, and -D (**177–180**) (*191, 195*). The sarcophytols-H, -O, -R, and -S (**172, 173, 175, 176**) (*191, 194*) are likely to arise by an analogous process occurring with the 7,8 double bond in sarcophytol-A (**126**).

Sarcophytol-M (**114**) (*191*), *i.e.* the enantiomer of cembrenol (**9**), nephthenol (**133**) (*191*), sinulariol-D (**136**) (*191*), the sarcophytols-C and -Q (**157, 192**) (*194, 195*) and sarcophytin-A (**142**) (*190*) are other structurally simple monools, diols, triols and epoxides present in Okinawaian *S. glaucum*.

Early studies of *S. glaucum* collected in the Red Sea led to the discovery of the toxin (+)-sarcophine (**209**), a 16,2-olide that constitutes up to 4% of the dry weight of the coral (*139, 196, 197*). Also present are two diastereoisomers (**211, 212**) of sarcophine (**209**), two isomeric deoxosarcophines (**162, 164**) and the alcohols **119** and **170** (*196*). It is noteworthy that the enantiomer of **162**, *i.e.* (–)-deoxosarcophine (**163**), has been reported as a constituent of several other *Sarcophyton* species (*142, 198–200*), and that the enantiomer of **209**, *i.e.* **210**, is present in *S. crassocaule* (*198*)**.

S. glaucum collected from Guam has yielded two isomeric isosarcophytoxides (**165, 166**) (*200*), while sarcoglaucol (**197**), a compound exhibiting interesting pharmacological properties, has been isolated from

---

* It follows from the crystal structure published for the *p*-bromophenylboronate derivative of sarcophytol-B (**182**) that the relative configuration of C-13 and C-14 is *R,R* (*201*). This is also the absolute configuration as determined by CD measurements (*194*). In our view, the formula for **182** drawn by KOBAYASHI *et al.* in *194* is correct.

** Because of insufficiencies and inconsistencies in some of the original publications, the reviewers are not confident that all structures and data relating to compounds **162–164** and **209–212** in Table 4 are fully correct.

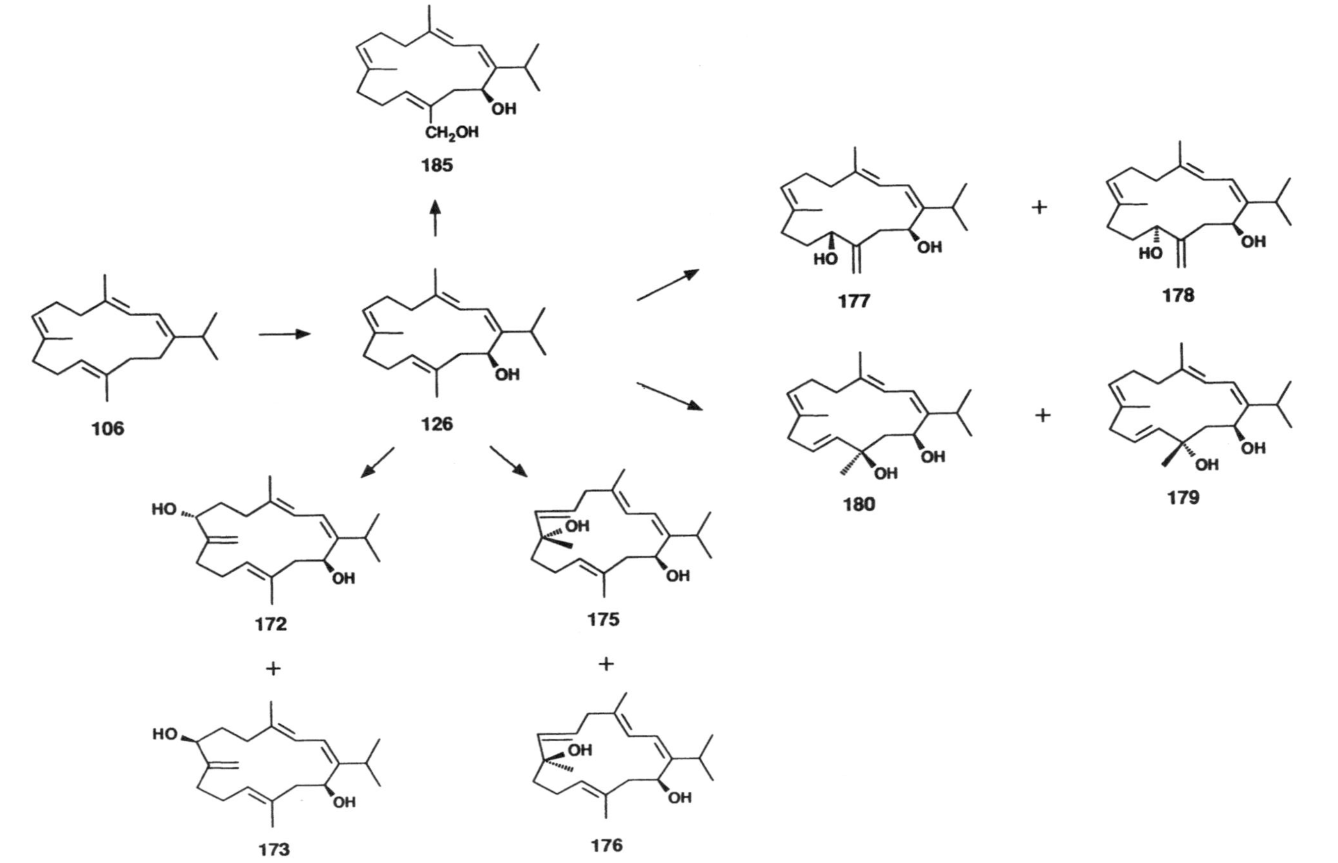

Scheme 13. Proposed biogenesis of the sarcophytols A (126), H (172), O (173), R (175), S (176), I (177), E (178), G (179), D (180) and P (185) from cembrene-C (106)

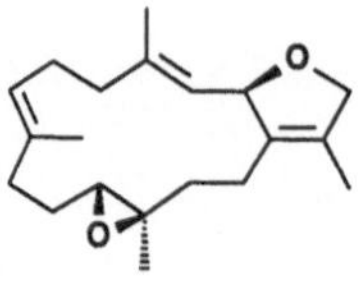

165

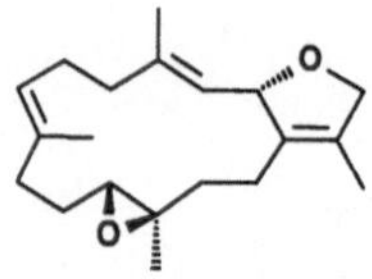

166

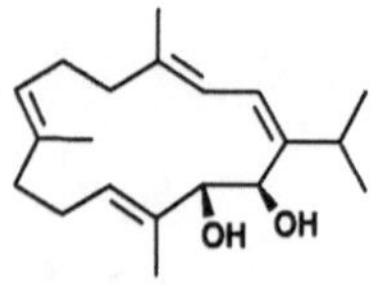

182

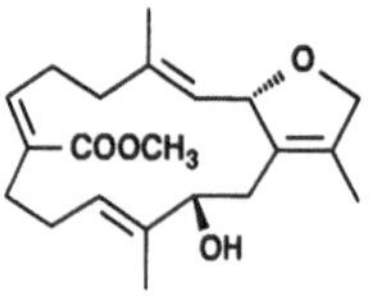

197

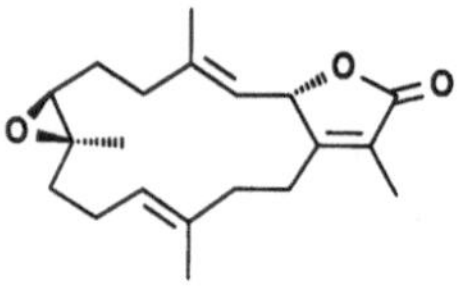

209

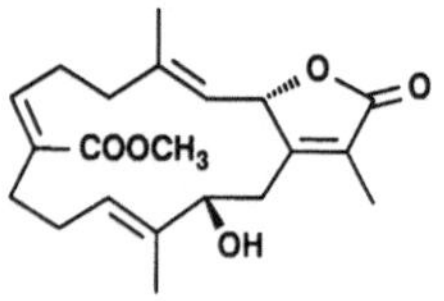

229

corals collected off the coast of Australia and New Guinea (*138, 202*). The Australian sample has also been found to contain the lactone (**229**) (*202*), a congener of sarcoglaucol (**197**).

Sarcophytolide (**291**), the corresponding (7R)-hydroxy derivative (**292**), and emblide (**294**) are three 20,8-olides isolated from *S. glaucum* (*203, 204*). Sarcophytolide (**291**) has also been found in *S. elegans* as has the structurally related ketoemblide (**293**) (*205*). Cembrene-C (**106**) has been suggested to be a biogenetic precursor of these compounds (*205*). As outlined in Scheme 14, oxidation of the methyl group at C-12 to a carboxylic acid group and epoxidation of the 7,8 double bond with formation of a (7R,8R)-epoxide (**329**) may be steps preceding the formation of **292**. This is then converted to sarcophytolide (**291**) and emblide (**294**).

As shown in Scheme 15, nephthenol (**133**), 3,4-epoxynephthenol (**150**) and decaryiol (**158**), all constituents of *S. decaryi*, are biogenetically closely related (*206*). Decaryiol (**158**) has also been detected in the eggs and coral tissue of *Lobophytum microlobulatum*, while its precursor 3,4-epoxynephthenol (**150**) has only been found in the eggs of this species (*143*).

5. *Sinularia Species*

Soft corals of the genus *Sinularia*, which are abundant in the Indo-Pacific coastal waters, contain α-methylene-lactones as the main cembranoids. Extensive studies of *S. mayi*, a species which is common in coral reefs of southern Japan, have resulted in the identification of a variety of

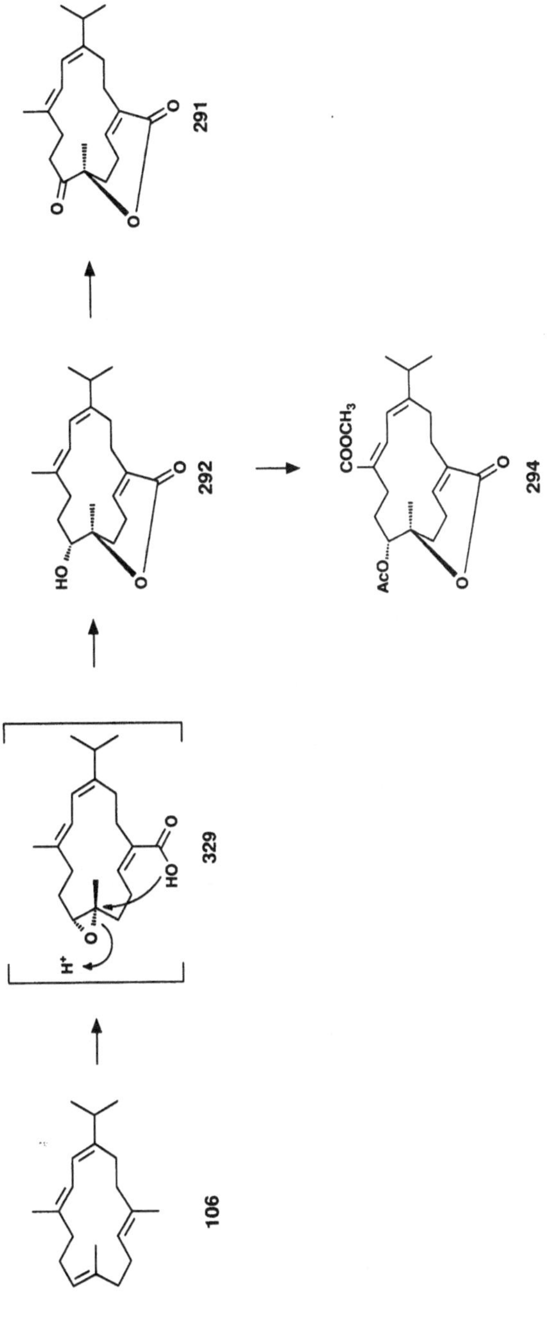

Scheme 14. Proposed biogenesis of compound **292**, sarcophytolide (**291**) and emblide (**294**) from cembrene-C (**106**)

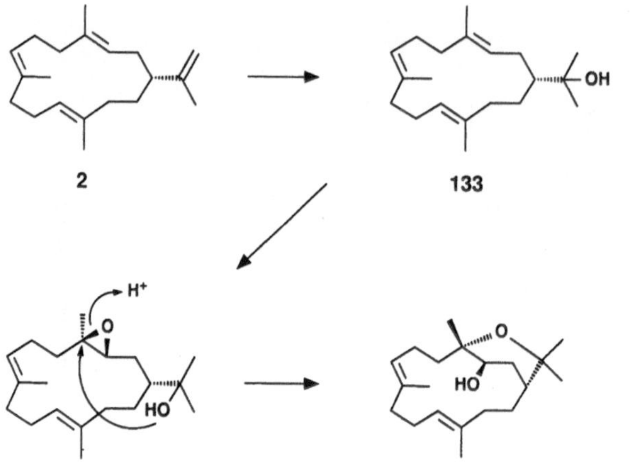

Scheme 15. Proposed biogenesis of nephthenol (**133**), 3,4-epoxynephthenol (**150**) and decaryiol (**158**) from cembrene-A (**2**)

cembran-16,2-olides (**201, 202, 206, 213, 216, 224–228, 295, 303**, and **305**) but also of a series of minor, structurally simple non-lactonic cembranoids (**106, 108, 109, 113, 115, 131, 132, 136, 143, 144, 148, 166, 169**, and **186**, see Table 4).

It is noteworthy that (−)-cembrene (**109**) (*207*) and mayol (**115**) (*207, 208*) belong to the β-series, while several of the other cembranic constituents of *S. mayi* belong to the α-series. This would imply that the cyclization of geranylgeranyl pyrophosphate is not fully stereoselective in this organism (*208*).

KOBAYASHI *et al.* (*208*) have suggested that the biogenesis of lactone **201** (*209*) takes place as illustrated in Scheme 16. Cembrene-A (**2**) undergoes epoxidation of the isopropenyl group. Subsequent hydrolysis affords sinulariol B (**186**) (*180*), which is converted to sinulariol D (**136**) (*208*) upon dehydration. This route to **136** is favoured over that involving allylic oxidation of C-16 in cembrene-A (**2**) because of the presence of both sinulariol A (**169**) and B (**186**) in *S. mayi* (*180*).

Sinulariol D (**136**) is converted to sinularic acid (**144**) *via* sinularial A (**143**) (*208*). Hydroxylation at C-2 and lactone formation complete one of the two proposed routes to **201**. The other route comprises introduction of a hydroxy group at C-2 in sinulariol D (**136**) to form sinulariol A (**169**) and subsequent oxidation.

Provided that the mayolides C and D (**226, 227**) (*210*) have the proper (11S,12R)-configuration, it seems likely that they arise *via* oxygenation of

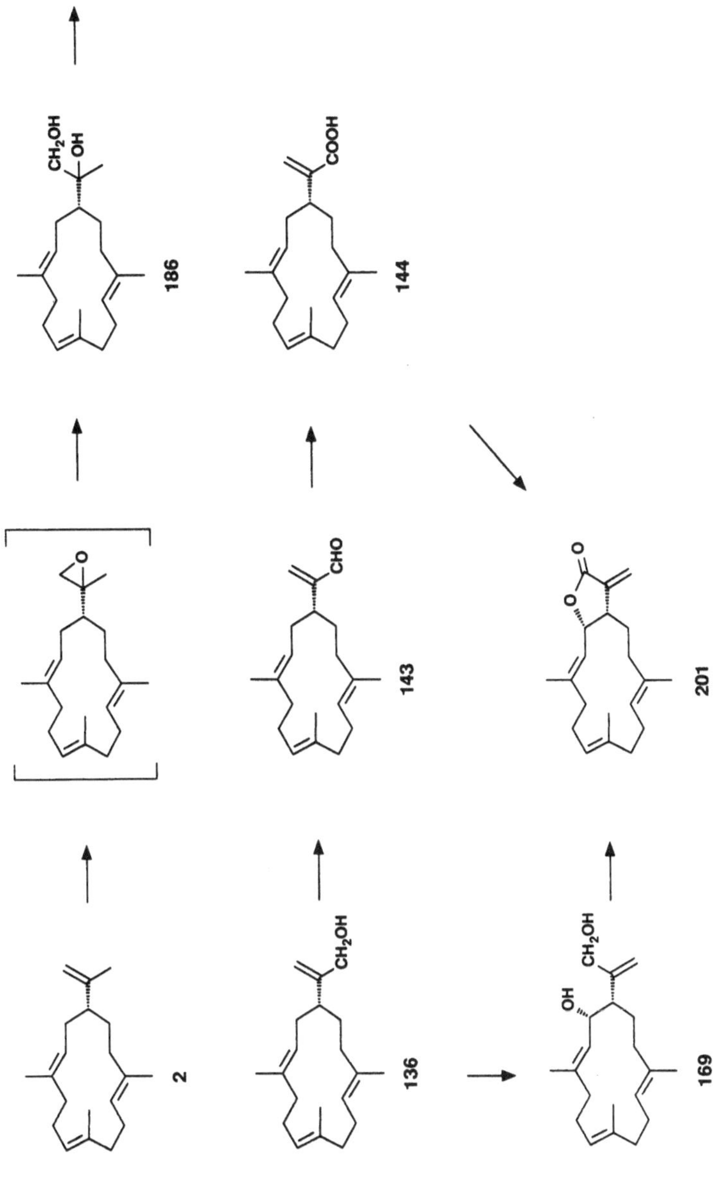

Scheme 16. Proposed biogenesis of the lactone **201** from cembrene-A (**2**) via sinulariol B (**186**), sinulariol D (**136**), sinularial A (**143**), sinularic acid (**144**) and sinulariol A (**169**)

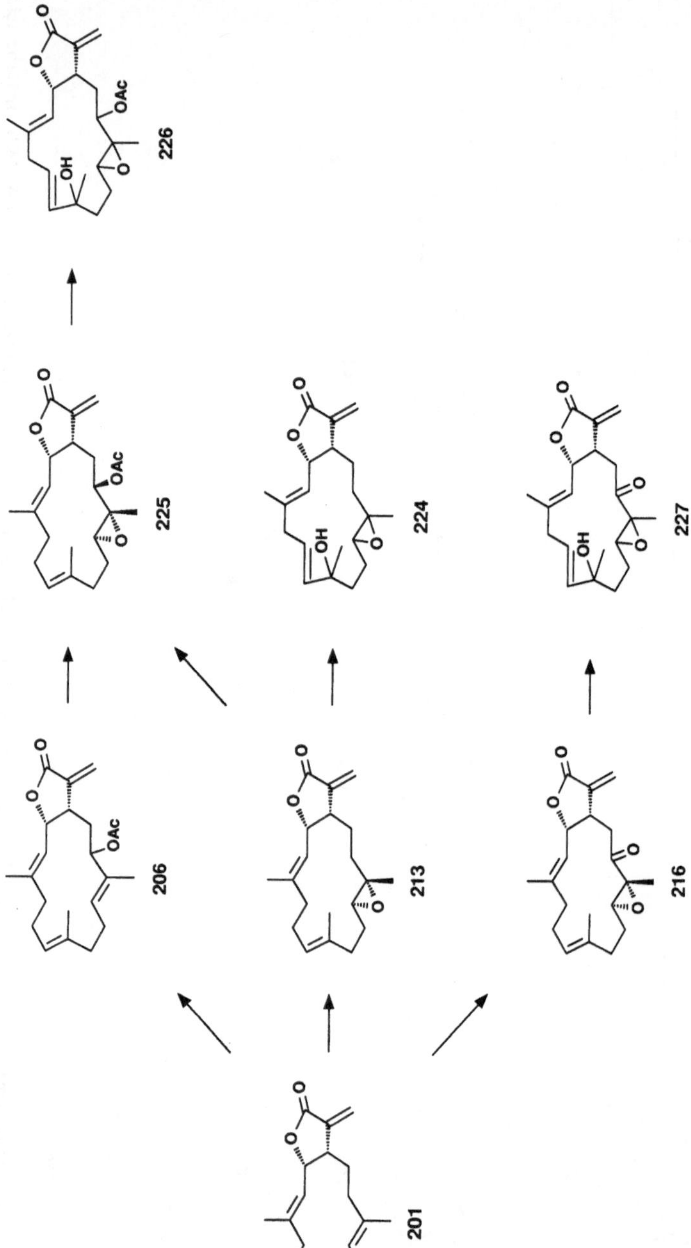

Scheme 17. Proposed biogenesis of the mayolides B (**224**), C (**226**) and D (**227**) from the lactone **201**

the 7,8 double bond in the lactones **225** and **216** (*180, 181*), respectively, either through the assistance of an enzyme or by the action of singlet oxygen. The lactone **225** in turn, may derive from **206** (*180*) and/or **213** (*180, 211*). Compound **213** may also serve as an intermediate in the biogenesis of mayolide B (**224**) (*210*), if the latter has the prerequisite (11S, 12S)-stereochemistry (Scheme 17).

To our knowledge, mayolide A (**295**) (*210*) is the only *seco*-cembranoid isolated from a marine organism. It is evidently formed by scission of the 12,13 bond in an appropriate cembranolide precursor.

S. flexibilis is a source of several well-known lactones, *e.g.*, the 16,3-olides sinularin (**232**) (*212, 213*) and dihydrosinularin (**233**) (*212, 213*) and the 16,12-olides sinulariolide (**259**) (*214, 215*) and a 6-hydroxy derivative thereof (**262**) (*216*), 11-episinulariolide (**260**) (*217*) and its monoacetate (**261**) (*4, 216*), 11-dehydrosinulariolide (**258**) (*216*) and the 8,11-epoxy bridged compound **263** (*217*). Being members of the α-series, these lactones are most likely derived from (1R)-cembrene-A (**2**) and its (3S,4S,11S,12S)-3,4:11,12-diepoxide (**161**), which are also present in *S. flexibilis* (*216*).

As suggested in Scheme 18, oxidation of the methyl group at C-15 and attack of the newly generated carboxylate on C-3 (a) or on C-12 (b) would give sinularin (**232**) or sinulariolide (**259**), respectively. The latter is oxidized to 11-dehydrosinulariolide (**258**), which, in turn, may be an intermediate in the biogenesis of 11-episinulariolide (**260**) and its acetate (**261**). Hydroxylation of the allylic C-6 position of sinulariolide (**259**) gives **262**. This compound may, however, be an artefact, since it has been found among the degradation products formed from sinulariolide (**259**) (*216*).

Sinulariolide (**259**) is an experimentally verified precursor of the 8,11-epoxide **263** (*217*). Thus, treatment of **259** with *m*-chloroperoxybenzoic acid gives the alcohol **330**, which is converted to **263** on acetylation (*cf.* Schemes 2 and 7 for similar processes).

Flexibilene (**302**), a hydrocarbon having a fifteen-membered carbocyclic ring, was first isolated from *S. flexibilis* (*213, 218*) and later from *S. conferta*, where it co-occurs with cembrene-A (**2**) and casbene (**331**) (*219*). The discovery of these three hydrocarbons in a single organism is of considerable interest, since they represent all three carbon skeletons expected to arise form the hypothetical intermediate **332** (Scheme 19) (*219*).

Six structurally closely related 18-nor-cembranoids (**296–301**) have been isolated from various *Sinularia* species (*220–223*). They are all 5,8-epoxy-bridged and possess oxo groups at C-3 and C-6. Five of them are 20,10-olides (**296–300**), while the remaining **301** is a methyl ester.

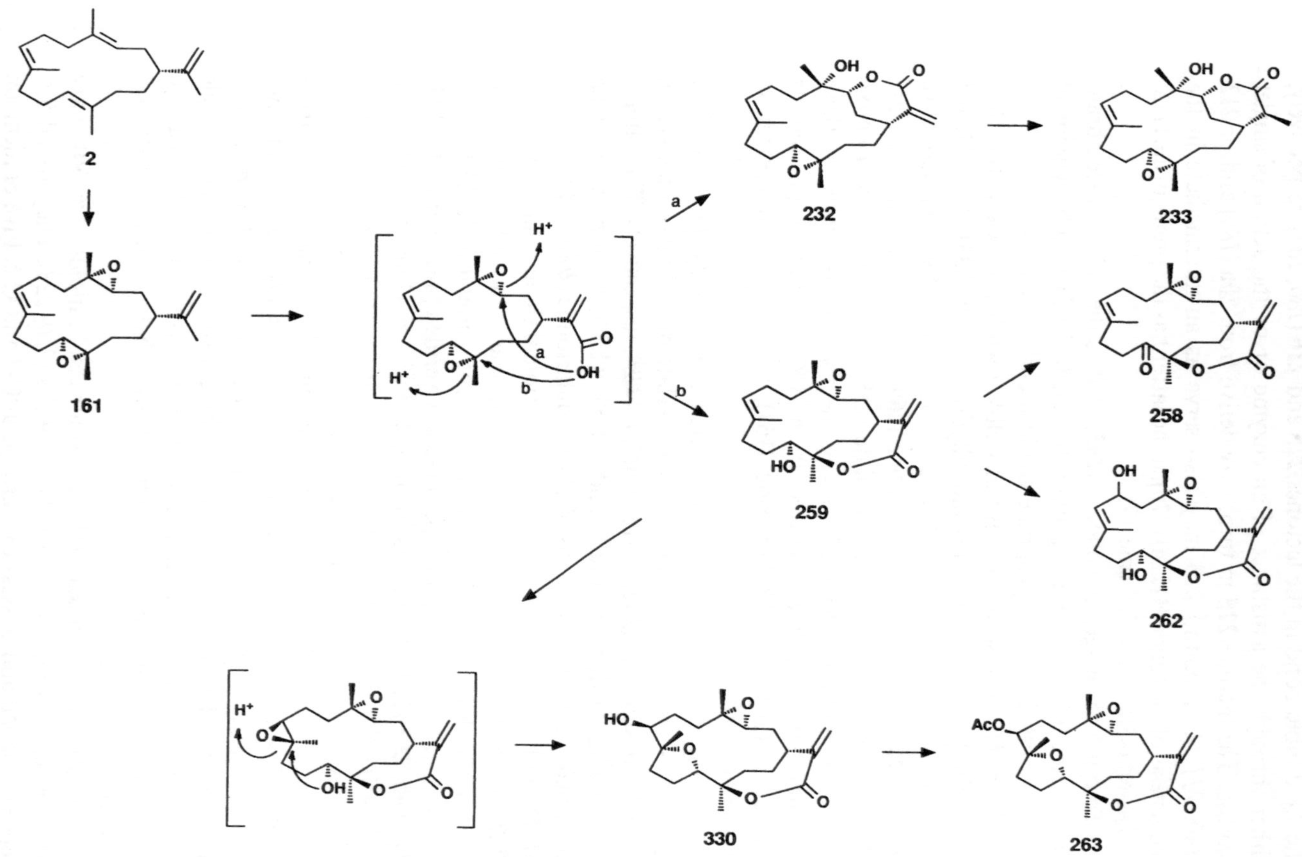

Scheme 18. Proposed biogenesis of sinularin (232), dihydrosinularin (233), sinulariolide (259), 11-dehydrosinulariolide (258), 6-hydroxysinulariolide (262) and compound 263 from cembrene-A (2)

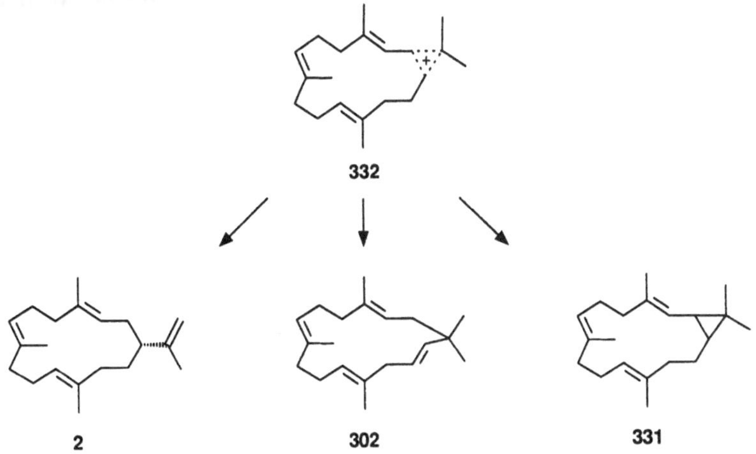

332

2   **302**   **331**

Scheme 19. Proposed biogenesis of cembrene-A (**2**), flexibilene (**302**) and casbene (**331**) from the hypothetical intermediate **332**

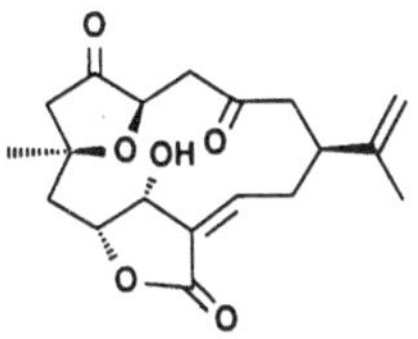

296 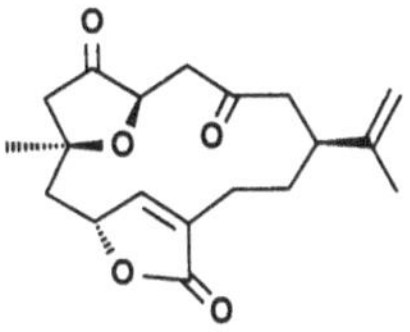   **297** 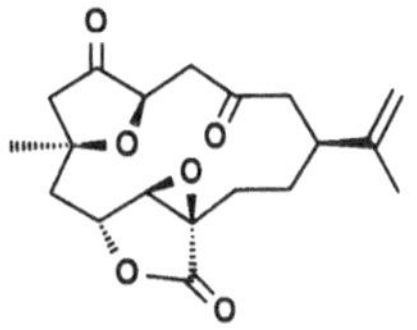   **298**

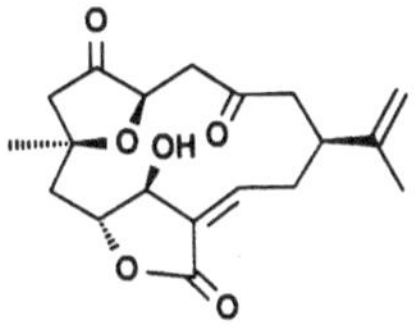

299 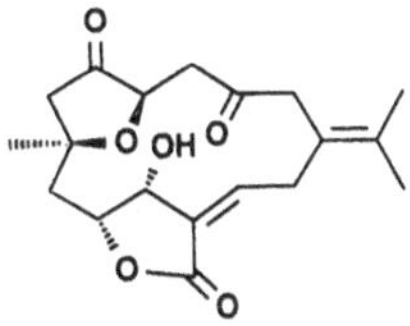   **300** 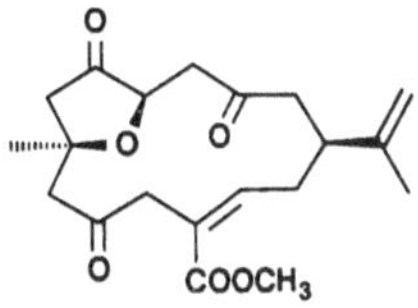   **301**

C. Stolonifera

Soft corals of the order *Stolonifera* are much less common in coral reefs than are alcyonaceans and gorgonaceans. Even though a wide array of terpenoids has been found within this order, cembranic metabolites have hitherto been encountered in only two genera, *Clavularia* and *Pachyclavularia* (family Clavulariidae).

Studies on *Clavularia koellikeri* have revealed the presence of five cytotoxic 16,2-olides of the β-series, the kericembrenolides A-E (**204, 205, 217–219**) (*224*). *Pachyclavularia violecea* is the source of four other cembranoids. These possess a 4,7-epoxy group and hydroxy/acetoxy substituents at C-2 and C-3. In compounds **198–200** the isopropyl group forms part of a furan group, while in pachyclavularolide (**257**) this has been transformed into a 16,14-olide group (*225, 226*).

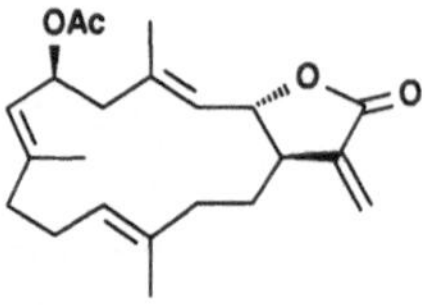

204

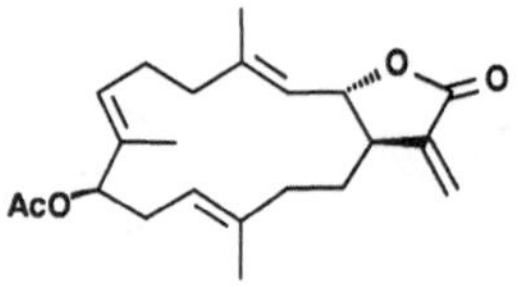

205

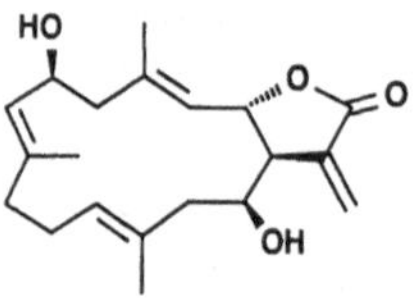

217

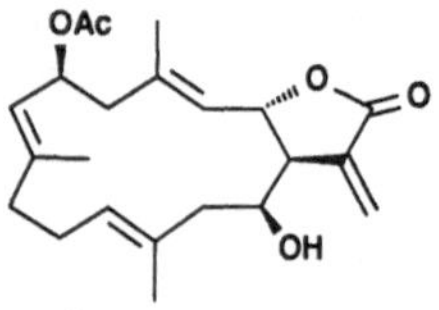

218

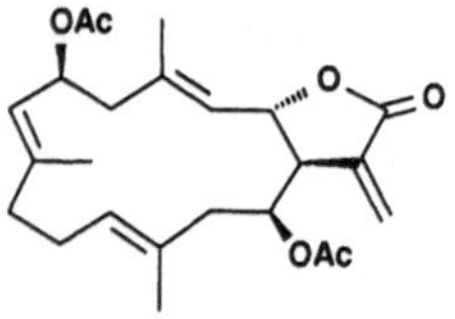

219

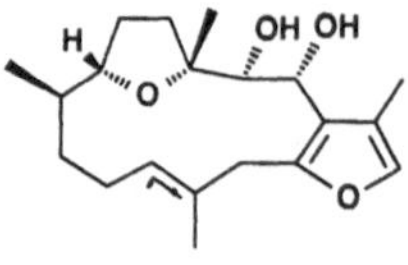

198

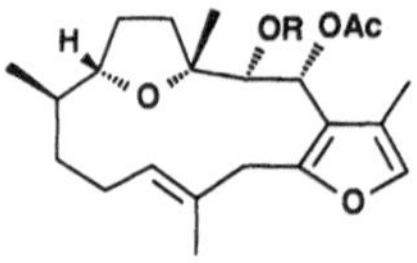

199 R=H
200 R=Ac

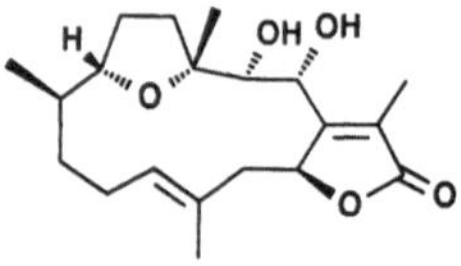

257

V. Pseudopteranoids from Marine Invertebrates

Pseudopterolide (**313**), a diterpenoid reported as a cytotoxic constituent of *Pseudopterogorgia acerosa* by FENICAL et al. in 1982 (*309*), was the first member of the pseudopterane group to be discovered in nature.

Since then the group has expanded and now comprises ten compounds, all being lactones (306–315).

The pseudopterane skeleton (D) is composed of a twelve-membered carbomonocyclic ring which is substituted by isopropyl groups at C-1 and C-7 and by methyl groups at C-4 and C-10. The biogenesis of the pseudopteranoids can, in principle, be accounted for by dimerization of two geranyl units. FENICAL et al. (309) have suggested, however, that because of the co-occurrence of cembranoids and pseudopteranoids in soft corals a more plausible mode of formation would involve ring contractions in appropriate cembranoid precursors.

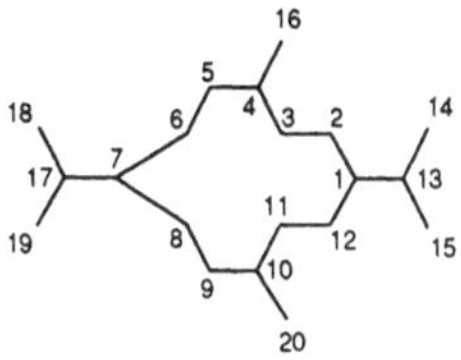

D

The nitrogen-containing tobagolide (315) has been isolated as a natural metabolite of *Pseudopterogorgia acerosa* (310, 317). It has been prepared from pseudopterolide (313) by treatment with dimethylamine in

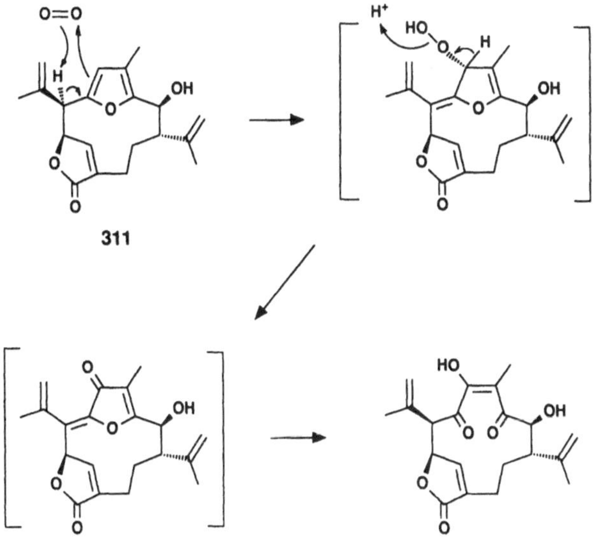

Scheme 20. Proposed mechanism for the conversion of kallolide A (**311**) to kallolide C (**309**) by a singlet oxygen reaction

Table 5. *Pseudopteranoids from Marine Invertebrates*

Structure no.	Compound Name	Source	X-ray and conformational studies	Biological activity
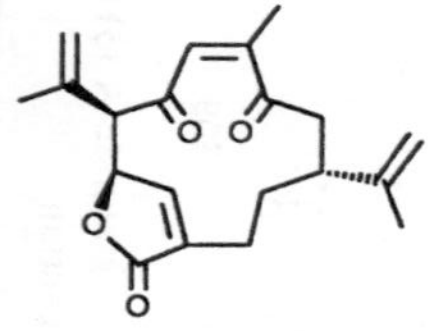 306	Gersemolide (1R*,4Z,7R*,8R*)-3,6-Dioxopseudoptera-4,9,13,17-tetraen-20,8-olide	*Gersemia rubiformis* (178, 179)	X-ray analysis of 306 (relative configuration) (178) MM calculations (178)	
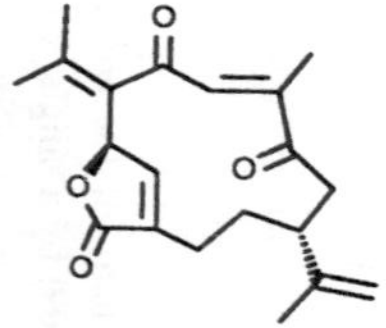 307	Isogersemolide A (1R*,4E,8R*)-3,6-Dioxopseudoptera-4,7(17),9,13-tetraen-20,8-olide	*Gersemia rubiformis* (179)		
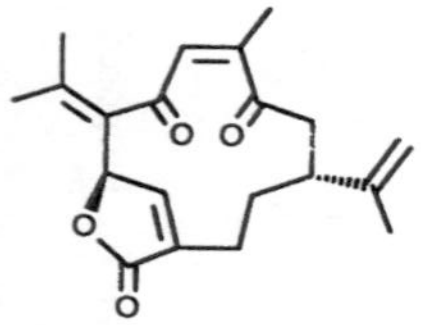 308	Isogersemolide B (1R*,4Z,8R*)-3,6-Dioxopseudoptera-4,7(17), 9,13-tetraen-20,8-olide	*Gersemia rubiformis* (179)		

Structure	Name	Source	Ref.	Remarks
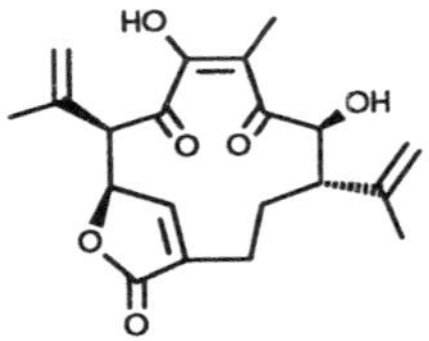 **309**	Kallolide C (1S*,2S*,4Z,7R*,8R*)-2,5-Dihydroxy-3,6-dioxo-pseudoptera-4,9,13,17-tetraen-20,8-olide	*Pseudopterogorgia kallos*	(311)	
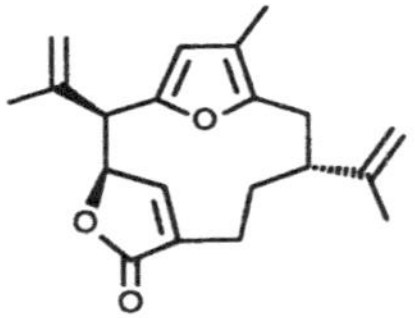 **310**	Kallolide B (1R*,7R*,8R*)-3,6-Epoxypseudoptera-3,5,9,13,17-pentaen-20,8-olide	*Pseudopterogorgia kallos*	(311)	
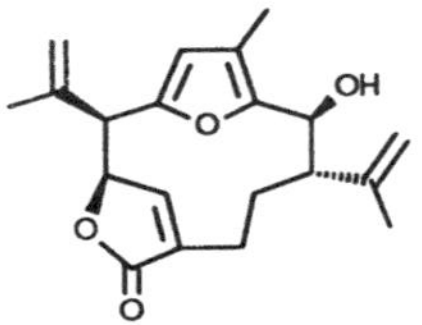 **311**	Kallolide A (1S*,2S*,7R*,8R*)-3,6-Epoxy-2-hydroxy-pseudoptera-3,5,9,13,17-pentaen-20,8-olide	*Pseudopterogorgia kallos*	(311)	Anti-inflamma-tory effect (311)
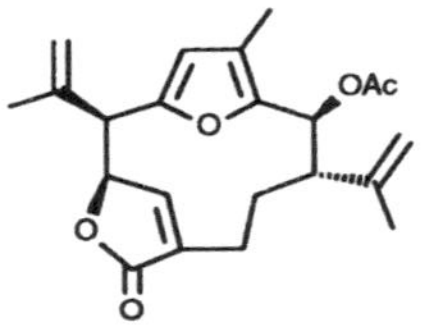 **312**	Kallolide A acetate (1S*,2S*,7R*,8R*)-2-Acetoxy-3,6-epoxy-pseudoptera-3,5,9,13,17-pentaen-20,8-olide	*Pseudopterogorgia kallos*	(311)	X-ray analysis of **312** (relative config-uration) (311)

Table 5 (*continued*)

Structure no.	Compound Name	Source	X-ray and conformational studies	Biological activity
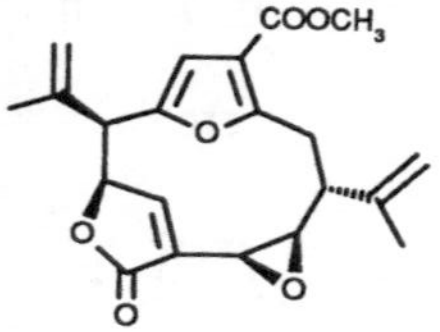 **313**	Pseudopterolide Methyl (1R,7R,8R,11S,12R)-3,6:11,12-Diepoxy-20,8-olidylpseudoptera-3,5,9,13,17-pentaen-16-oate	*Pseudopterogorgia acerosa* (309)	X-ray analysis of the urethane derivative of **313** (absolute configuration) (309)	Cytotoxic (309)
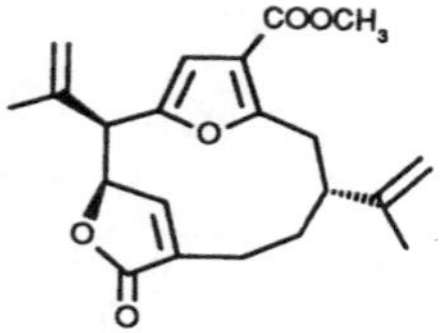 **314**	Desoxypseudopterolide Methyl (1R*,7R*,8R*)-3,6-Epoxy-20,8-olidyl-pseudoptera-3,5,9,13,17-pentaen-16-oate	*Pseudopterogorgia acerosa* (302, 317)		
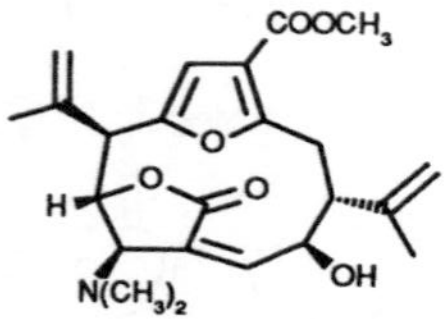 **315**	Tobagolide Methyl (1R,7R,8S,9R,10Z,12R)-9-N-Dimethylamino-3,6-epoxy-12-hydroxy-20,8-olidylpseudoptera-3,5,10,13,17-pentaen-16-oate	*Pseudopterogorgia acerosa* (310, 317)		

References, pp. 276–294

tetrahydrofuran, one driving force for the reaction being the release of strain on opening of the 11,12-epoxide group (*302*).

Kallolide A (**311**), another furan-containing pseudopteranolide, exhibits marked anti-inflammatory activity. It has been isolated along with the related **309**, **310**, and **312** from *Pseudopterogorgia kallos* (*311*). Kallolide A (**311**) has been converted to kallolide C (**309**) by singlet oxygen in a manner that has been proposed to take place as illustrated in Scheme 20 by an unprecedented ene reaction and subsequent fragmentation (*311*).

Gersemolide (**306**) and its congeners **307** and **308** have been obtained from *Gersemia rubiformis*, an alcyonacean containing both pseudopteranolides and cembranolides (*178, 179*). It is noteworthy that gersemolide (**306**) is directly related to epilophodione (**266**) (also present in this coral) by ring contraction.

Addendum

It is our intention in this addendum to review articles that were not available or have appeared after completion of the original manuscript. Several deal with the isolation of new cembranoids from marine organisms; relevant data are summarized in Table 6. In view of recent findings that diterpenoids of the cubitane class are most likely generated by ring contraction of appropriate cembranic precursors (*319*), we have now also included a section on naturally occurring cubitanoids.

Cubitanoids from Insects and Marine Organisms

Cubitene (**344**) was the first compound of the cubitane class reported as a natural product. It was isolated from the defensive secretions released from the frontal glands of soldiers of the East African termite *Cubitermes umbratus* (*321*). Cubitanoids, *i.e.* the calyculones A–G (**345–351**), have later been found in the Caribbean gorgonian *Eunicea calyculata* (*278, 319*).

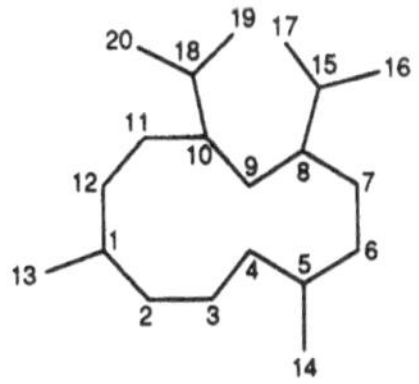

E

Table 6. Cembranoids from Marine Invertebrates

Structure no.	Compound Name	Source	X-ray and spectral studies	Biological activity
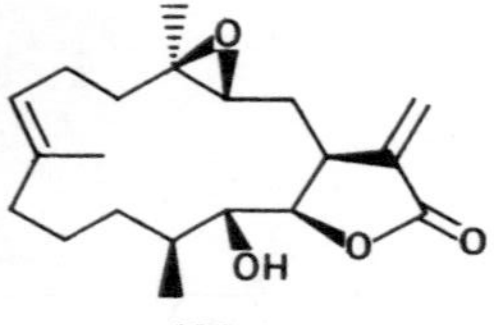 325	Eupalmerin (1S*,3R*,4R*,7E,12S*,13R*,14R*)-3,4-Epoxy-13-hydroxycembra-7,15(17)-dien-16,14-olide	Eunicea mammosa	(318)	Cytotoxic activity against CHO-K1 cells: ED$_{50}$ 5.46 μg/ml
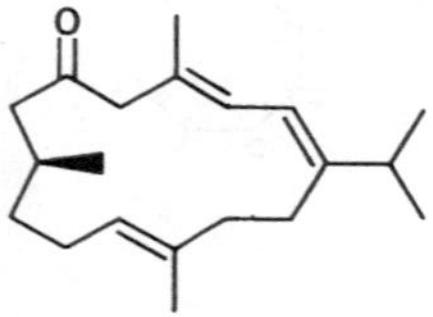 333	(1E,3E,8S*,11E)-Cembra-1,3,11-trien-6-one	Eunicea calyculata	(319)	
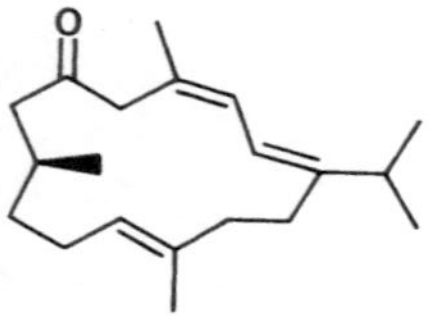 334	(1Z,3Z,8S*,11E)-Cembra-1,3,11-trien-6-one	Eunicea calyculata	(319)	

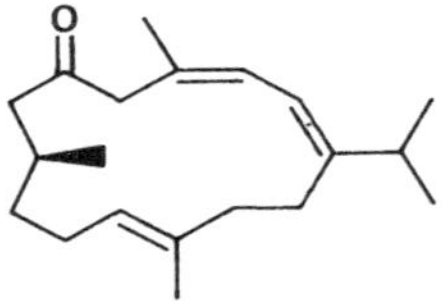

Cytotoxic activity against HCT-116 cells at 27 µg/ml (316)

(1E,3Z,8S*,11E)-Cembra-1,3,11-trien-6-one

Eunicea calyculata (319)

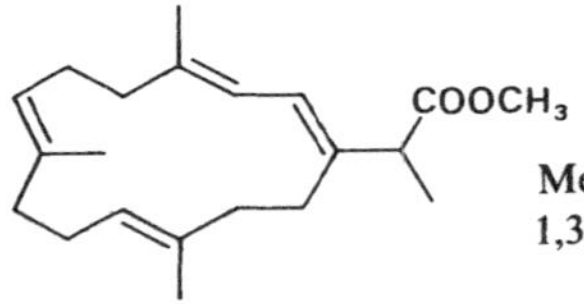

335

Methyl(1E,3E,7E,11E)-Cembra-1,3,7,11-tetraen-16-oate

Sinularia mayi (316)

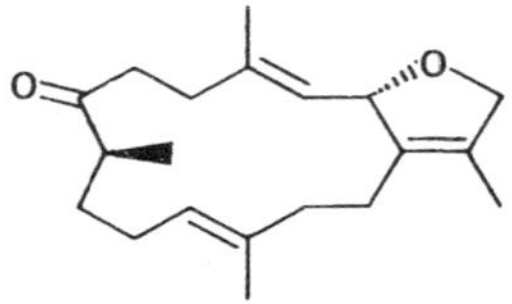

COOCH₃

336

(2S,3E,8S,11E)-2,16-Epoxycembra-1(15),3,11-trien-6-one

Sarcophyton species (313)

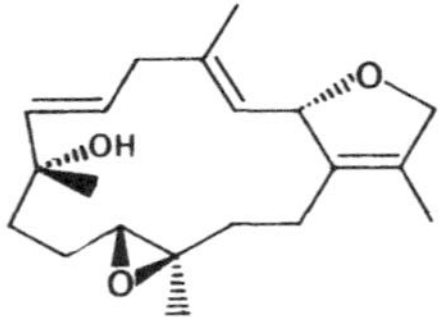

337

8-Hydroxyisosarcophytoxide-6-ene (2S*,3E,6E,8R*,11R*,12R*)-2,16:11,12-Diepoxycembra-1(15),3,6-trien-8-ol

Sinularia mayi (316)

338

Table 6 (continued)

Structure no.	Compound Name	Source	X-ray and spectral studies	Biological activity
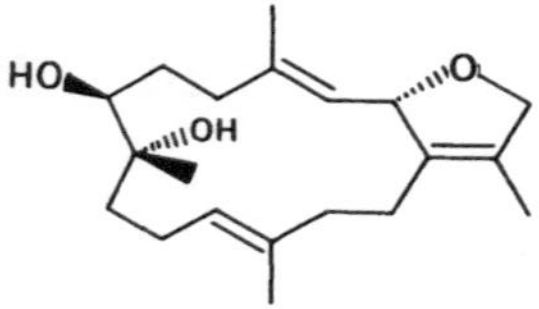 339	(2S,3E,7S,8R,11E)-2,16-Epoxycembra-1(15),3,11-trien-7,8-diol	Sarcophyton species	(313)	
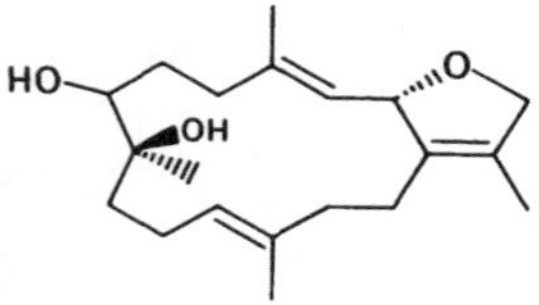 340	Sarcophytonin C (2S,3E,8S,11E)-2,16-Epoxycembra-1(15),3,11-trien-7,8-diol	Sarcophyton species	(313)	
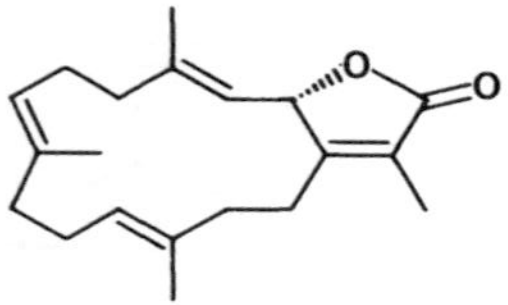 341	Sarcophytonin B (2S,3E,7E,11E)-Cembra-1(15),3,7,11-tetraen-16,2-olide	Sarcophyton species	(313)	

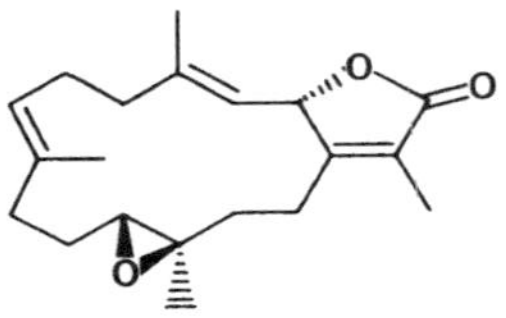

Cytotoxic activity
against HCT-116 cells
at 64 µg/ml (316)

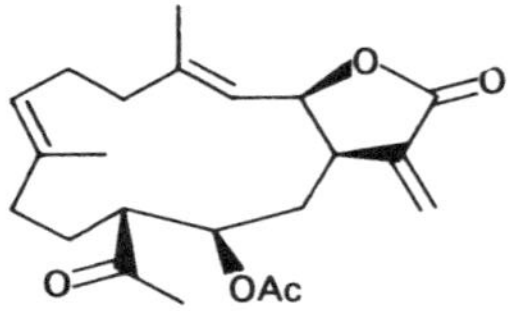

(316)

X-ray analysis of **343**
(relative configura-
tion) (320)

(320)

Sinularia mayi

Lobophytum species

(+)-Isosarcophine
(2S,3E,7E,11R,12R)-11,12-
Epoxycembra-1(15),3,7-trien-16,2-olide

Chilobolide A
(1R*,2E,6E,10R*,11R*,13R*)-11-
Acetoxy-10-acetyl-3,7-dimethyl-14-
methylen-16-oxabicyclo[11.3.0]
hexadeca-2,6-dien-15-one

342

343

Table 7. *Cubitanoids from Insects and Marine Invertebrates*

Structure no.	Compound	Name	Source	X-ray and spectral studies	Synthesis
	Cubitene (4E,8S*,10S*,12E) Cubita-4,12,15,18-tetraene		Cubitermes umbratus (321)	X-ray analysis of 344 (relative configuration) (321)	(±)-344 (323)
	Calyculone D (1S*,4E,8Z,10S*)-Cubita-4,8,18-trien-6-one		Eunicea calyculata (319)		
	Calyculone E (1S*,4E,8Z,10R*)-Cubita-4,8,18-trien-6-one		Eunicea calyculata (319)		

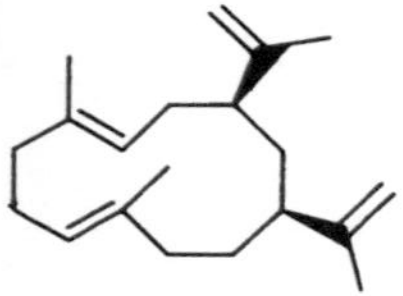

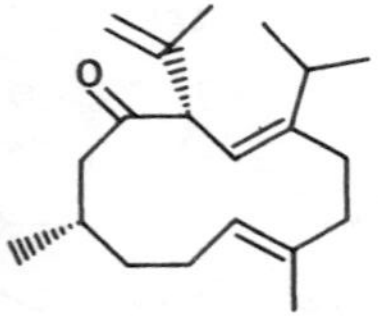

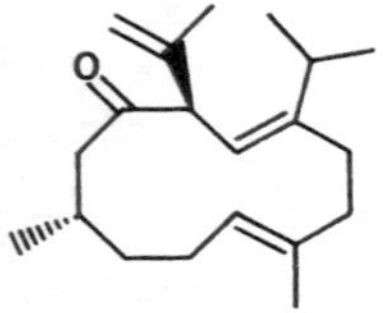

344

345

346

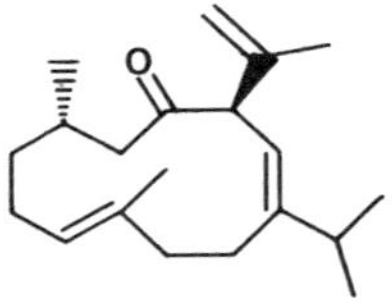

Calyculone F
(1S*,4E,8E,10R*)-Cubita-4,8,18-trien-6-one

Eunicea calyculata

(319)

Calyculone G
(1S*,4E,8E,10S*)-Cubita-4,8,18-trien-6-one

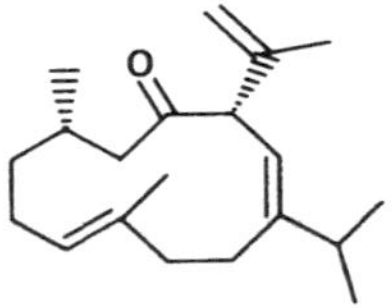

Eunicea calyculata

(319)

Calyculone A
(1S*,4R*,5R*,8E,10S*)-4,5-Epoxycubita-8,18-dien-11-one

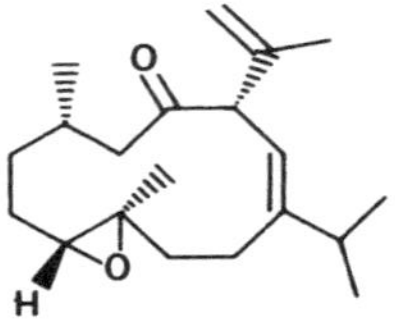

Eunicea calyculata

(278)

X-ray analysis of 349 (relative configuration) (278)

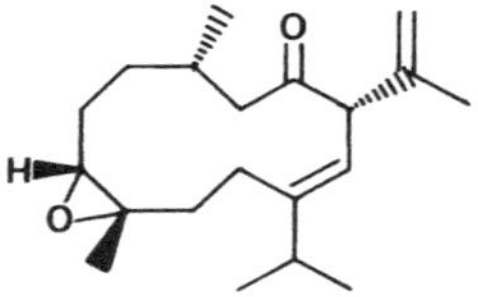

Calyculone B
(1S*,4R*,5S*,8E,10S*)-4,5-Epoxycubita-8,18-dien-11-one

Eunicea calyculata

(278)

347

348

349

350

Table 7 *(continued)*

| Structure no. | Compound | | Source | X-ray and spectral studies | Synthesis |
	Name			
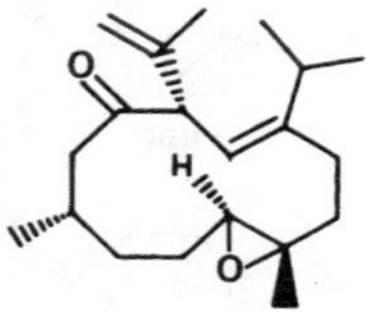 351	Calyculone C (1S*,4S*,5S*,8Z,10S*)-4,5-Epoxycubita-8,18-dien-11-one		*Eunicea calyculata*	(278)	

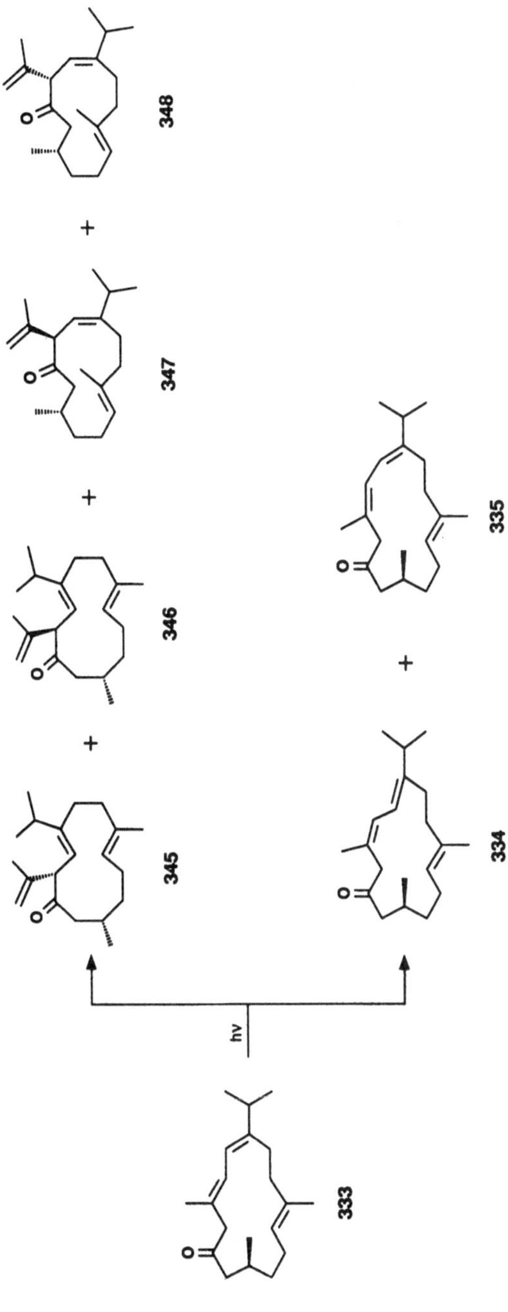

Scheme 21. Formation of compounds **334**, **335**, and **345–348** by photochemical treatment of compound **333**

The cubitane skeleton (**E**) is based on a twelve-membered carbo-monocyclic ring possessing methyl substituents at C-1 and C-5 and isopropyl substituents at C-8 and C-10. The biosynthesis of these irregular diterpenoids has been the subject of discussion and different routes have been proposed in the past (*321, 322*). Strong support for the view that cembranoids do play a role as intermediates was, however, recently presented by FENICAL *et al.* (*319*) who found that the calyculones D–G (**345–348**) co-occur with similarly functionalized cembranoids (**333–335**) in the same *Eunicea* species and that one of these cembratrienones (**333**) undergoes a photochemically induced 1,3 acyl migration (and double bond isomerization) with the formation of the aforementioned calyculones. As shown in Scheme 21, the cembratrienones **334** and **335** were also obtained as products in this experiment.

Acknowledgements

We are grateful to Dr. Arne Björnberg and Professor Curt R. Enzell for their interest in this work and to Ms. Gabriella Huss and Ms. Elsa Moreau for typing this manuscript.

References

1. DAUBEN, W.G., W.E. THIESSEN, and P.R. RESNICK: Cembrene, a 14-Membered Ring Diterpene Hydrocarbon. J. Amer. Chem. Soc. **84**, 2015–2016 (1962).

2. KOBAYASHI, H., and S. AKIYOSHI: Thunbergene, a Macrocyclic Diterpene. Bull. Chem. Soc. Japan **35**, 1044–1045 (1962).

3. ROBERTS, D.L., and R.L. ROWLAND: Macrocyclic Diterpenes, α- and β-4,8,13-Duvatriene-1,3-diols from Tobacco. J. Organ. Chem. **27**, 3989–3995 (1962).

4. WEINHEIMER, A.J., C.W.J. CHANG, and J.A. MATSON: Naturally Occurring Cembranes. Fortschr. Chem. organ. Naturstoffe **36**, 285–387 (1979).

5. TIUS, M.A.: Synthesis of Cembranes and Cembranolides. Chem. Rev. **88**, 719–732 (1988).

6. DAUBEN, W.G., W.E. THIESSEN, and P.R. RESNICK: Cembrene, a Fourteen-Membered Ring Diterpehe Hydrocarbon. J. Organ. Chem. **30**, 1693–1698 (1965).

7. DREW, M.G.B., D.H. TEMPLETON, and A. ZALKIN: The Crystal and Molecular Structure of Cembrene. Acta Crystallogr. **B25**, 261–267 (1969).

8. SHMIDT, E.N., and V.A. PENTEGOVA: Chemical Composition of *Picea* Oleoresins. Diterpenoid from *Picea obovata* Oleoresin. Khim. Prir. Soedin. 769–770 (1970).

9. SHMIDT, E.N., L.E. CHUPAKHINA, and V.A. PENTEGOVA: Diterpenoids of the Oleoresins of Three Species of the Genus *Larix: Larix sibirica, Larix sukaczawii* and *Larix czekanovskii.* Izv. Sib. Otd. Akad. Nauk. SSSR, Ser. Khim. Nauk. 173–175 (1975).

10. FUJISE, Y., I. MARUTA, S. ITO, and T. NOZOE: Chemical Constituents of Essential Oil of *Chamaecyparis obtusa* (Sieb. et Zucc.). Chem. Pharm. Bull. (Japan) **12**, 991 (1964).

11. ERDTMAN, H., B. KIMLAND, T. NORIN, and P.J.L. DANIELS: The Constituents of the

"Pocket Resin" from Douglas Fir *Pseudotsuga menziesii* (Mirb.) Franco. Acta Chem. Scand. **22**, 930–942 (1968).

12. KIMLAND, B., and T. NORIN: Thunbergol, a New Macrocyclic Diterpene Alcohol. Acta Chem. Scand. **22**, 943–948 (1968).

13. PENTEGOVA, V.A., and N.K. KASHTANOVA: Diterpene Hydrocarbons of the Resin of *Pinus sibirica* R. Mayr. Khim. Prir. Soedin., 223–224 (1965).

14. PATIL, V.D., U.R. NAYAK, and S. DEV: Chemistry of Ayurvedic Crude Drugs-II. *Guggulu* (Resin from *Commiphora mukul*)-2: Diterpenoid Constituents. Tetrahedron **29**, 341–348 (1973).

15. SHMIDT, E.N., N.K. KASHTANOVA, and V.A. PENTEGOVA: Neocembrene, a New Diterpenic Hydrocarbon from *Picea obovata* and *Pinus koraiensis*. Khim. Prir. Soedin. 694–698 (1970).

16. KASTHANOVA, N.K., A.I. LISINA, and V.A. PENTEGOVA: The New Diterpenes, Isocembrene and Isocembrol, from Oleoresin of *Pinus sibirica*. Khim. Prir. Soedin. 52–53 (1968).

17. WAHLBERG, I., I. WALLIN, C. NARBONNE, T. NISHIDA, and C.R. ENZELL: Note on the Stereostructures of Thunbergol (Isocembrol) and 4-Epiisocembrol. Acta Chem. Scand. **B35**, 65–68 (1981).

18. RALDUGIN, V.A., and V.A. PENTEGOVA: New Diterpenoid Components of the Oleoresin of *Pinus koraiensis*. Khim. Prir. Soedin. 174 (1976).

19. — —: 4-Epiisocembrol, a New Diterpenoid from Oleoresins of *Pinus koraiensis* and *Pinus sibirica*. Khim. Prir. Soedin. 669–670 (1971).

20. LISINA, A.I., A.I. REZVUKHIN, and V.A. PENTEGOVA: Composition of the Neutral Part of the Resin *Pinus sibirica*. II. Oxygen Containing Compounds of the High Boiling Neutral Part of the Cedar Resin. Khim. Prir. Soedin. 250–256 (1965).

21. RALDUGIN, V.A., and V.A. PENTEGOVA: Absolute Configuration of Isocembrol and Its Stereospecific Synthesis from Cembrene. Khim. Prir. Soedin. 577–578 (1977).

22. ZDERO, C., F. BOHLMANN, R.M. KING, and H. ROBINSON: Sesquiterpene Lactones and other Constituents from Australian *Helipterum* Species. Phytochem. **28**, 517–526 (1989).

23. SATO, A., M. KURABAYASHI, A. OGISO, and H. KUWANO: Poilaneic Acid, a Cembranoid Diterpene from *Croton poilanei*. Phytochem. **20**, 1915–1918 (1981).

24. PRASAD, R.S., and S. DEV: Chemistry of Ayurvedic Crude Drugs-IV *Guggulu* (Resin from *Commiphora mukul*)-4. Absolute Stereochemistry of Mukulol. Tetrahedron **32**, 1437–1441 (1976).

25. RÜCKER, G.: Über monocyclische Diterpene aus dem indischen Guggul-Harz (*Commiphora mukul*). Arch. Pharm. **305**, 486–493 (1972).

26. RALDUGIN, V.A., O.B. SHELEPINA, I.P. SEKATSIS, A.I. REZVUKHIN, and V.A. PENTEGOVA: Configuration of the C-3 Double Bond and Partial Synthesis of Allylcembrol. Khim. Prir. Soedin. 108–109 (1976).

27. TOUBIANA, R., M.J. TOUBIANA, A.T. MCPHAIL, R.W. MILLER, and K. TORI: Structures and Conformations of the Fourteen-membered Ring Diterpene Ovatodiolide and Its Acid Cyclization Product: Nuclear Overhauser Effect Studies in Solution and X-Ray Crystal Structure Analyses of Ovatodiolide and Ovatodiolic Acid. J. Chem. Soc. Perkin II, 1881–1889 (1976).

28. MANCHAND, P.S., and J.F. BLOUNT: Stereostructures of the Macrocyclic Diterpenoids Ovatodiolide and Isoovatodiolide. J. Organ. Chem. **42**, 3824–3828 (1977).

29. PURUSHOTHAMAN, K.K., R. BHIMA RAO, and K. KALYANI: Ovatodiolide and Anisomelic Acid, Two Diterpenoid Lactones from *Anisomeles malabarica* R.Br. Indian J. Chem. **13**, 1357–1358 (1975).

30. IMMER, H., J. POLONSKY, R. TOUBIANA, and H.D. AN: Structure of Ovatodiolide, a Macrocyclic Diterpene Isolated from *Anisomeles ovata*. Tetrahedron **21**, 2117–2131 (1965).

31. KOSELA, S., E.L. GHISALBERTI, P.R. JEFFERIES, B.W. SKELTON, and A.H. WHITE: Unsaturated Cembrane Acids from *Cleome viscosa* L. (Capparidaceae). Austral. J. Chem. **38**, 1365–1370 (1985).

32. JENTE, R., J. JAKUPOVIC, and G.A. OLATUNJI: A Cembranoid Diterpene from *Cleome viscosa*. Phytochem. **29**, 666–667 (1990).

33. GHISALBERTI, E.L., P.R. JEFFERIES, and T.A. MORI: The Chemistry of *Eremophila* spp. XXV. New Cembrene Derivatives from *E. dempsteri*, *E. platycalyx and E. fraseri*. Austral. J. Chem. **39**, 1703–1710 (1986).

34. COATES, P., E.L. GHISALBERTI, and P.R. JEFFERIES: The Chemistry of *Eremophila* spp. VIII. A Cembrenetriol from *E. clarkei*. Austral. J. Chem. **30**, 2717–2721 (1977).

35. GHISALBERTI, E.L., P.R. JEFFERIES, T.A. MORI, V.A. PATRICK, and A.H. WHITE: The Chemistry of *Eremophila* spp. XIX. New Cembrane Diterpenes from *E. granitica* and *E. abietina*. Austral. J. Chem. **36**, 1187–1196 (1983).

36. GHISALBERTI, E.L., P.R. JEFFERIES, J.R. KNOX, and P.N. SHEPPARD: The Chemistry of *Eremophila* spp. VII. An Epoxycembradienol from *Eremophila georgei*. Tetrahedron **33**, 3301–3303 (1977).

37. GHISALBERTI, E.L., P.R. JEFFERIES, and G.M. PROUDFOOT: The Chemistry of *Eremophila* spp. XV. New Acyclic Diterpenes from *Eremophila* spp. Austral. J. Chem. **34**, 1491–1499 (1981).

38. FORCELLESE, M.L., R. NICOLETTI, and U. PETROSSI: The Structure of Isoincensole-oxide. Tetrahedron **28**, 325–331 (1972).

39. KLEIN, E., and H. OBERMANN: (*S*)-1-Isopropyl-4,8,12-trimethylcyclotetradeca-3*E*,7*E*, 11*E*-trien-1-ol, ein neues Cembrenol aus dem Ätherischen Öl von Olibanum. Tetrahedron Letters 349–352 (1978).

40. NICOLETTI, R., and M.L. FORCELLESE: The Structure of Incensole-oxide. Tetrahedron **24**, 6519–6525 (1968).

41. CORSANO, S., and R. NICOLETTI: The Structure of Incensole. Tetrahedron **23**, 1977–1984 (1967).

42. FORCELLESE, M.L., R. NICOLETTI, and C. SANTARELLI: The Revised Structure of Isoincensole-oxide. Tetrahedron Letters 3783–3786 (1973).

43. BOSCARELLI, A., E. GIGLIO, and C. QUAGLIATA: Structure and Conformation of Incensole Oxide. Acta Crystallogr. **B37**, 744–746 (1981).

44. STRAPPAGHETTI, G., G. PROIETTI, S. CORSANO, and I. GRGURINA: Synthesis of Incensole. Bioorgan. Chem. **11**, 1–3 (1982).

45. KOBAYASHI, H., and S. AKIYOSHI: Terpenoids. VI. The Structure of Thunbergene. Bull. Chem. Soc. Japan **36**, 823–826 (1963).

46. SHMIDT, E.N., A.I. LISINA, and V.A. PENTEGOVA: The Neutral Part of the Oleoresin of Siberian Larch (*Larix sibirica*). Chem. Abstr. **62**, 14732f (1965).

47. DAUBEN, W.G., G.H. BEASLEY, M.D. BROADHURST, B. MULLER, D.J. PEPPARD, P. PESNELLE, and C. SUTER: A Synthesis of Cembrene. A 14-Membered Ring Diterpene. J. Amer. Chem. Soc. **96**, 4724–4726 (1974).

48. — — — — — — —: A Synthesis of (±)-Cembrene, a Fourteen-Membered Ring Diterpene. J. Amer. Chem. Soc. **97**, 4973–4980 (1975).

49. KATO, T., T. KOBAYASHI, T. KUMAGAI, and Y. KITAHARA: Cyclization of Polyenes. XVIII. Selective Synthesis of Cembrene. Synth. Commun. **6**, 365–369 (1976).

50. FARKAS, I., and H. PFANDER: 182. Neue Synthese von (−)-(*R*)-Cembren A, Synthese von (−)-(*R*)-Cembrenen und (+)-(*S*)-Cembren. Helv. Chim. Acta **73**, 1980–1985 (1990).

51. KODAMA, M., Y. MATSUKI, and S. ITO: Syntheses of Macrocyclic Terpenoids by Intramolecular Cyclization I. (±)-Cembrene-A, a Termite Trail Pheromone, and (±)-Nephthenol. Tetrahedron Letters 3065–3068 (1975).

52. KITAHARA, Y., T. KATO, T. KOBAYASHI, and B.P. MOORE: Cyclization of Polyenes XVII. Synthesis and Pheromone Activity of DL-Neocembrene. Chem. Letters 219–222 (1976).

53. TAKAYANAGI, H., T. UYEHARA, and T. KATO: Alternative Synthetic Route to the Cembrene Skeleton. J. Chem. Soc. Chem. Commun. 359–360 (1978).

54. VIG, O.P., R. NANDA, R. GAUBA, and S.K. PURI: Synthesis of (±)-Cembrene-A. Indian J. Chem. **24B**, 918–922 (1985).

55. KATO T., M. SUZUKI, T. KOBAYASHI, and B.P. MOORE: Synthesis and Pheromone Activities of Optically Active Neocembrenes and Their Geometrical Isomers, (E,Z,E)- and (E,E,Z)-Neocembrenes. J. Organ. Chem. **45**, 1126–1130 (1980).

56. SCHWABE, R., I. FARKAS, and H. PFANDER: Synthese von (−)-(R)-Nephthenol und (−)-R-Cembren A. Helv. Chim. Acta **71**, 292–297 (1988).

57. KATO, T., C. KABUTO, K.H. KIM, H. TAKAYANAGI, T. UEYHARA, and Y. KITAHARA: Cyclization of Polyenes. XXV. Conformational Study of Cembrene Type Diterpenes by X-Ray Analysis. Chem. Letters 827–830 (1977).

58. KATO, T., T. KOBAYASHI, and Y. KITAHARA: Cyclization of Polyenes XVI. Biogenetic Type Synthesis of Cembrene Type Compounds. Tetrahedron Letters 3299–3302 (1975).

59. COX, N.J.G., G. PATTENDEN, and S.D. MILLS: Radical Macrocyclizations in Synthesis. A New Approach to Mukulol and Marine Cembranolide Lactones. Tetrahedron Letters **30**, 621–624 (1989).

60. GAMOV, N.S., M.A. CHIRKOVA, T.F. TITOVA, V.A. RALDUGIN, and V.A. PENTEGOVA: Labdane and Cembrane Diterpenoids from Picea ajanensis Resin. Khim. Prir. Soedin. 178–181 (1981).

61. KATO, T., M. SUZUKI, M. TAKAHASHI, and Y. KITAHARA: Cyclization of Polyenes XXII. Synthesis and Stereochemistry of Thunbergol. Chem. Letters 465–466 (1977).

62. ASTLES, P.C., and E.J. THOMAS: Cembranoid Synthesis Using Ketophosphonate-Aldehyde Cyclization: Syntheses of Thunbergols and α- and β-Cembra-2,7,11-triene-4,6-diols. Synlett 42–45 (1989).

63. PARDHY, R.S., and S.C. BHATTACHARYYA: Structure of Serratol, a New Diterpene Cembranoid Alcohol from Boswellia serrata Roxb. Indian J. Chem. **16B**, 171–173 (1978).

64. SUZUKI, M., A. SHIMADA, and T. KATO: Cyclization of Polyenes. XXX. Synthesis of 2-Hydroxy- and 1,2-Dehydro-7,8-oxidonephtenols and Cembrenol. Chem. Letters 759–762 (1978).

65. GACS-BAITZ, E., L. RADICS, G. FARDELLA, and S. CORSANO: ^{13}C NMR Study of Some 14-Membered Macrocyclic Diterpenes. J. Chem. Research (M) 1701–1709 (1978).

66. KATO, T., C.C. YEN, T. KOBAYASHI, and Y. KITAHARA: Cyclization of Polyenes XXI. Synthesis of DL-Incensole. Chem. Letters 1191–1192 (1976).

67. KATO, T., C.C. YEN, T. UYEHARA, and Y. KITAHARA: Cyclization of Polyenes XXIII. Synthesis and Stereochemistry of Isoincensole-oxide. Chem. Letters 565–568 (1977).

68. MASLEN, E.N., C.L. RASTON, and A.H. WHITE: Crystal Structure of an Epoxycembra-dienol, 3,15-Epoxy-4-Hydroxycembra-7(Z),11(Z)-diene. Tetrahedron **33**, 3305–3311 (1977).

69. ———: Crystal Structure of (Z)-Cembr-4-ene-15,19,20-triol. Austral. J. Chem. **30**, 2723–2727 (1977).

70. MARSHALL, J.A., and B.S. DeHOFF: Stereoselective Total Synthesis of the Cembranolide Diterpene Anisomelic Acid. Tetrahedron Letters **27**, 4873–4876 (1986).

71. MARSHALL, J.A., and B.S. DEHOFF: Cembranolide Total Synthesis. Anisomelic Acid. Tetrahedron **43**, 4849–4860 (1987).

72. SPRINGER, J.P., J. CLARDY, R.H. COX, H.G. CUTLER, and R.J. COLE: The Structure of a New Type of Plant Growth Inhibitor Extracted from Immature Tobacco Leaves. Tetrahedron Letters, 2737–2740 (1975).

73. AASEN, A.J., N. JUNKER, and C.R. ENZELL: Tobacco Chemistry 36. Absolute Configuration of Tobacco Thunberganoids. Tetrahedron Letters 2607–2610 (1975).

74. WAHLBERG, I., I. WALLIN, C. NARBONNE, T. NISHIDA, C.R. ENZELL, and J.-E. BERG: Tobacco Chemistry. 55. Three New Cembranoids from Greek Tobacco. The Stereochemistry of (1S,2E,4S,6R,7E,11E)-2,7,11-Cembratriene-4,6-diol. Acta Chem. Scand. **B36**, 147–153 (1982).

75. KEENE, C.K., and G.J. WAGNER: Direct Demonstration of Duvatrienediol Biosynthesis in Glandular Heads of Tobacco Trichomes. Plant Physiol. **79**, 1026–1032 (1985).

76. SEVERSON, R.F., R.F. ARRENDALE, O.T. CHORTYK, A.W. JOHNSON, D.M. JACKSON, G.R. GWYNN, J.F. CHAPLIN, and M.G. STEPHENSON: Quantitation of the Major Cuticular Components from Green Leaf of Different Tobacco Types. J. Agric. Food Chem. **32**, 566–570 (1984).

77. CROMBIE, L., D. MCNAMARA, D.F. FIRTH, S. SMITH, and P.C. BEVAN: Biosynthetic Precursors for α- and β-Cembrenediol Formation in Tobacco. Phytochem. **27**, 1685–1693 (1988).

78. BEGLEY, M.J., L. CROMBIE, D. MCNAMARA, D.F. FIRTH, S. SMITH, and P.C. BEVAN: Cembranediols in the Curing of Tobacco. X-Ray Crystal Structures of β-Cembrenediol and α-Cembreneketol. Phytochem. **27**, 1695–1703 (1988).

79. SHEPHERD, C.J., and M. MANDRYK: Germination of Conidia of *Peronospora tabacina* Adam (II. Germination *in vivo*). Austral. J. Biol. Sci. **16**, 77–87 (1963).

80. JOHNSON, A.W., and R.F. SEVERSON: Leaf Surface Chemistry of Tobacco Budworm Resistant Tobacco. J. Agric. Entomol. **1**, 23–32 (1984).

81. SEVERSON, R.F., G.R. GWYNN, J.F. CHAPLIN, and J.D. MILES: Leaf Trichome Exudate Associated with Insect Resistance in *Nicotiana tabacum* L. Tobacco Science **27**, 82–83 (1983).

82. CUTLER, H.G., W.W. REID, and J. DELETANG: Plant Growth Inhibiting Properties of Diterpenes from Tobacco. Plant Cell Physiol. **18**, 711–714 (1977).

83. KOSEKI, K., F. SAITO, N. KAWASHIMA, and M. NOMA: New Cembranoic Diterpene with IAA Inhibitory Activity from *Nicotiana tabacum*. Agric. Biol. Chem. **50**, 1917–1918 (1986).

84. SAITO, Y., H. TAKIZAWA, S. KONISHI, D. YOSHIDA, and S. MIZUSAKI: Identification of Cembratriene-4,6-diol as Antitumor-Promoting Agent from Cigarette Smoke Condensate. Carcinogenesis **6**, 1189–1194 (1985).

85. SAITO, Y., H. NISHINO, D. YOSHIDA, S. MIZUSAKI, and A. OHNISHI: Inhibition of 12-O-Tetradecanoylphorbol-13-Acetate-Stimulated $^{32}P_i$ Incorporation into Phospholipids and Protein Phosphorylation by 2,7,11-Cembratriene-4,6-diol, an Antitumor-Promoting Agent. Oncology **45**, 122–126 (1988).

86. SAITO, Y., Y. TSUJINO, H. KANEKO, D. YOSHIDA, and S. MIZUSAKI: Inhibitory Effects of Cembratriene-4,6-diol Derivatives on the Induction of Epstein-Barr Virus Early Antigen by 12-O-Tetradecanoylphorbol-13-acetate. Agric. Biol. Chem. **51**, 941–943 (1987).

87. WAHLBERG, I., and C.R. ENZELL: Tobacco Cembranoids. Beitr. Tabakforsch. **12**, 93–104 (1984).

88. — —: Tobacco Isoprenoids. Nat. Prod. Rep. **4**, 237–276 (1987).

89. WAHLBERG, I., K. NORDFORS, C. VOGT, T. NISHIDA, and C.R. ENZELL: Tobacco Chemistry. 60. Five New Hydroperoxycembratrienediols from Tobacco. Acta Chem. Scand. **B37**, 653–656 (1983).

90. WAHLBERG, I., R. ARNDT, I. WALLIN, C. VOGT, T. NISHIDA, and C.R. ENZELL: Tobacco Chemistry. 59. Six New Cembratrienetriols from Tobacco. Acta Chem. Scand. **B38**, 21–30 (1984).

91. SINNWELL, V., V. HEEMANN, A.-M. BYLOV, W. HASS, C. KAHRE, and F. SEEHOFER: A New Cembranoid from Tobacco, IV. Z. Naturforsch. **39c**, 1023–1026 (1984).

92. KINZER, G.W., T.F. PAGE, and R.R. JOHNSON: Structure of Two Solanone Precursors from Tobacco. J. Organ. Chem. **31**, 1797–1800 (1966).

93. WAHLBERG, I., R. ARNDT, T. NISHIDA, and C.R. ENZELL: Tobacco Chemistry. 63. Syntheses and Stereostructures of Six Tobacco *Seco*-Cembranoids. Acta Chem. Scand. **B40**, 123–134 (1986).

94. OLSSON, E., A.-M. EKLUND, and I. WAHLBERG: Tobacco Chemistry. 72. Five New Cembratrienetriols from Tobacco. Acta Chem. Scand. **45**, 92–98 (1991).

95. WAHLBERG, I., I. FORSBLOM, C. VOGT, A.-M. EKLUND, T. NISHIDA, C.R. ENZELL, and J.-E. BERG: Tobacco Chemistry. 62. Five New Cembranoids from Tobacco. J. Organ. Chem. **50**, 4527–4538 (1985).

96. NISHIDA, T., I. WAHLBERG, K. NORDFORS, C. VOGT, and C.R. ENZELL: Application of 2D-NMR Spectroscopy in the Structural Determination of a New Tobacco Cembranoid. Tetrahedron Letters **25**, 1299–1302 (1984).

97. ARNDT, R., J.-E. BERG, and I. WAHLBERG: Tobacco Chemistry. 71. Structure Determination and Biomimetic Studies of Five New Tobacco Cembranoids. Acta Chem. Scand. **44**, 814–825 (1990).

98. BEHR, D., I. WAHLBERG, T. NISHIDA, C.R. ENZELL, J.-E. BERG, and A.-M. PILOTTI: Tobacco Chemistry. 51. New Cembranic Diterpenoids from Greek Tobacco. Acta Chem. Scand. **B34**, 195–202 (1980).

99. VOGT, C., J.-E. BERG, and I. WAHLBERG: To be published.

100. WAHLBERG, I., D. BEHR, A.-M. EKLUND, T. NISHIDA, C.R. ENZELL, and J.-E. BERG: Tobacco Chemistry. 54. (1*S*,2*E*,4*S*,6*E*,8*S*,11*R*,12*S*)-8,11-Epoxy-2,6-cembradiene-4,12-diol, a New Constituent of Greek Tobacco. Acta Chem. Scand. **B36**, 37–41 (1982).

101. WAHLBERG, I., A.-M. EKLUND, C. VOGT, C.R. ENZELL, and J.-E. BERG: Tobacco Chemistry. 65. Two New 7,8-Epoxycembranoids from Tobacco. Acta Chem. Scand. **B40**, 855–860 (1986).

102. ARNDT, R., I. WAHLBERG, C.R. ENZELL, and J.-E. BERG: Tobacco Chemistry. 68. Structure Determination and Biomimetic Syntheses of Two New Tobacco Cembranoids. Acta Chem. Scand. **B42**, 294–302 (1988).

103. OLSSON, E., and I. WAHLBERG: To be published.

104. FORSBLOM, I., J.-E. BERG, and I. WAHLBERG: To be published.

105. YAMAZAKI, Y., and Y. MIKAMI: Cembratrienetriol and Production thereof. Patent JP 62-126146, 1987.

106. — —: 2,7,11-Cembratriene-4,6,20-triol, Production thereof and Flavor and Taste Improver Consisting of Said Compound for Tobacco. Patent JP 62-234037, 1987.

107. ZANE, A.: 4,8,13-Duvatriene-1-ol-3-one and 11-Isopropyl-4,8-dimethyl-3,7,12-pentadecatriene-2,14-dione Isomers from *Nicotiana tabacum*. Phytochem. **12**, 731–732 (1973).

108. COURTNEY, J.L., and S. MACDONALD: A New C20 α,β-Unsaturated Aldehyde (3,7,13-Trimethyl-10-isopropyl-2,6,11,13-tetradecatetraen-1-al) (I) from Tobacco. Tetrahedron Letters 459–466 (1967).

109. ROWLAND, R.L., and D.L. ROBERTS: Macrocyclic Diterpenes Isolated from Tobacco. α- and β-3,8,13-Duvatriene-1,5-diols. J. Organ. Chem. **28**, 1165–1169 (1963).

110. WAHLBERG, I., D. BEHR, A.-M. EKLUND, T. NISHIDA, C.R. ENZELL, and J.-E. BERG: Tobacco Chemistry. 56. The Stereochemistries of the Tobacco Diterpenoids: The (1*S*,2*E*,4*S*,6*E*,8*S*,11*E*)- and (1*S*,2*E*,4*R*,6*E*,8*S*,11*E*)-2,6,11-Cembratriene-4,8-diols. Acid-induced Transformations of Cembratrienediols. Acta Chem. Scand. **B36**, 443–449 (1982).

111. LLOYD, R.A., C.W. MILLER, D.L. ROBERTS, J.A. GILES, J.P. DICKERSON, N.H. NELSON, C.E. RIX, and P.H. AYERS: Flue-Cured Tobacco Flavor. I. Essence and Essential Oil Components. Tobacco Science, **20**, 40–48 (1976).

112. BYLOV, A.-M., U. BRÜMMER, W. HASS, F. SEEHOFER, V. HEEMANN, and V. SINNWELL: New Cembranoids from Tobacco, II. Z. Naturforsch. **38c**, 515–516 (1983).

113. TAKAGI, Y., T. FUJIMORI, H. KANEKO, and K. KATO: Cembrene, from Japanese Domestic Tobacco, *Nicotiana tabacum* cv. Suifu. Agric. Biol. Chem. **44**, 467–468 (1980).

114. BRUEMMER, U., C. PAULSEN, G. SPREMBERG, F. SEEHOFER, V. HEEMANN, and V. SINNWELL: New Cembranoids from Burley Tobacco. Z. Naturforsch. **36c**, 1077–1080 (1981).

115. ROBERTS, D.L., and W.A. ROHDE: Isolation and Identification of Flavor Components of Burley Tobacco. Tobacco Science **16**, 107–112 (1972).

116. HEEMANN, V., A.-M. BYLOV, U. BRÜMMER, W. HASS, and F. SEEHOFER: 3,7,11,15-Cembratetraen-6-ol, a New Cembranoid from Tobacco, III. Z. Naturforsch. **38c**, 517–518 (1983).

117. MARSHALL, J.A., E.D. ROBINSON, and J. LEBRETON: Synthesis of the Tumor Inhibitory Tobacco Constituents α- and β-2,7,11-Cembratriene-4,6-diol by Diastereoselective [2,3] Wittig Ring Contraction. J. Organ. Chem. **55**, 227–239 (1990).

118. MARSHALL, J.A., E.D. ROBINSON, and R.D. ADAMS: Stereoselective Total Synthesis of β-2,7,11-Cembratriene-4,6-diol (β-CBT), a Tumor Inhibitory Constituent of Tobacco Smoke, Tetrahedron Letters **29**, 4913–4916 (1988).

119. CUTLER, H.G., and R.J. COLE: Properties of a Plant Growth Inhibitor Extracted from Immature Tobacco Leaves. Plant Cell Physiol. **15**, 19–28 (1974).

120. CUTLER, H.G.: A Growth Inhibitor from Young Expanding Tobacco Leaves. Science **170**, 856–857 (1970).

121. OKAMOTO, H., and D. YOSHIDA: Inhibitor for Aldose Reductase. Patent JP 63-68524, 1988.

122. MARSHALL, J.A., and E.D. ROBINSON: Enantioselective Total Synthesis of (+)-α-2,7,11-Cembratriene-4,6-diol (α-CBT). Tetrahedron Letters **30**, 1055–1058 (1989).

123. ROWLAND, R.L., A. RODGMAN, J.N. SCHUMACHER, D.L. ROBERTS, L.C. COOK, and W.E. WALKER: Macrocyclic Diterpene Hydroxy Ethers from Tobacco and Cigarette Smoke. J. Organ. Chem. **29**, 16–21 (1964).

124. BEHR, D., I. WAHLBERG, A.J. AASEN, T. NISHIDA, C.R. ENZELL, J.-E. BERG, and A.-M. PILOTTI: Tobacco Chemistry. 44. (1*S*,2*E*,4*R*,6*E*,8*R*,11*S*,12*R*)- and (1*S*,2*E*,4*S*,6*E*,8*R*,11*S*,12*R*)-8,11-Epoxy-2,6-thunbergadiene-4,12-diol. Two New Diterpenoids of Greek Tobacco. Acta Chem. Scand. **B32**, 221–227 (1978).

125. AASEN, A.J., Å. PILOTTI, C.R. ENZELL, J.-E. BERG, and A.-M. PILOTTI: Tobacco Chemistry. 31. (1*S*,4*S*,8*R*,11*S*,12*R*)-8,12-Epoxy-2*E*,6*E*-thunbergadiene-4,11-diol, a New Constituent of Greek Tobacco. Acta Chem. Scand. **B30**, 999–1000 (1976).

126. NOMA, M., Y. KOSEKI, and S. KUBO: (2*E*,6*E*)-2,6,12(20)-Cembratriene-4,8,11-triol as Tobacco Flavor Improver. Jpn. Kokai Tokkyo Koho JP 61,225,144. Chem. Abstr. **106**, 116738h (1987).

127. WAHLBERG I., C. VOGT, A.-M. EKLUND, and C.R. ENZELL: Tobacco Chemistry. 67. Two New 20-Norcembranoids from Tobacco. Acta Chem. Scand. **B41**, 749–753 (1987).

128. BIRCH, A.J., W.V. BROWN, J.E.T. CORRIE, and B.P. MOORE: Neocembrene-A, a Termite Trail Pheromone. J. Chem. Soc. Perkin I, 2653–2658 (1972).

129. PRESTWICH, G.D., R.W. JONES, and M.S. COLLINS: Terpene Biosynthesis by Nasute Termite Soldiers (Isoptera: Nasutitermitinae). Insect Biochem. **11**, 331–336 (1981).

130. MCDOWELL, P.G., and G.W. OLOO: Isolation, Identification, and Biological Activity of Trail-Following Pheromone of Termite *Trinervitermes bettonianus* (Sjöstedt) (Termitidae: Nasutitermitinae). J. Chem. Ecol. **10**, 835–851 (1984).

131. EDWARDS, J.P., and J. CHAMBERS: Identification and Source of a Queen-specific Chemical in the Pharaoh's Ant, *Monomorium pharaonis* (L). J. Chem. Ecol. **10**, 1731–1747 (1984).

132. WIEMER, D.F., J. MEINWALD, G.D. PRESTWICH, and I. MIURA: Cembrene A and (3Z)-Cembrene A: Diterpenes from a Termite Soldier (Isoptera: Termitidae; Termitinae). J. Organ. Chem. **44**, 3950–3952 (1979).

133. PRESTWICH, G.D.: Interspecific Variation of Diterpene Composition of *Cubitermes* Soldier Defense Secretion. J. Chem. Ecol. **10**, 1219–1231 (1984).

134. SHIMADA, K., M. KODAMA, and S. ITO: Synthesis of Macrocyclic Terpenoids by Intramolecular Cyclization. VI. Synthesis of 3Z-Cembrene A and Cembrenene. Tetrahedron Letters **22**, 4275–4276 (1981).

135. WENDER, P.A., and D.A. HOLT: Macroexpansion Methodology. 3. Eight-Step Synthesis of (−)-(3Z)-Cembrene A. J. Amer. Chem. Soc. **107**, 7771–7772 (1985).

136. KOKKE, W.C.M.C., S. EPSTEIN, S.A. LOOK, G.H. RAU, W. FENICAL, and C. DJERASSI: On the Origin of Terpenes in Symbiotic Associations between Marine Invertebrates and Algae (Zooxanthellae). J. Biol. Chem. **259**, 8168–8173 (1984).

137. RICE, J.R., C. PAPASTEPHANOU, and D.G. ANDERSON: Isolation, Localization and Biosynthesis of Crassin Acetate in *Pseudoplexaura porosa* (Houttuyn). Biol. Bull. **138**, 334–343 (1970).

138. ALBERICCI, M., J.C. BRAEKMAN, D. DALOZE, B. TURSCH, J.P. DECLERCQ, G. GERMAIN, and M. VAN MEERSSCHE: Chemical Studies of Marine Invertebrates. XXXV. Sarcoglaucol, a Novel Cembrane Diterpene from the Soft Coral *Sarcophyton glaucum* (Coelenterata, Octocorallia). Bull. soc. chim. Belges. **87**, 487–492 (1978).

139. NÉEMAN, I., L. FISHELSON, and Y. KASHMAN: Sarcophine – a New Toxin from the Soft Coral *Sarcophyton glaucum* (Alcyonaria). Toxicon **12**, 593–598 (1974).

140. UCHIO, Y., S. EGUCHI, J. KURAMOTO, M. NAKAYAMA, and T. HASE: Denticulatolide, an Ichthyotoxic Peroxide-Containing Cembranolide from the Soft Coral *Lobophytum denticulatum*. Tetrahedron Letters **26**, 4487–4490 (1985).

141. TURSCH, B., J.C. BRAEKMAN, D. DALOZE, H. DEDEURWAERDER, and R. KARLSSON: Chemical Studies of Marine Invertebrates. XXXI. Crassolide, a Highly Oxygenated Diterpene from the Soft Coral *Lobophytum crassum* (Coelenterata, Octocorallia, Alcyonacea). Bull. soc. chim. Belges. **87**, 75–81 (1978).

142. TURSCH, B.: Some Recent Developments in the Chemistry of Alcyonaceans. Pure & Applied Chem. **48**, 1–6 (1976).

143. COLL J.C., B.F. BOWDEN, G.M. KÖNIG, R. BRASLAU, and I.R. PRICE: Studies of Australian Soft Corals. XXXX. The Natural Products Chemistry of Alcyonacean Soft Corals with Special Reference to the Genus *Lobophytum*. Bull. soc. chim. Belges. **95**, 815–834 (1986).

144. COLL, J.C., B.F. BOWDEN, A. HEATON, P.J. SCHEUER, M.K.W. LI, J. CLARDY, G.K. SCHULTE, and J. FINER-MOORE: Structures and Possible Functions of Epoxypukalide

and Pukalide. Diterpenes Associated with Eggs of Sinularian Soft Corals (Cnidaria, Anthozoa, Octacorallia, Alcyonacea, Alcyoniidae). J. Chem. Ecol. **15**, 1177–1191 (1989).

145. FAULKNER, D.J.: Marine Natural Products: Metabolites of Marine Invertebrates. Nat. Prod. Rep. **1**, 551–598 (1984).

146. —: Marine Natural Products. Nat. Prod. Rep. **3**, 1–33 (1986).

147. —: Marine Natural Products. Nat. Prod. Rep. **4**, 539–576 (1987).

148. —: Marine Natural Products. Nat. Prod. Rep. **5**, 613–663 (1988).

149. —: Marine Natural Products. Nat. Prod. Rep. **7**, 269–309 (1990).

150. TURSCH, B., J.C. BRAEKMAN, D. DALOZE, and M. KAISIN: Terpenoids from Coelenterates. Marine Natural Products. Chemical and Biological Perspectives. (P. J. SCHEUER, Ed.) New York: Academic Press **2**, 247–296 (1978).

151. FENICAL, W.: Diterpenoids. Marine Natural Products. Chemical and Biological Perspectives. (P.J. SCHEUER, Ed.) New York: Academic Press **2**, 173–245 (1978).

152. KASHMAN Y., A. GROWEISS, S. CARMELY, Z. KINAMONI, D. CZARKIE, and M. ROTEM: Recent Research in Marine Natural Products from the Red Sea. Pure & Appl. Chem. **54**, 1995–2010 (1982).

153. KREBS, H.C.: Recent Developments in the Field of Marine Natural Products with Emphasis on Biologically Active Compounds. Fortschr. Chem. organ. Naturstoffe **49**, 151–363 (1986).

154. MORALES, J.J., J.R. ESPINA, and A.D. RODRIGUEZ: The Structure of Euniolide, a New Cembranoid Diterpene from the Caribbean Gorgonians *Eunicea succinea* and *Eunicea mammosa*. Tetrahedron **46**, 5889–5894 (1990).

155. GOPICHAND, Y., L.S. CIERESZKO, F.J. SCHMITZ, D. SWITZNER, A. RAHMAN, M.B. HOSSAIN, and D. VAN DER HELM: Further Studies of the Terpenoid Content in the Gorgonian *Eunicea succinea*: 12,13-Bisepieupalmerin, a New Cembranolide. J. Nat. Prod. **47**, 607–614 (1984).

156. WEINHEIMER, A.J., R.E. MIDDLEBROOK, J.O. BLEDSOE, W.E. MARSICO, and T.K.B. KARNS: Eunicin, an Oxa-bridged Cembranolide of Marine Origin. Chem. Commun. 384–385 (1968).

157. CIERESZKO, L.S., D.S. SIFFORD, and A.J. WEINHEIMER: Chemistry of Coelenterates. I. Occurrence of Terpenoid Compounds in Gorgonians. Ann. New York Acad. Sci. **90**, 917–919 (1960).

158. SANDUJA R., G.S. LINZ, M. ALAM, A.J. WEINHEIMER, G.E. MARTIN, and E.L. EZELL: Two-Dimensional NMR Studies of Marine Natural Products. IV. Isolation of the Cembranoid Diterpene Jeunicin from the Mollusc *Planaxis sulcatus*: Assignment of the Proton and Carbon NMR Spectra by Two-Dimensional Techniques. J. Heterocyclic Chem. **23**, 529–535 (1986).

159. VAN DER HELM, D., E.L. ENWALL, A.J. WEINHEIMER, T.K.B. KARNS, and L.S. CIERESZKO: *p*-Iodobenzoate of Jeunicin. Acta Crystallogr. **B32**, 1558–1560 (1976).

160. GROSS, R.A.: Cueunicin and Cueunicin Acetate – Two New Marine Cembranolides. II. Structure and PMR Spectra of Natural Products. The Association of High Field Vinyl Methyl Signals with *Trans* Double Bond Geometry in Germacrene Derivatives. III. New Marine Diterpenoids. Dissertation, University of Oklahoma, USA, 1974.

161. REHM, S.J.: I. Eupalmerin Acetate: A New Marine Epoxy Cembranolide. II. *Eunicea palmeri* Bayer: An Ambivalent Species. Dissertation, University of Oklahoma, Norman, 1971.

162. CIERESZKO, L.S., and T.K.B. KARNS: In: Biology and Geology of Coral Reef; JONES, O.A. and ENDEAU, R. Eds. New York: Academic Press, 183–203 (1973).

163. CHANG, C.Y.: Dissertation, University of Oklahoma, USA., 1977.

164. CHANG, C.Y., L.S. CIERESZKO, M.B. HOSSAIN, and D. VAN DER HELM: Structure of Pounicln. Acta Crystallogr. **B36**, 731–733 (1980).

165. WEINHEIMER, A.J., J.A. MATSON, D. VAN DER HELM, and M. POLING: Marine Anticancer Agents: Asperdiol, a Cembranoid from the Gorgonians, *Eunicea asperula* and *E. tourneforti*. Tetrahedron Letters 1295–1298 (1977).

166. BANDURRAGA, M.M., B. MCKITTRICK, W. FENICAL, E. ARNOLD, and J. CLARDY: Diketone Cembrenolides from the Pacific Gorgonian *Lophogorgia alba*. Tetrahedron **38**, 305–310 (1982).

167. FENICAL, W., R.K. OKUDA, M.M. BANDURRAGA, P. CULVER and R.S. JACOBS: Lophotoxin: A Novel Neuromuscular Toxin from Pacific Sea Whips of the Genus *Lophogorgia*. Science **212**, 1512–1514 (1981).

168. KSEBATI, M.B., L.S. CIERESZKO, and F.J. SCHMITZ: 11β,12β-Epoxypukalide, a Furanocembranolide from the Gorgonian *Leptogorgia setacea*. J. Nat. Prod. **47**, 1009–1012 (1984).

169. LI, M.K.W.: Dissertation, University of Hawaii at Manoa, Honolulu, Hawaii, U.S.A., 1986.

170. EALICK, S.E., D. VAN DER HELM, R.A. GROSS, A.J. WEINHEIMER, and L.S. CIERESZKO: The Structure and Absolute Configuration at Low Temperature of the Acetate Derivative of Plexaurolone, a Marine Cembranoid. Acta Crystallogr. **B36**, 1901–1907 (1980).

171. CHAN, W.R., W.F. TINTO, P.S. MANCHAND, L.J. TODARO, and L.S. CIERESZKO: New Cembranoids from *Plexaura*. Tetrahedron **45**, 103–106 (1989).

172. WEINHEIMER, A.J., and J.A. MATSON: Crassin Acetate, the Principal Antineoplastic Agent in Four Gorgonians of the *Pseudoplexaura* Genus. Lloydia, **38**, 378–382 (1975).

173. MARSHALL, J.A., L.J. KARAS, and M.J. COGHLAN: Interconversion of Cembranolide δ- and γ-Lactones: Synthesis of the C-1 Epimer of Isolobophytolide. J. Organ. Chem. **47**, 699–701 (1982).

174. FENICAL, W.: Marine Soft Corals of the Genus *Pseudopterogorgia*: a Resource for Novel Anti-inflammatory Diterpenoids. J. Nat. Prod. **50**, 1001–1008 (1987).

175. WRIGHT, A.E., N.S. BURRES, and G.K. SCHULTE: Cytotoxic Cembranoids from the Gorgonian *Pseudopterogorgia bipinnata*. Tetrahedron Letters **30**, 3491–3494 (1989).

176. D'AMBROSIO, M., D. FABBRI, A. GUERRIERO, and F. PIETRA: 7. Coralloidolide A and Coralloidolide B, the First Cembranoids from a Mediterranean Organism, the Alcyonacean *Alcyonium coralloides*. Helv. Chim. Acta **70**, 63–70 (1987).

177. D'AMBROSIO, M., A. GUERRIERO, and F. PIETRA: 176. Novel Cembranolides (Coralloidolide D and E) and a 3,7-Cyclized Cembranolide (Coralloidolide C) from the Mediterranean Coral *Alcyonium coralloides*. Helv. Chim. Acta **72**, 1590–1596 (1989).

178. WILLIAMS, D., R.J. ANDERSEN, G.D. VAN DUYNE, and J. CLARDY: Cembrane and Pseudopterane Diterpenes from the Soft Coral *Gersemia rubiformis*. J. Organ. Chem. **52**, 332–335 (1987).

179. WILLIAMS, D.E., R.J. ANDERSEN, J.F. KINGSTON, and A.G. FALLIS: Minor Metabolites of the Cold Water Soft Coral *Gersemia rubiformis*. Canad. J. Chem. **66**, 2928–2934 (1988).

180. KOBAYASHI, M., T. ISHIZAKA, N. MIURA, and H. MITSUHASHI: Marine Terpenes and Terpenoids. III. Isolation and Structures of Two Cembrane Diols from the Soft Coral *Sinularia mayi*. Chem. Pharm. Bull (Japan) **35**, 2314–2318 (1987).

181. KUSUMI, T., I. OHTANI, Y. INOUYE, and H. KAKISAWA: Absolute Configurations of Cytotoxic Marine Cembranolides; Consideration of Mosher's Method. Tetrahedron Letters **29**, 4731–4734 (1988).

182. UCHIO, Y., J. TOYOTA, H. NOZAKI, M. NAKAYAMA, Y. NISHIZONO, and T. HASE: Lobohedleolide and (7Z)-Lobohedleolide, New Cembranolides from the Soft Coral *Lobophytum hedleyi* Whitelegge. Tetrahedron Letters **22**, 4089–4092 (1981).

183. TURSCH, B., J.C. BRAEKMAN, D. DALOZE, M. HERIN, and R. KARLSSON: Chemical Studies of Marine Invertebrates. X. Lobophytolide, a New Cembranolide Diterpene from the Soft Coral *Lobophytum cristagalli* (Coelenterata, Octocorallia, Alcyonacea). Tetrahedron Letters 3769–3772 (1974).

184. FRINCKE, J.M., D.E. MCINTYRE, and D.J. FAULKNER: Deoxosarcophine from a Soft Coral, *Sarcophyton* Sp. Tetrahedron Letters **21**, 735–738 (1980).

185. KASHMAN, Y., and A. GROWEISS: Lobolide: A New Epoxy Cembranolide from Marine Origin. Tetrahedron Letters 1159–1160 (1977).

186. KASHMAN, Y., S. CARMELY, and A. GROWEISS: Further Cembranoid Derivatives from the Red Sea Soft Corals *Alcyonium flaccidum* and *Lobophytum crassum*. J. Organ. Chem. **46**, 3592–3596 (1981).

187. KINAMONI, Z., A. GROWEISS, S. CARMELY, Y. KASHMAN, and Y. LOYA: Several New Cembranoid Diterpenes from Three Soft Corals of the Red Sea. Tetrahedron **39**, 1643–1648 (1983).

188. BOWDEN, B.F., J.A. BRITTLE, J.C. COLL, N. LIYANAGE, S.J. MITCHELL, and G.J. STOKIE: Studies of Australian Soft Corals. VI. A New Cembranolide Diterpene from the Soft Coral *Lobophytum crassum* (Coelenterata, Anthozoa, Octocorallia, Alcyonacea). Tetrahedron Letters 3661–3662 (1977).

189. AHOND, A., B.F. BOWDEN, J.C. COLL, J.-D. FOURNERON, and S.J. MITCHELL: Studies of Australian Soft Corals. XII. Further Cembranolide Diterpenes from *Lobophytum crassospiculatum* and a Correction of a Previous Stereochemical Assignment. Austral. J. Chem. **32**, 1273–1280 (1979).

190. KOBAYASHI, M., T. NAKAGAWA, and H. MITSUHASHI: Marine Terpenes and Terpenoids. I. Structures of Four Cembrane-Type Diterpenes; Sarcophytol-A, Sarcophytol-A Acetate, Sarcophytol-B, and Sarcophytonin-A, from the Soft Coral, *Sarcophyton glaucum*. Chem. Pharm. Bull. (Japan) **27**, 2382–2387 (1979).

191. KOBAYASHI, M., and K. OSABE: Marine Terpenes and Terpenoids. VII. Minor Cembranoid Derivatives, Structurally Related to the Potent Anti-Tumor-Promoter Sarcophytol A, from the Soft Coral *Sarcophyton glaucum*. Chem. Pharm. Bull. (Japan) **37**, 631–636 (1989).

192. FUJIKI, H., M. SUGANUMA, H. SUGURI, S. YOSHIZAWA, K. TAKAGI, and M. KOBAYASHI: Sarcophytols A and B Inhibit Tumor Promotion by Teleocidin in Two-Stage Carcinogenesis in Mouse Skin. J. Cancer Res. Clin. Oncol. **115**, 25–28 (1989).

193. KOBAYASHI, M., and E. NAKANO: Stereochemical Course of the Transannular Cyclization, in Chloroform, of Epoxycembranoids Derived from the Geometrical Isomers of (14S)-14-Hydroxy-1,3,7,11-cembratetraenes. J. Organ. Chem. **55**, 1947–1951 (1990).

194. KOBAYASHI, M., T. IESAKA, and E. NAKANO: Marine Terpenes and Terpenoids. IX. Structures of Six New Cembranoids, Sarcophytols F, K, P, Q, R and S, from the Soft Coral *Sarcophyton glaucum*. Chem. Pharm. Bull. (Japan) **37**, 2053–2057 (1989).

195. NAKAGAWA, T., M. KOBAYASHI, K. HAYASHI, and H. MITSUHASHI: Marine Terpenes and Terpenoids. II. Structures of Three Cembrane-Type Diterpenes, Sarcophytol-C, Sarcophytol-D, and Sarcophytol-E, from the Soft Coral, *Sarcophyton glaucum* Q. et G. Chem. Pharm. Bull. (Japan) **29**, 82–87 (1981).

196. KASHMAN, Y., E. ZADOCK, and I. NÉEMAN: Some New Cembrane Derivatives of Marine Origin. Tetrahedron **30**, 3615–3620 (1974).

197. BERNSTEIN, J., U. SHMEULI, E. ZADOCK, Y. KASHMAN, and I. NÉEMAN: Sarcophine, a New Epoxy Cembranolide from Marine Origin. Tetrahedron **30**, 2817–2824 (1974).

198. BOWDEN, B.F., J.C. COLL, and B.J. MITCHELL: Studies of Australian Soft Corals. XVIII. Further Cembranoid Diterpenes from Soft Corals of the Genus *Sarcophyton*. Austral. J. Chem. **33**, 879–884 (1980).

199. BOWDEN, B.F., J.C. COLL, W. HICKS, R. KAZLAUSKAS, and S.J. MITCHELL: Studies of Australian Soft Corals. X. The Isolation of Epoxyisoneocembrene-A from *Sinularia grayi* and Isoneocembrene-A from *Sarcophyton ehrenbergi*. Austral. J. Chem. **31**, 2707–2712 (1978).

200. BOWDEN, B.F., J.C. COLL, A. HEATON, and G. KÖNIG: The structures of Four Isomeric Dihydrofuran-Containing Cembranoid Diterpenes from Several Species of Soft Coral. J. Nat. Prod. **50**, 650–659 (1987).

201. MCMURRY, J.E., J.G. RICO, and Y. SHIH: Synthesis and Stereochemistry of Sarcophytol B: An Anticancer Cembranoid. Tetrahedron Letters **30**, 1173–1176 (1989).

202. KAZLAUSKAS, R., J.A. BAIRD-LAMBERT, P.T. MURPHY, and R.J. WELLS: Two New Cembrane Diterpenes from a Soft Coral (*Sarcophyton* Species). Austral. J. Chem. **35**, 61–68 (1982).

203. BOWDEN, B.F., J.C. COLL, and R.H. WILLIS: Studies of Australian Soft Corals. XXVII. Two Novel Diterpenes from *Sarcophyton glaucum*. Austral. J. Chem. **35**, 621–627 (1982).

204. TOTH, J.A., B.J. BURRESON, P.J. SCHEUER, J. FINER-MOORE and J. CLARDY: Emblide, a New Polyfunctional Cembranolide from the Soft Coral *Sarcophyton glaucum*. Tetrahedron **36**, 1307–1309 (1980).

205. UCHIO, Y., M. NITTA, M. NAKAYAMA, T. IWAGAWA, and T. HASE: Ketoemblide and Sarcophytolide, Two New Cembranolides with ε-Lactone Function from the Soft Coral *Sarcophyta elegans*. Chem. Letters 613–616 (1983).

206. CARMELY, S., A. GROWEISS, and Y. KASHMAN: Decaryiol, a New Cembrane Diterpene from the Marine Soft Coral *Sarcophyton decaryi*. J. Organ. Chem. **46**, 4279–4284 (1981).

207. UCHIO, Y., H. NABEYA, M. NAKAYAMA, S. HAYASHI, and T. HASE: Cembrenene and Mayol, Two New Cembranoid Diterpenes from the Soft Coral *Sinularia mayi*. Tetrahedron Letters **22**, 1689–1690 (1981).

208. KOBAYASHI, M., and T. HAMAGUCHI: Marine Terpenes and Terpenoids. VI. Isolation of Several Plausible Precursors of Marine Cembranolides from the Soft Coral, *Sinularia mayi*. Chem. Pharm. Bull. (Japan) **36**, 3780–3786 (1988).

209. UCHIO, Y., S. EGUCHI, M. NAKAYAMA and T. HASE: The Isolation of Two Simple γ-Lactonic Cembranolides from the Soft Coral *Sinularia mayi*. Chem. Letters 277–278 (1982).

210. KOBAYASHI, M.: Marine Terpenes and Terpenoids. IV. Isolation of New Cembranoid and Secocembranoid Lactones from the Soft Coral *Sinularia mayi*. Chem. Pharm. Bull. (Japan) **36**, 488–494 (1988).

211. KOBAYASHI, M., and T. HAMAGUCHI: Marine Terpenes and Terpenoids. X. Acid-Catalyzed Transannular Cyclization of 11,12-Epoxycembranolide. Chem. Pharm. Bull. (Japan) **38**, 664–668 (1990).

212. WEINHEIMER, A.J., J.A. MATSON, M.B. HOSSAIN, and D. VAN DER HELM: Marine Anticancer Agents: Sinularin and Dihydrosinularin, New Cembranolides from the Soft Coral, *Sinularia flexibilis*. Tetrahedron Letters 2923–2926 (1977).

213. KAZLAUSKAS, R., P.T. MURPHY, R.J. WELLS, P. SCHÖNHOLZER and J.C. COLL: Cembranoid Constituents from an Australian Collection of the Soft Coral *Sinularia flexibilis*. Austral. J. Chem. **31**, 1817–1824 (1978).

214. TURSCH, B., J.C. BRAEKMAN, D. DALOZE, M. HERIN, R. KARLSSON, and D. LOSMAN: Chemical Studies of Marine Invertebrates. XI. Sinulariolide, a New Cembranolide

Diterpene from the Soft Coral *Sinularia flexibilis* (Coelenterata, Octocorallia, Alcyonacea). Tetrahedron **31**, 129–133 (1975).

215. KARLSSON, R.: The Structure and Absolute Configuration of Sinulariolide, a Cembranolide Diterpene. Acta Crystallogr. **B33**, 2027–2031 (1977).

216. HERIN, M., and B. TURSCH: Chemical Studies of Marine Invertebrates. XXIV. Minor Cembrane Diterpenes from the Soft Coral *Sinularia flexibilis* (Coelenterata, Octocorallia). Bull. soc. chim. Belges. **85**, 707–719 (1976).

217. MORI, K., S. SUZUKI, K. IGUCHI, and Y. YAMADA: 8,11-Epoxy Bridged Cembranolide Diterpene from the Soft Coral *Sinularia flexibilis*. Chem. Letters 1515–1516 (1983).

218. HERIN, M., M. COLIN and B. TURSCH: Chemical Studies of Marine Invertebrates. XXV. Flexibilene, an Unprecedented Fifteen-membered Ring Diterpene Hydrocarbon from the Soft Coral *Sinularia flexibilis* (Coelenterata, Octocorallia, Alcyonacea). Bull. soc. chim. Belges. **85**, 801–803 (1976).

219. SCHMITZ, F.J., Y. GOPICHAND, D.P. MICHAUD, R.S. PRASAD, S. REMALEY, M.B. HOSSAIN, A. RAHMAN, P.K. SENGUPTA, and D. VAN DER HELM: Recent Developments in Research on Metabolites from Caribbean Marine Invertebrates. Pure & Applied Chem. **51**, 853–865 (1981).

220. SATO, A., W. FENICAL, Z. QI-TAI and J. CLARDY: Norcembrene Diterpenoids from Pacific Soft-Corals of the Genus *Sinularia* (Alcyonacea; Octocorallia). Tetrahedron **41**, 4303–4308 (1985).

221. BOWDEN, B.F., J.C. COLL, S.J. MITCHELL, J. MULDER, and G.J. STOKIE: Studies of Australian Soft Corals. IX. A Novel Nor-Diterpene from the Soft Coral *Sinularia leptoclados*. Austral. J. Chem. **31**, 2049–2056 (1978).

222. LAKSHMI, V., and F.J. SCHMITZ: Metabolites from Two Soft Corals from Guam: *Sinularia leptoclados* and *Sinularia gyrosa*. J. Nat. Prod. **49**, 728–730 (1986).

223. LONG, K., and Y. LIN: Studies of the Chemical Constituents of Chinese Soft Corals. (IV). Zhongshan Daxue Xuebao, Ziran Kexueban 98–104 (1981).

224. KOBAYASHI, M., B. WHA SON, Y. KYOGOKU, and I. KITAGAWA: Kericembrenolides A, B, C, D and E, Five New Cytotoxic Cembrenolides from the Okinawan Soft Coral *Clavularia koellikeri*. Chem. Pharm. Bull. (Japan) **34**, 2306–2309 (1986).

225. BOWDEN, B.F., J.C. COLL, S.J. MITCHELL, C.L. RASTON, G.J. STOKIE, and A.H. WHITE: Studies of Australian Soft Corals. XV. The Structure of Pachyclavulariadiol, a Novel Furano-Diterpene from *Pachyclavularia violacea*. Austral. J. Chem. **32**, 2265–2274 (1979).

226. INMAN, W., and P. CREWS: The Structure and Conformational Properties of a Cembranolide Diterpene from *Clavularia violacea*. J. Organ. Chem. **54**, 2526–2529 (1989).

227. BLACKMAN, A.J., B.F. BOWDEN, J.C. COLL, B. FRICK, M. MAHENDRAN, and S.J. MITCHELL: Studies of Australian Soft Corals. XXIX. Several New Cembranoid Diterpenes from *Nephthea brassica* and Related Diterpenes from a *Sarcophyton* Species. Austral. J. Chem. **35**, 1873–1880 (1982).

228. VANDERAH, D.J., N. RUTLEDGE, F.J. SCHMITZ, and L.S. CIERESZKO: Marine Natural Products: Cembrene-A and Cembrene-C from Soft Coral, *Nephthea* Sp. J. Organ. Chem. **43**, 1614–1616 (1978).

229. RAVI, B.N., and D.J. FAULKNER: Cembranoid Diterpenes from a South Pacific Soft Coral. J. Organ. Chem. **43**, 2127–2131 (1978).

230. GUERRIERO, A., M. D'AMBROSIO, and F. PIETRA: 30. Isolation of the Cembranoid Preverecynarmin Alongside Some Briaranes, the Verecynarmins, from Both the Nudibranch Mollusc *Armina maculata* and the Octocoral *Veretillum cynomorium* of the East Pyrenean Mediterranean Sea. Helv. Chim. Acta **73**, 277–283 (1990).

231. BOWDEN, B.F., J.C. COLL, and D,M TAPIOLAS. Studies of Australian Soft Corals. XXXIII. New Cembranoid Diterpenes from a *Lobophytum* Species. Austral. J. Chem. **36**, 2289–2295 (1983).
232. BOWDEN, B.F., J.C. COLL, and S.J. MITCHELL: Studies of Australian Soft Corals. XXI. A New Sesquiterpene from *Nephthea chabrolii* and an Investigation of the Common Clam *Tridacna maxima*. Austral. J. Chem. **33**, 1833–1839 (1980).
233. POET, S.E., and B.N. RAVI: Three New Diterpenes from a Soft Coral *Nepthea* Species. Austral. J. Chem. **35**, 77–83 (1982).
234. HUANG, J., J. LI, and T. ZHONG: Isolation and Identification of Some Chemical Constituents from a Soft Coral (*Clavularia* sp.) Huaxue Xuebao **43**, 199–202 (1985).
235. AOKI, M., T. UYEHARA, T. KATO, K. KABUTO, and S. YAMAGUCHI: Preparation of Enantiomerically Pure 1-Hydroxy-Neocembrenes to Determine the Unsolved Absolute Configuration of Cembrenoids. Chem. Letters 1121–1124 (1983).
236. UCHIO, Y., M. NITTA, H. NOZAKI, M. NAKAYAMA, T. IWAGAWA, and T. HASE: 10-Oxo-, 10-Hydroxy-, and 10-Methoxycembrenes from the Soft Coral *Sarcophyta elegans*. Chem. Letters 1719–1720 (1983).
237. AOKI, M., T. KATO, Y. UCHIO, M. NAKAYAMA, and M. KODAMA: Characterization of 13-Hydroxyneocembrene from Soft Corals. Bull. Chem. Soc. Japan **58**, 779–780 (1985).
238. BOWDEN, B.F., J.C. COLL, and I.M. VASILESCU: Studies of Australian Soft Corals. XLVI. New Diterpenes from a *Briareum* Species (Anthozoa, Octocorallia, Gorgonacea). Austral. J. Chem. **42**, 1705–1726 (1989).
239. BOWDEN, B.F., J.C. COLL, S.J. MITCHELL, and R. KAZLAUSKAS: Studies of Australian Soft Corals. XXIV. Two Cembranoid Diterpenes from the Soft Coral *Sinularia facile*. Austral. J. Chem. **34**, 1551–1556 (1981).
240. BOWDEN, B.F., J.C. COLL, and A.D. WRIGHT: Studies of Australian Soft Corals. XLIV. New Diterpenes from *Sinularia polydactyla* (Coelenterata, Anthozoa, Octocorallia). Austral. J. Chem. **42**, 757–763 (1989).
241. SULEIMENOVA, A.M., A.I. KALINOVSKII, V.A. RALDUGIN, S.A. SHEVTSOV, I.Y. BAGRYANSKAYA, Y.V. GATILOV, T.A. KUZNETSOVA, and G. B. ELYAKOV: New Derivatives of (−)-Neocembrene from the Soft Coral *Sarcophyton trocheliophorum*. Chem. Nat. Comp. **24**, 453–458 (1988).
242. KUZNETSOVA, T.A., A.M. POPOV, I.G. AGAFONOVA, A.M. SULEIMENOVA, and G.B. ELYAKOV: Physiological Activity of Diterpenoids from the Soft Coral *Sarcophyton trocheliophorum*. Khim. Prir. Soedin. 137–139 (1989).
243. KOBAYASHI, M., K. KONDO, K. OSABE, and H. MITSUHASHI: Marine Terpenes and Terpenoids. V. Oxidation of Sarcophytol A, a Potent Anti-Tumor-Promoter from the Soft Coral *Sarcophyton glaucum*. Chem. Pharm. Bull. (Japan) **36**, 2331–2341 (1988).
244. KOBAYASHI, M., K. KOBAYASHI, M. NOMURA, and H. MUNAKATA: Conformational Study of the Cembranoid Sarcophytol A, a Potent Anti-Tumor-Promoter. Chem. Pharm. Bull. (Japan) **38**, 815–817 (1990).
245. TAKAYANAGI, H., Y. KITANO, and Y. MORINAKA: Stereo- and Enantioselective Total Synthesis of Sarcophytol-A. Tetrahedron Letters **31**, 3317–3320 (1990).
246. KASHMAN, Y., M. BODNER, Y. LOYA, and Y. BENAYAHU: Cembranolides from Marine Origin (Red Sea), Survey, and Isolation of a New Sinulariolide Derivative. Israel. J. Chem. **16**, 1–3 (1977).
247. TURSCH, B., J.C. BRAEKMAN, and D. DALOZE: Chemical Studies of Marine Invertebrates - XIII. 2-Hydroxynephtenol, a Novel Cembrane Diterpene from the Soft Coral *Litophyton viridis* (Coelenterata, Octocorallia, Alcyonacea). Bull. soc. chim. Belges **84**, 767–774 (1975).

248. SCHMITZ, F.J., D.J. VANDERAH, and L.S. CIERESZKO: Marine Natural Products: Nephthenol and Epoxynephthenol Acetate, Cembrene Derivatives from a Soft Coral. J. Chem. Soc. Chem. Commun. 407–408 (1974).

249. GROWEISS, A., Y. KASHMAN, D. J. VANDERAH, B. TURSCH, P. CORNET, J.C. BRAEKMAN, and D. DALOZE: The Structure of Trocheliophorol, a Cembranoid Diterpene from Soft Corals of the Genus *Sarcophyton*. Bull. soc. chim. Belges **87**, 277–283 (1978).

250. COLL, J.C., G.B. HAWES, N. LIYANAGE, W. OBERHÄNSLI, and R.J. WELLS: Studies of Australian Soft Corals. I. A New Cembrenoid Diterpene from a *Sarcophyton* Species. Austral. J. Chem. **30**, 1305–1309 (1977).

251. COLL, J.C., B.F. BOWDEN, D.M. TAPIOLAS, R.H. WILLIS, P. DJURA, M. STREAMER and L. TROTT: Studies of Australian Soft Corals. XXXV. The Terpenoid Chemistry of Soft Corals and its Implications. Tetrahedron **41**, 1085–1092 (1985).

252. KOBAYASHI, J., Y. OHIZUMI, H. NAKAMURA, T. YAMAKADO, T. MATSUZAKI, and Y. HIRATA: Ca-Antagonistic Substance from Soft Coral of the Genus *Sarcophyton*. Experientia **39**, 67–69 (1983).

253. ESTRELLA, D.J., and R.S. JACOBS: Deoxosarcophine (DXS), a Potentiator of Skeletal Muscle Contraction. Fed. Proc. **43**, 586 (1984).

254. BOWDEN, B.F., and J.C. COLL: Studies of Australian Soft Corals. XLV. Epoxidation Reactions of Cembranoid Diterpenes: Stereochemical Outcomes. Heterocycles **28**, 669–672 (1989).

255. BOWDEN, B.F., J.C. COLL, S.J. MITCHELL, and G.J. STOKIE: Studies of Australian Soft Corals. XI. Two New Cembranoid Diterpenes from a *Sarcophyton* Species. Austral. J. Chem. **32**, 653–659 (1979).

256. KODAMA, M., K. SHIMADA, and S. ITO: Syntheses of Macrocyclic Terpenoids by Intramolecular Cyclization. III. Model Reactions for the Synthesis of Cembrene and Thunbergol Derivatives. Tetrahedron Letters 2763–2764 (1977).

257. LINZ, G.S., R. SANDUJA, A.J. WEINHEIMER, M. ALAM, and G.E. MARTIN: Applications of COSY and Homonuclear Relay 2D-NMR in the Determination of the Structure of a New Cembrane Isolated from the Mollusc *Planaxis sulcatus*. Tetrahedron Letters **27**, 4833–4836 (1986).

258. MARTIN, G.E., J.A. MATSON, and A.J. WEINHEIMER: ^{13}C NMR Studies of Marine Natural Products II. Total Assignment of the ^{13}C NMR Spectrum of Asperdiol. Tetrahedron Letters 2195–2198 (1979).

259. AOKI, M., Y. TOOYAMA, T. UYEHARA, and T. KATO: Synthesis of ($\pm$)-Asperdiol, a Marine Anticancer Cembrenoid. Tetrahedron Letters **24**, 2267–2270 (1983).

260. STILL, W.C., and D. MOBILIO: Synthesis of Asperdiol. J. Organ. Chem. **48**, 4785–4786 (1983).

261. KATO, T., M. AOKI, and T. UYEHARA: Cyclization of Polyenes. 46. Synthesis of ($\pm$)-Asperdiol, an Anticancer Cembrenoid. J. Organ. Chem. **52**, 1803–1810 (1987).

262. TIUS, M.A., and A. FAUQ: Total Synthesis of (−)-Asperdiol. J. Amer. Chem. Soc. **108**, 6389–6391 (1986).

263. COLL, J.C., S.J. MITCHELL, and G.J. STOKIE: Studies of Australian Soft Corals. II. A Novel Cembrenoid Diterpene from *Lobophytum michaelae*. Austral. J. Chem. **30**, 1859–1863 (1977).

264. AOKI, M., T. UYEHARA, and T. KATO: Synthesis of Cembra-3E,7E,11E,15(17)-tetraen-cis-16,2-olide. Chem. Letters 695–698 (1984).

265. MARSHALL, J.A., and S.L. CROOKS: Stereoselective Synthesis of Cembranolides via Conjugate Addition to Cycloalkynones. Tetrahedron Letters **28**, 5081–5082 (1987).

266. MARSHALL, J.A., S.L. CROOKS, and B.S. DEHOFF: Cembranolide Total Synthesis. Macrocyclization of (α-Alkoxyallyl)stannane-Acetylenic Aldehydes as a Route to Cembrane Lactones. J. Organ. Chem. **53**, 1616–1623 (1988).

267. MARSHALL, J.A., and W.Y, GUNG: Stereoselective Total Synthesis of Cembranolides through Cyclization of a Homochiral (α-Alkoxyallyl)stannane Precursor. Tetrahedron Letters **29**, 3899–3902 (1988).

268. KODAMA, M., T. TAKAHASHI, and S. ITO: Synthesis of Macrocyclic Terpenoids by Intramolecular Cyclization. VIII. Synthesis of Cembra-3E,7E,11E,15(17)-tetraentrans-16,2-olide. Tetrahedron Letters **23**, 5175–5176 (1982).

269. LIU, Y.-X., L.-M. ZENG, and K.N. TRUEBLOOD: Sarcophinone at 115 K: A Cembranolide Diterpenoid from the South China Sea. Acta Crystallogr. **C42**, 373–376 (1986).

270. BOWDEN, B.F., J.C. COLL, L.M. ENGELHARDT, G.V. MEEHAN, G.G. PEGG, D.M. TAPIOLAS, A.H. WHITE, and R.H. WILLIS: Studies of Australian Soft Corals. XXXVII. The Structure Determination of Two Cembranolide Diterpenes from Soft Corals of the Genus *Efflatounaria* (Coelenterata, Octocorallia, Alcyonacea). Austral. J. Chem. **39**, 123–135 (1986).

271. ERMAN, A., and I. NÉEMAN: Inhibition of Phosphofructokinase by the Toxic Cembranolide Sarcophine Isolated from the Soft-Bodied Coral *Sarcophyton glaucum*. Toxicon **15**, 207–215 (1977).

272. YAMADA, Y., S. SUZUKI, K. IGUCHI, H. KIKUCHI, Y. TSUKITANI, and H. HORIAI: Studies on Marine Natural Products. III. Two New Cembranolides from the Soft Coral *Lobophytum pauciflorum* (Ehrenberg). Chem. Pharm. Bull. (Japan) **28**, 2035–2038 (1980).

273. BOWDEN, B.F., J.C. COLL, S.J. MITCHELL, and G.J. STOKIE: Studies of Australian Soft Corals. VII. Two New Diterpenes from an Unknown Species of Soft Coral (Genus *Lobophytum*). Austral. J. Chem. **31**, 1303–1312 (1978).

274. BOWDEN, B.F., J.C. COLL, M.S.L. DE COSTA, M.F. MACKAY, M. MAHENDRAN, E.D. DE SILVA, and R.H. WILLIS: The Structure Determination of a New Cembranolide Diterpene from the Soft Coral *Lobophytum cristigalli* (Coelenterata, Octocorallia, Alcyonacea). Austral. J. Chem. **37**, 545–552 (1984).

275. BURNS, K.P., R. KAZLAUSKAS, P.T. MURPHY, R.J. WELLS, and P. SCHÖNHOLZER: Two Cembranes from *Cespitularia* Species (Soft Coral). Austral. J. Chem. **35**, 85–94 (1982).

276. SU, J., Z. JIAN, and K. LONG: Studies on the Chinese Soft Corals. Zhongshan Daxue Xuebao, Ziran Kexueban 89–92 (1982).

277. FUKAZAWA, Y., S. USUI, and Y. UCHIO: Conformational Study of the Cembranolide Diterpene Denticulatolide by Molecular Mechanics Method. Tetrahedron Letters **27**, 1825–1828 (1986).

278. LOOK, S.A., W. FENICAL, Q. ZHENG, and J. CLARDY: Calyculones, New Cubitane Diterpenoids from the Caribbean Gorgonian Octocoral *Eunica calyculata*. J. Organ. Chem. **49**, 1417–1423 (1984).

279. LEBIODA, L., J.A. MARSHALL, and M.J. COGHLAN: Structure of the Thermodynamic Benzenethiol 1,4-Adduct of the Cembranolide Crassin Alcohol, $C_{26}H_{36}O_4S$. Acta Crystallogr. **C40**, 1191–1193 (1984).

280. HOSSAIN, M.B., and D. VAN DER HELM: The Crystal Structure of Crassin p-Iodobenzoate. Rec. trav. chim. Pays-Bas **88**, 1413–1419 (1969).

281. MARTIN, G.E., J.A. MATSON, J.C. TURLEY, and A.J. WEINHEIMER: ^{13}C NMR Studies of Marine Natural Products. 1. Use of the SESFORD Technique in the Total ^{13}C NMR Assignment of Crassin Acetate. J. Amer. Chem. Soc. **101**, 1888–1890 (1979).

282. MCMURRY, J.E., and R.G. DUSHIN: Total Synthesis of (±)-Crassin by Titanium-Induced Pinacol Coupling. J. Amer. Chem. Soc. **111**, 8928–8929 (1989).

283. MCMURRY, J.E., and R.G. DUSHIN: Total Synthesis of (±)-Isolobophytolide and (±)-Crassin by Titanium-Induced Carbonyl Coupling. J. Amer. Chem. Soc. **112**, 6942–6949 (1990).

284. DAUBEN, W.G., T.-Z. WANG, and R.W. STEPHENS: Total Synthesis of ($\pm$)-Crassin Acetate Methyl Ether. Tetrahedron Letters 31, 2393–2396 (1990).

285. LEE, W.Y., S.A. MACKO, and L.S. CIERESZKO: Toxic Effects of Cembranolides Derived from Octocorals on the Rotifer *Brachionus plicatilis* Müller and the Amphipod *Parhyale Hawaiensis* (Dana). J. Exp. Mar. Biol. Ecol. 54, 91–96 (1981).

286. CIERESZKO, L.S., and R.R.L. GUILLARD: The Influence of Some Cembranolides from Gorgonian Corals on Motility of Marine Flagellates. J. Exp. Mar. Biol. Ecol. 127, 205–210 (1989).

287. HOSSAIN, M.B., D. VAN DER HELM, J.A. MATSON, and A.J. WEINHEIMER: The Molecular Structures and Absolute Configurations of Two New Marine Cembranolides: Sinularin and Dihydrosinularin. Acta Crystallogr. B35, 660–666 (1979).

288. GAMPE, R.T., A.J. WEINHEIMER, and G.E. MARTIN: ^{13}C NMR Studies of Marine Natural Products. A Further Investigation into the Spin-Lattice Relaxation Behavior of the Cembranoid Diterpene Sinularin. Spectroscopy Letters 15, 47–55 (1982).

289. NORTON, R.S., and R. KAZLAUSKAS: ^{13}C NMR-Study of Flexibilide, an Anti-inflammatory Agent from a Soft Coral. Experientia 36, 276–278 (1980).

290. SANDUJA, R., S.K. SANDUJA, A.J. WEINHEIMER, M. ALAM, and G.E. MARTIN: Isolation of the Cembranolide Diterpenes Dihydrosinularin and 11-Epi-Sinulariolide from the Marine Mollusk *Planaxis sulcatus*. J. Nat. Prod. 49, 718–719 (1986).

291. MARSHALL, J.A., R.C. ANDREWS, and L. LEBIODA: Synthetic Studies on Cembranolides. Stereoselective Total Synthesis of Isolobophytolide. J. Organ. Chem. 52, 2378–2388 (1987).

292. MARSHALL, J.A., and R.C. ANDREWS: Stereoselective Total Synthesis of ($\pm$)-Isolobophytolide, a Marine Cembranolide Natural Product. Tetrahedron Letters 27, 5197–5200 (1986).

293. KARLSSON, R.: Lobophytolide, a Cembranolide Diterpene. Acta Crystallogr. B33, 2032–2034 (1977).

294. WASYLYK, J.M., and M. ALAM: Isolation and Identification of a New Cembranoid Diterpene from the Tunicate *Styela plicata*. J. Nat. Prod. 52, 1360–1362 (1989).

295. FONTAN, L.A., W.Y. YOSHIDA, and A.D. RODRIGUEZ: Application of Two-Dimensional NMR Spectroscopy in the Structural Determination of Marine Natural Products. Total Structural Assignment of the Cembranoid Diterpene Eupalmerin Acetate through the Use of Two-Dimensional ^{1}H-^{1}H, ^{1}H-^{13}C, and ^{13}C-^{13}C Chemical Shift Correlation Spectroscopy. J. Organ. Chem. 55, 4956–4960 (1990).

296. EALICK, S.E., D. VAN DER HELM, and A.J. WEINHEIMER: The Molecular Structures and Absolute Configurations of Eupalmerin Acetate and Eupalmerin Acetate Dibromide at Low Temperature. Acta Crystallogr. B31, 1618–1626 (1975).

297. VAN DER HELM, D., S.E. EALICK, and A.J. WEINHEIMER: 15(R)-Acetoxy-6(S),10(S)-Dibromo-3a(S),4,7,8,11,12,13,14(S),15,15a(R)-decahydro-6,10,14-trimethyl-3-methylene-5(R),9(R)-epoxycyclotetradeca[b]2-furanone, $C_{22}H_{32}Br_2O_5$. Cryst. Struct. Commun. 3, 167–171 (1974).

298. HOSSAIN, M.B., A.F. NICHOLAS, and D. VAN DER HELM: The Molecular Structure of Eunicin Iodoacetate. Chem. Commun. 385–386 (1968).

299. GAMPE, R.T., M. ALAM, A.J. WEINHEIMER, G.E. MARTIN, J.A. MATSON, M.R. WILLCOTT, R.R. INNERS, and R.E. HURD: Total Assignment of the ^{13}C NMR Spectrum of the Cembranoid Diterpene Eunicin through the Use of Two-Dimensional Proton-Carbon Chemical Shift Correlation. J. Amer. Chem. Soc. 106, 1823–1826 (1984).

300. SEN GUPTA, P.K., M.B. HOSSAIN, and D. VAN DER HELM: Structure of Cueunicin Acetate. Acta Crystallogr. C42, 434–436 (1986).

301. WILLIAMS, D.E., and R.J. ANDERSEN: Terpenoid Metabolites from Skin Extracts of the Dendronotid Nudibranch *Tochuina tetraquetra.* Canad. J. Chem. **65**, 2244–2247 (1987).

302. PAQUETTE, L.A., and P.C. ASTLES: Furanocembranolide Interconversions. Transformation of Pseudopterolide into Tobagolide and its Reversal. Tetrahedron Letters **31**, 6505–6508 (1990).

303. CULVER, P., W. FENICAL, and P. TAYLOR: Lophotoxin Irreversibly Inactivates the Nicotinic Acetylcholine Receptor by Preferential Association at One of the Two Primary Agonist Sites. J. Biol. Chem. **259**, 3763–3770 (1984).

304. MISSAKIAN, M.G., B.J. BURRESON and P.J. SCHEUER: Pukalide, a Furanocembranolide from the Soft Coral *Sinularia abrupta.* Tetrahedron **31**, 2513–2515 (1975).

305. NAGAOKA, H., M. IWASHIMA, H. ABE, and Y. YAMADA: Total Synthesis of (+)-Mayolide A: The Absolute Configuration of Mayolide A. Tetrahedron Letters **30**, 5911–5914 (1989).

306. MCMURRY, J.E.: Titanium-Induced Dicarbonyl-Coupling Reactions. Accounts Chem. Res. **16**, 405–411 (1983).

307. YAMADA, Y., S. SUZUKI, K. IGUCHI, K. HOSAKA, H. KIKUCHI, Y. TSUKITANI, H. HORIAI, and F. SHIBAYAMA: Studies on Marine Natural Products. I. 13-Membered Carbocyclic Cembranolide Diterpenes from the Soft Coral *Lobophytum pauciflorum* (Ehrenberg). Chem. Pharm. Bull. (Japan) **27**, 2394–2397 (1979).

308. YAMADA, Y., S. SUZUKI, K. IGUCHI, H. KIKUCHI, Y. TSUKITANI, H. HORIAI, and F. SHIBAYAMA: Studies on Marine Natural Products. IV. The Stereochemistry of 13-Membered Carbocyclic Cembranolide Diterpenes from the Soft Coral *Lobophytum pauciflorum* (Ehrenberg). Tetrahedron Letters **21**, 3911–3914 (1980).

309. BANDURRAGA, M.M., W. FENICAL, S.F. DONOVAN, and J. CLARDY: Pseudopterolide, an Irregular Diterpenoid with Unusual Cytotoxic Properties from the Caribbean Sea Whip *Pseudopterogorgia acerosa* (Pallas) (Gorgonacea). J. Amer. Chem. Soc. **104**, 6463–6465 (1982).

310. TINTO, W.F., W.R. CHAN, W.F. REYNOLDS, and S. MCLEAN: Tobagolide, a New Pseudopterane Diterpenoid of the Octocoral *Pseudopterogorgia acerosa.* Tetrahedron Letters **31**, 465–468 (1990).

311. LOOK, S.A., M.T. BURCH, W. FENICAL, Z. QI-TAI, and J. CLARDY: Kallolide A, a New Antiinflammatory Diterpenoid, and Related Lactones from the Caribbean Octocoral *Pseudopterogorgia kallos* (Bielschowsky). J. Organ. Chem. **50**, 5741–5746 (1985).

312. MCMURRY, J., J.R. MATZ, K.L. KEES, and P.A. BOCK: Synthesis of Flexibilene, a Naturally Occurring 15-Membered-Ring Diterpene. Tetrahedron Letters **23**, 1777–1780 (1982).

313. KOBAYASHI, M., and T. HIRASE: Marine Terpenes and Terpenoids. XI. Structures of New Dihydrofuranocembranoids Isolated from a *Sarcophyton* sp. Soft Coral of Okinawa. Chem. Pharm. Bull. (Japan) **38**, 2442–2445 (1990).

314. MAO, J., Y. LI, Z. HOU, and X. LIANG: Total Synthesis of Cembrene-C. Chin. Science Bull. **35**, 2020–2021 (1990).

315. COLL, J.C., D.M. TAPIOLAS, B.F. BOWDEN, L. WEBB, and H. MARSH: Transformation of Soft Coral (Coelenterata: Octocorallia) Terpenes by *Ovula ovum* (Mollusca: Prosobranchia). Marine Biol. **74**, 35–40 (1983).

316. KUSUMI, T., K. YAMADA, M.O. ISHITSUKA, Y. FUJITA, and H. KAKISAWA: New Cembranoids from the Okinawan Soft Coral *Sinularia mayi.* Chem. Letters 1315–1318 (1990).

317. CHAN, W.R., W.F. TINTO, R.S. LAYDOO, P.S. MANCHAND, W.F. REYNOLDS, and S. MCLEAN: Cembrane and Pseudopterane Diterpenoids of the Octocoral *Pseudopterogorgia acerosa.* J. Organ. Chem. **56**, 1773–1776 (1991).

318. FONTAN, L.A., and A.D. RODRIGUEZ: Isolation of Eupalmerin, a Minor Cembranoid Diterpene from the Caribbean Gorgonian *Eunicea mammosa*. J. Nat. Prod. **54**, 298–301 (1991).

319. SHIN, J., and W. FENICAL: New Diterpenoids from the Caribbean Gorgonian *Eunicea calyculata*. Photochemical Interconversion of the Cembrene and Cubitene Skeletons. J. Organ. Chem. **56**, 1227–1233 (1991).

320. HAN, Y., Q. YANG, G. LI, Y. TANG, and J. SU: The Crystal Structure of a New Diterpene Lactone – Chilobolide A. Huaxue Xuebao **46**, 1055–1061 (1988).

321. PRESTWICH, G.D., D.F. WIEMER, J. MEINWALD, and J. CLARDY: Cubitene: An Irregular Twelve-Membered-Ring Diterpene from a Termite Solider. J. Amer. Chem. Soc. **100**, 2560–2561 (1978).

322. PRESTWICH, G.D.: Chemical Defense by Termite Soliders. J. Chem. Ecol. **5**, 459–480 (1979).

323. KODAMA, M., T. TAKAHASHI, T. KOJIMA, and S. ITO: Synthesis of Macrocyclic Terpenoids by Intramolecular Cyclization VII. Total Synthesis of (±)-Cubitene. Tetrahedron Letters **23**, 3397–3400 (1982).

(Received February 14, 1991)

Author Index

Page numbers printed in *italics* refer to References

Subject Index

Fortschritte der Chemie organischer Naturstoffe

Progress in the Chemistry of Organic Natural Products

Volume 58:

1991. 64 figures. VII, 343 pages. Cloth DM 280,–, öS 1960,–.
ISBN 3-211-82265-8

Contents: J. A. Robinson: Chemical and Biochemical Aspects of Polyether-Ionophore Antibiotic Biosynthesis. – R. D. H. Murray: Naturally Occurring Plant Coumarins.

Volume 57:

1991. 26 figures and 2 plates. X, 212 pages. DM 210,–, öS 1470,–.
ISBN 3-211-82245-3

Contents: P. Metzger, C. Largeau, E. Casadevall: Lipids and Macromolecular Lipids of the Hydrocarbon-rich Microalga *Botryococcus braunii*. Chemical Structure and Biosynthesis. Geochemical and Biotechnological Importance. – D. P. Chakraborty and S. Roy: Carbazole Alkaloids III. – G. R. Pettit: The Bryostatins.

Volume 56:

1991. 8 figures. X, 188 pages. Cloth DM 220,–, öS 1540,–.
ISBN 3-211-82188-0

Contents: J. Asselineau: Bacterial Lipids Containing Amino Acids or Peptides Linked by Amine Bonds. – J. Kagan: Naturally Occurring Di- and Trithiophenes.

Volume 55:

1989. 41 figures. X, 208 pages. Cloth DM 190,–, öS 1330,–.
ISBN 3-211-82087-6

Contents: M. T. Davies-Coleman and D. E. A. Rivett: Naturally Occurring 6-substituted 5,6-dihydro-α-pyrones – K. Krohn: Building Blocks for the Total Synthesis of Anthracyclinones – M. Lounasmaa and J. Galambos: Indole Alkaloid Production in Catharanthus Roseus Cell Suspension Cultures – C. E. James, L. Hough, and R. Khan: Sucrose and Its Derivatives.

Volume 54:

1988. VII, 353 pages. Cloth DM 320,-, öS 2240,-.
ISBN 3-211-82086-8

Contents: T. Murakami and N. Tanaka: Occurrence, Structure and Taxonomic Implications of Fern Constituents.

Volume 53:

1988. 72 figures. VIII, 311 pages. Cloth DM 275,-, öS 1930,-.
ISBN 3-211-82074-4

Contents: L. F. Alves: Chemical Ecology and the Social Behavior of Animals – T. Nomura: Phenolic Compounds of the Mulberry Tree and Related Plants – A. Chimiak and M. J. Milewska: N-Hydroxyamino Acids and Their Derivatives.

Volume 52:

1987. 65 figures. VIII, 224 pages. Cloth DM 210,-, öS 1470,-.
ISBN 3-211-81989-4

Contents: U. Weiss, L. Merlini, and G. Nasini: Naturally Occurring Perylenequinones – H. Achenbach: The Pigments of the Flexirubin-Type. A Novel Class of Natural Products – T. Goto: Structure, Stability and Color Variation of Natural Anthocyanins – P. Bhattacharyya and D. P. Chakraborty: Carbazole Alkaloids.

All Volumes and Cumulative Index 1–20 available

Price reduction for subscribers: 10%

Special reduced price (20% reduction) for the complete Series Vols. 1–59 incl. the Cumulative Index to Vols. 1–20

Springer-Verlag Wien New York

Sachsenplatz 4–6, A-1201 Wien
175 Fifth Avenue, New York, NY 10010, U.S.A.
Heidelberger Platz 3, D-1000 Berlin 33
37-3, Hongo 3-chome, Bunkyo-ku, Tokyo 113, Japan